AF786035

Computer Graphics: Systems and Applications

Managing Editor: J. L. Encarnação

Editors: K. Bø J. D. Foley R. A. Guedj
P. J. W. ten Hagen F. R. A. Hopgood M. Hosaka
M. Lucas A. G. Requicha

T. Yagiu

Modeling Design Objects and Processes

With 82 Figures

Springer-Verlag
Berlin Heidelberg New York
London Paris Tokyo
HongKong Barcelona
Budapest

Takaaki Yagiu
Nihon Unisys, Ltd.
2-17-51 Akasaka
Minato-ku
Tokyo 107
Japan

ISBN-13: 978-3-642-84422-5 e-ISBN-13: 978-3-642-84420-1
DOI: 10.1007/978-3-642-84420-1

Library of Congress Cataloging-in-Publication Data
Yagiu, Takaaki, 1935- Modeling design objekts and processes/Takaaki Yagiu.
p. cm. – (Symbolic computation. Computer graphics – systems and applications)
Includes bibliographical references and index.
ISBN-13: 978-3-642-84422-5
1. Computer-aided design. I. Title. II. Series. TA345.Y34 1991 620' .0042'0285–dc20
91-2206 CIP

This work is subject to copyright. All rights are reserved, whether the whole or part of the material is concerned, specifically the rights of translation, reprinting, reuse of illustrations, recitation, broadcasting, reproduction on microfilms or in other ways, and storage in data banks. Duplication of this publication or parts thereof is only permitted under the provisions of the German Copyright Law of September 9, 1965, in its current version, and a copyright fee must always be paid. Violations fall under the prosecution act of the German Copyright Law.

© Springer-Verlag Berlin Heidelberg 1991
Softcover reprint of the hardcover 1st edition 1991
45/3140-543210 – Printed on acid-free paper

Preface

A little more than a decade ago my colleagues and I faced the necessity for providing a database management system which might commonly serve a number of different types of computer aided design applications at different manufacturing enterprises. We evaluated some wellknown cases of conceptual models and commercially available DBMSs, and found none fully meeting the requirements. Yet the analysis of them led us to the development of what we named the Logical Structure Management System (LMS). Syntactically the LMS language is somewhat similar to ALPHA by E. F. Codd. The underlying conceptual model is entirely different from that of the relational model, however. LMS has been since put into practical use, meanwhile a further effort in search of a sound theoretical base and a concrete linguistic framework for true product modeling together with comparative studies of various approaches has been made. Here, the term product modeling is used to signify the construction of informational models of design objects and design processes in which it must be possible to include not a fixed set of attributes and relations, such as geometry, physical properties, part-of hierarchy, etc., but whatever aspects of design designers may desire to be included.

The purpose of this book is to present the major results of the said effort, which are primarily of a theoretical or conceptual nature. Following the introduction (Chap. 0) the body of the book is divided into three parts: requirements of CAD databases (Chap. 1), criticism of past and current methods (Chaps. 2 and 4), and the proposal of a linguistic framework (Chaps. 5–8). Chapter 3, which gives a succinct description of first order predicate logic, is to prepare those readers who are not familiar with the subject as well as to summarize the important results which are used in later discussions. The last chapter (Chap. 9) contains an overview of LMS and our experiences of applying it in production environments.

As indicated above, this book is heavily theory-oriented, for which it will be appropriate to offer an appology here. First, it is desirable or even necessary to focus our attention on the conceptual level and put aside implementation issues if we want to discuss conceptual frameworks as such. Second, it might take another decade if we wait for an implementation to be completed and practical experiences to be acquired. Third, one of my fundamental theses is that modeling design objects which exist as physical entities and serve human purposes through manifestations of their physical nature should entail an inherently different approach than that of business data, which is hardly ever taken account of in the literature on databases or conceptual models and can be clearly understood only through in-depth theoretical analyses of the respective universes of discourse.

This last point made me dare to expound rather lengthy philosophical arguments (Chap. 5) at the risk of taxing readers with intricate, yet superficial and misguided, discussions. It might be argued that, even at best, philosophy can contribute little to enriching the technical substance of any specific discipline. I do believe, however, that philosophical analyses are greatly beneficial or even mandatory for evaluating different positions within the discipline in question by illuminating what may underlie or be assumed implicitly in each of these positions.

I owe much to many of my colleagues at Nihon Unisys for the comparative evaluations of alternate approaches and the implementation of LMS in the early years of the last decade. I wish to express my sincere gratitude to Prof. M. Hosaka whose advice and encouragement led me to write this book. Special thanks are due to Mrs. M. Horiuchi who did all the laborious works of typing and re-typing during the preparation of the manuscript for the past years, and Mr. J. A. Ross of Springer-Verlag for his careful copy-editing.

March, 1991 T. Yagiu

Contents

Chapter 0
Introduction

0.1 Primary Organization of a CAD System

A computer aided design system that not only edits drawings but supports designers and engineers in truly creative and synthetic works has been a long-dreamt dream since the era of SKETCHPAD [Sut] and General Motors' DAC [Jac]. Lack of mainframes, mass storage devices, graphic terminals and communication links at reasonable costs and yet with required functionality and performance had been among several major obstacles to realizing the dream. When these pieces of hardware finally became available in the late 1970s, however, we immediately faced a host of other kinds of problem such as the following:

- definition of user-friendly interface between the user and the system,
- balanced distribution of load and functionality to local (graphic) and central processors where the balance is taken with respect to the response, cost and number of terminals the system can support,
- specification of the structure of graphic entities and manipulations on them in accordance with the said distribution of processing,
- construction and management of the database which describes full complexities of design objects (artifacts) in a coherent and yet open-endedly expandable way, thus lending itself to access by diverse users for diversified purposes,
- development of appropriate mathematical representations for spatial objects (curves, surfaces, solid bodies, form features, etc.) and of efficient algorithms for carrying out the geometric calculations involving them,
- devising a supervisory mechanism which monitors the functioning of various modules and controls the information flow between them including among others independently developed application programs.

Generally, the solution to these problems is embodied in an organized collection of software components as depicted in Fig. 0.1.

While these components are closely interrelated, it must be possible to change any one of them as freely as required and propagate changes to others naturally and methodically. This means that they should not only be written as well-structured programs but be preferably implemented as in-

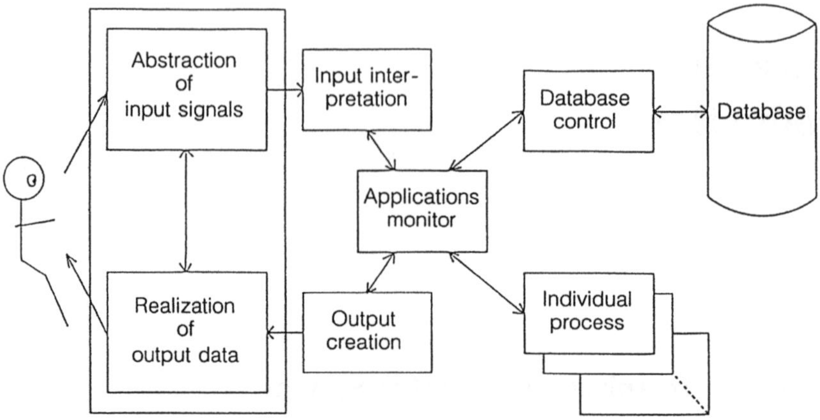

Fig. 0.1. Primary organization of a CAD system

stances of typified or generic pieces of software. Genericity in turn implies the existence of certain assumptions at a more abstract or conceptual level, and if the rationale for such assumptions is pursued, we will end up in a theory or 'world view' for each component concerning the universe of discourse it deals with. For example, the command interpreter may as well be generated from a meta-interpreter which accepts any set of commands conforming to a specific syntactic and semantic set of rules. By letting it be so, the system builder is committed to a specific view as to what class of languages will allow human users to dialogue with the system in a manner appropriate for them. Another example is the case of database management where the schema is described in a language with its vocabulary and grammar having established in advance the conceptual framework in which the world is to be seen. It is very important to clearly recognize the significance of this aspect of CAD software since it will lead us to the position that each component be treated as a self-sustaining discipline with its own methods and theories.

0.2 Current State of Integration and Distribution

The organization in Fig. 0.1 is termed 'primary' since that of a CAD system in practical use is far more complicated. The complication is multidimensional, that is, due to integration and distribution along the temporal (phasic), vertical (hierarchical) and horizontal (spatial) divisions of the whole work of design. While it is difficult to express this situation fully and precisely in a single diagram, Fig. 0.2 shows an exemplary case where a few subsystems on mainframes and many on workstations coexist.

It should be noted that quite often a CAD system is developed gradually over many years, and at each addition of a new subsystem or replacement of an old one, a key technical issue is the transfer of data from one sub-

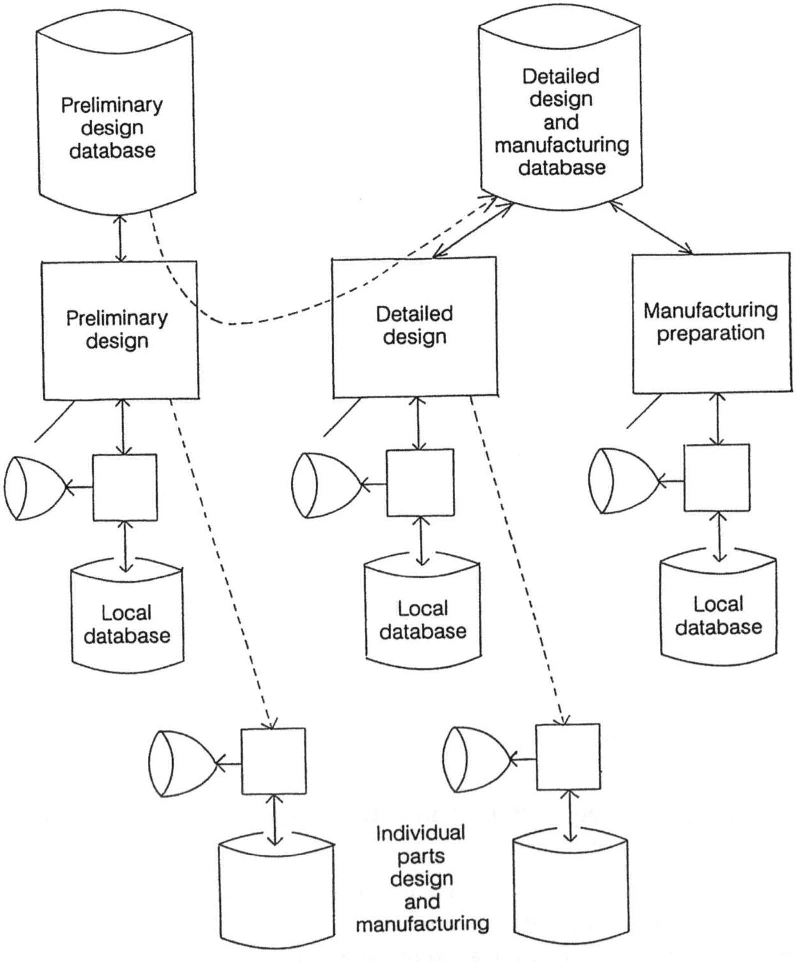

Fig. 0.2. Integration and distribution of CAD systems

system to another or the sharing of data among them. In some cases (e.g., between detailed design and preparation for manufacturing in Fig. 0.2), a common database is maintained through which the output of any subsystem is fully available to others. In other cases (e.g., between preliminary and detailed designs), databases are more loosely connected and the subsystem following in time requests the preceding subsystem for whatever data it may need. In still others (e.g., between detailed design and individual part design/manufacturing), batches of data are transferred downstream in a fixed and standard (e.g., STEP [ISO]) format. Even within a subsystem, databases may possibly be distributed among different processors (e.g., central and local) necessitating rather frequent exchange of data for updating and integrity maintenance. While there rarely exists a single and fully integrated database that serves all purposes and intentions, these examples together clearly indicate the existence of close relations among separate databases, hence the need for cohering data at an abstract, not to say physical, level.

0.3 Roles of a Database

The roles and functions that distinguish a database from a mere collection of data may be summarized as follows:

(1) The sheer volume of stored data raises serious technical problems concerning cost, efficiency and complexity.

(2) A database serves multiple users with multiple purposes. Moreover, it must anticipate additions of applications currently unknown or undefined, and be adaptable accordingly if these actually occur without causing any serious impact on existing ones.

(3) It is 'permanent' in that it continues to exist and be of service for an indefinitely long period of time.

(4) It is structured in accordance with a set of general rules and statements (schemas) which reflects the fundamental understanding by the database designer of the universe of discourse in question. In addition, it allows the definition of a specific view through which it is to be presented to and accessed by a specific user or application.

(5) Accessing (that is, storing and retrieving data to and from) the database is done through a precisely defined and rather restricted interface, and is guided by the structure and view.

(6) Often, facilities are provided so that end-users may directly inquire into the state of a database and obtain answers in a form friendly and easily understandable to them. This man-system interface is quite possibly related to but not necessarily the same as that between application program and database manager as in (5) above.

A large population of databases are found in the business transaction processing class of applications, typical examples including reservation systems, bank account balancing, personnel information management, etc. In this class, a database plays the role of a sort of general ledger. More specifically:

- it accumulates all the pertinent pieces of information, which are relatively static,
- it is self-contained in that nothing external but the information system itself generates the demand for a specific piece to be deemed pertinent, hence contained in it,
- these pieces are instances of a small number of strictly formatted collections of typed items (records) and relations among them, and

- the schema is specified in terms of these generic entities.

Of design/manufacturing applications, production control systems belong to this class with databases comprising such information as bill of materials, inventory levels, shop-floor work loads, order backlogs, etc., and sharing these properties. In the case of CAD, however, the database exhibits substantially different characteristics, namely:

- it is markedly dynamic, evolving from the initial and almost empty state to the final and full complexity which describes a finished design during a relatively short period of time,
- it responds to the demand to represent those entities and phenomena that lie outside any information system and carry profound significance for human designers, including shapes, form features, dimensional tolerances, motions, causal relationships, alternate plans, versions and a lot more, but
- this significance cannot naturally or easily be captured in rigidly formatted records but requires far richer and more flexible representational apparatus, and
- the representations inevitably entail redundancies (for example, dependencies among coordinate values and explicit relationships of a geometric entity to its boundaries), hence integrity constraints which are both justified and imposed by the facts lying outside the abstract world of information and 'directly observable'.

With these characteristics, a CAD database is intended to appear as something that behaves just like the real design object to the eyes of human designers. In the recent terminology, this something is called a product model [Lil].

0.4 Toward Product Modeling

Even within the limited context of information processing, the word 'model' or 'modeling' is used frequently and in different senses. Let us then take a look at a few of the more important cases and try to further characterize a 'product model' through comparison with them.

First, we have the classical notion of a 'mathematical model', which is the formulation of one or more specific aspects of the subject matter as a set of statements in well-established branches of mathematics. Different models for different aspects of the same subject may or may not be mutually related. An example is the vibration equation to (partially) describe the dynamic properties of a mechanical artifact, say a bridge. It may be based on a coarse approximation consisting of a series of springs connecting concentrated masses or a more precise 'geometric model' defining shapes of constituent pieces and relations among them. A geometric model is generally considered as forming a prototype or nucleus of a product model. It may also be considered as a kind of mathematical model, but perhaps not quite without reservation since we do not have a single paradigmatic method for treating finitely bounded spatial objects.

Second, a 'data model' provides a linguistic framework in which the database schema is described and accesses are implemented. It ranks at one level more abstract than a product model, hence might better be called a 'metamodel' in that a large number of databases from diversified universes of discourse are constructed under the same data model.

Third, against a formal theory comprising symbols, inference rules and axioms which are meaningless by themselves, any of its interpretations in terms of a concrete set, self-mapping functions and predicates with all the axioms holding true is simply called a 'model' of that theory. While an actual mathematical treatise as a whole is rarely presented in a purely formal fashion, there exists a distinct tendency to rely on logical inferences in several branches of computer science, such as automatic theorem proving, logic programming, deductive databases, etc. On the other hand, databases have traditionally been dealt with as structured sets of elements where general statements serve as integrity constraints rather than as axioms. A fundamental question then is whether a product model should be a formal theory or an interpretation ([Nic], [Gal], [Rei]).

Perhaps a physical mock-up is intuitively the closest to the idea of a product model as far as the shape, visual impression and (to a certain extent) motion are concerned. This comparison, though seemingly superficial, points to a profound problem of how concrete/physical and abstract/conceptual models may come close to and yet differ from each other.

It will not be irrelevant if we take a pause here and refer to the dictionary explication of the word 'model' to better understand its diversified uses and their shared meaning element. It is said that a model is:

- a structural design or pattern
- a 3-dimensional representation of something existing in nature, constructed or or to be constructed
- an image in clay, wax or the like to be reproduced in more durable material
- a person or thing regarded as worthy of imitation : something perfect of its kind
- a person or thing that serves as a pattern or source of inspiration for an artist or writer
- an archetype
- an organism whose appearance a mimic imitates (zoological)
- one who is employed to display clothes or other merchandise (a mannequin)
- a specific type or design of clothing or product (as a car or airplane)
- a description, a collection of statistical data, or an analogy used to help visualize, often in a simplified way, something that cannot be directly observed (as an atom)
- a theoretical projection in detail of a possible system of human relationships

- a system of postulates, data and inferences presented as a mathematical description of an entity or state of affairs
- a system of things and relations satisfying a set of axioms so that the axioms can be interpreted as true statements about the system

(abridged quotations from [Ran], [Sho], [Web])

Despite the diversity, we can clearly observe the existence of common elements constituting the act of modeling, that is:

- things and/or facts to be modeled,
- material for models to be made of,
- functions to be performed by models, and
- tools, techniques and/or methods for modeling.

Let us now return to the concept of a product model and try to define it by concretely specifying these elements.

(1) A product model models an entire product of arbitrary kind together with all of its parts and pieces that appears primarily in discrete manufacturing enviroments. It takes into consideration any attributes and functions of these parts and pieces and relations among them that are desired by designers and capable of being modeled on computers. These include among others such entities and phenomena as mentioned in the last paragraph of the preceding section.

(2) A product model, like any info-logical construct, is made of active elements (procedures) and passive elements (data). They comprise primitive symbols and higher level constructs created by the repetitive (and often recursive) applications of binding rules to those at lower levels.

(3) That a product model behaves just like a real object particularly implies the following:

(3.1) Designers can see and inspect the shape to whatever degree of detail they may desire. Moreover, they can demand that such analytic properties as curvatures, center of gravity, and cross sections be readily available even when not directly observable in physical objects.

(3.2) They can feel visual (and possibly in the future, other kinds of) sensations such as colors, shades, reflections, etc., which are as close to reality as possible.

(3.3) They can simulate the motion and movement of various parts and verify their proper functioning as to path, velocity, acceleration, smoothness, clearance and so on.

(3.4) They can perform more sophisticated simulation such as structural, thermal, electromagnetic, fluid-mechanical and other kinds of scientific/engineering analysis.

(3.5) Adequate geometric and topological information is available and contributes to finding the way to assemble and disassemble the product from and into sub-assemblies and parts respectively.

(3.6) Likewise, it is possible to plan the fabrication of each part, for example, to transmit data for designing dies and molds, and to determine the sequence of machining, select machine tools, jigs, fixtures and cutters, determine cutting conditions and calculate cutter paths. Finally, the finished part measurements can be matched against the product model to verify the precision of fabrication.

(4) How can we establish methods, tools and techniques to implement the product models described above? For the moment, we should remain content not to be too specific and systematic, since answering the question is indeed one of the major purposes of this book. In addition, the current problem is a particular case, probably more difficult than others, of those regarding the genenal methodology of data modeling which has been under intensive study for well over a decade but is still not settled. A recent trend is to incorporate the results into logic programming, data abstraction and artificial intelligence, thus making the problem more sophisticated and interesting but yet more difficult to resolve. Bearing these in mind, let us tentatively state that:

(4.1) The principal method is the division of the whole task of modeling into two aspects (static and dynamic), two levels (type/variable and instance) and three realms (conceptual, external and internal). The conceptual model represents the universe of discourse in abstract terms which are independent of application specificities and implementation issues (such as storage and efficiency). These concerns are left respectively to the external and internal models which are derived from the conceptual model. The schema describes the static aspect at generic level. It defines types, attributes and functions of entities and relations among them, and prescribes rules and constraints in the form of general statements. Generality here means statements' referring to one or more types, each of which is represented by a variable. A collection of specifications for unitary state transitions called events or transactions describes the dynamic aspect. Each unit asserts pre- and post-conditions around the transition, which is composed of a sequence of primitive manipulations. These are also at generic level. Instantiating entity, attribute and relation types to give a database which observes the prescribed rules and constraints, and implementing the primitive manipulations, completes the modeling.

(4.2) The primary and conventional tools are the language facilities for schema description, view (sub-schema) definition and primitive manipulation, and a set of manipulation programs. The transaction (event) specification language is for more abstract and macroscopic modeling of dynamism with which the model's behavior is implemented in levels of procedural abstractions. In case a logical or set-theoretic approach is taken to form general statements in the schema, the corresponding manipulative tools incorporate an inference or set-theoretic calculation engine.

(4.3) The basic technique is the use of interpreted predicate logic, possibly of higher order and/or augmented by modality. On the other hand, a variety of more specialized techniques have been developed and extensively used. These include:

- directed graph structures such as seen in graphics packages in the 1960s,
- classical data models, that is, hierarchical, CODASYL, and relational models,
- M. Minsky's theory of frames,
- special binary relations from the field of artificial intelligence and the subfield of semantic networks, that is, classification (*instance-of*), generalization (*is-a*) and aggregation (*part-of*),
- algebraic specifications of abstract data types,
- object oriented programming,
- Horn logic and its extensions, and so on.

The basic technique, predicate-logical or not, is therefore intended to be:
(4.3.1) simple and abstract so that it may be free of undesirable dependencies on excessive specificities and attain conceptual economy,
(4.3.2) rich and powerful so that it may accomplish what these specialized techniques accomplish, and
(4.3.3) general and well structured so that it may be applied to a wide range of modeling tasks and, further, naturally incorporate or interface individually developed models such as geometric, finite element, and simulation models.

It is of course arguable whether predicate logic is really qualified in the light of these goals, and this will actually be argued in succeeding chapters.
A final remark in concluding the section is that we have skirted issues concerning several important kinds of models, particularly those just mentioned, because first, we are primarily interested in providing a platform or general framework for product modeling rather than discussing specific aspects, and second, treatments of specific models require and deserve separate volumes, hence are beyond the scope of the present book.

0.5 The Subject Matter

The subject of this book is the search for a conceptual framework or metamodel for modeling design objects and design processes. A (partial) object model is the combination of a schema and a database as one of its instances, and a process model is the process of enlarging and enriching the database from the initial to the final state in which all the design requirements are met. Note that we have chosen to use the phrase 'design object' instead of 'product'. The reason is twofold. First, the targets of modeling may include not only an entire product but all the parts, pieces, aspects, abstract concepts or whatever designers may recognize as entities and deal with during the design process, and second, we want to emphasize the duality of static and dynamic, or operand and operator.

The state of the database is moved one step forward at each step of the design process. But the direction of move is generally not unique nor deterministic, hence its decision requires some extra-algorithmic instructions be given. While these instructions may depend more or less on designers' intuition, imagination and/or value judgement, the system is also intended to collect and organize pieces of knowledge so that it may support more objective and intelligent decision making. This organized body of knowledge or 'knowledge base' together with the mechanism for exploiting it is considered to embody the process model. We thus come to Fig. 0.3 which is a revision of Fig. 0.1 with the encircled portion indicating the present subject matter.

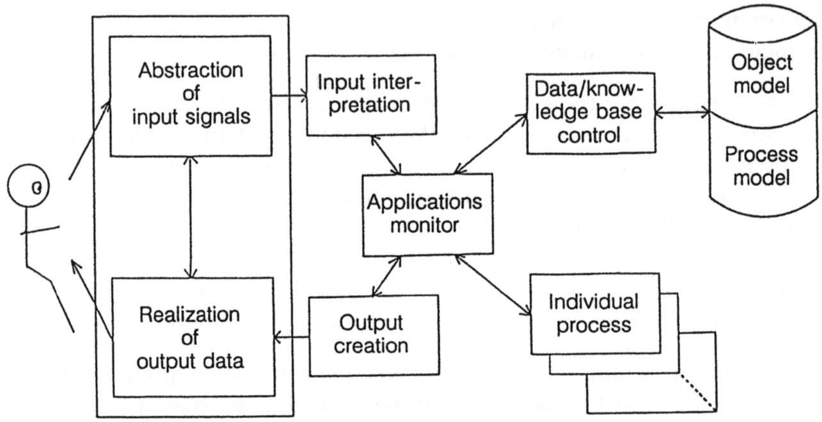

Fig. 0.3. Primary organization of a CAD system (revised)

The indeterminacy of the design process implies that there may exist more than one design solution and that designers might go astray on a path leading to a dead end from which they must backtrack to the point where they erred. The stored knowledge is again expected to support them in properly deciding how to backtrack. It is also expected that with an accumulation of knowledge and experiences, the system can prune unnecessary or undesirable branches. Figure 0.4 shows what has just been stated. It is among the most fundamental aspects of the design process which must be made manifest in the behavior of the process model.

It is emphasized that we envisage the framework (metamodel) as being applicable not only to specific types of products but to a general class of objects and processes which we observe in design activities in discrete manufacturing environments. Also, we will concentrate on the conceptual realm rather than on the external or internal realms. Therefore, we will not discuss implementation issues as such, though efficiencies attract our attention and will be considered so far as they relate to the representational capability and naturalness of the framework.

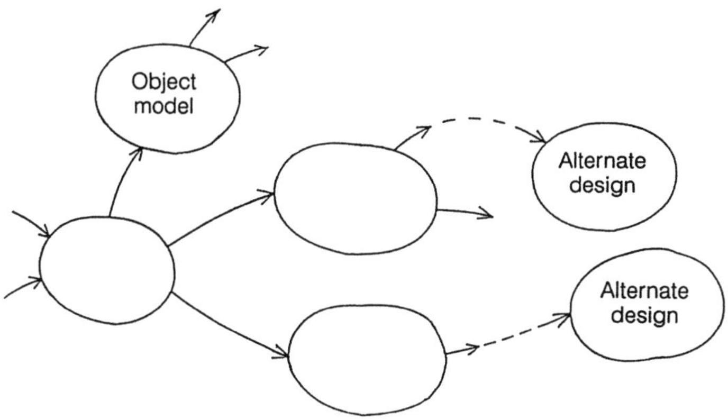

Fig. 0.4. The indeterminacy of the design process

0.6 An Example

It will be appropriate for us here to present a simple illustrative example so that readers may clearly understand the concepts and terminology we have been dealing with. What is to be modeled is the class of tetrahedral parts (Fig. 0.5). The function of the model is to represent shape. Specifically, we are interested in (a simplified version of) the so-called *winged-edge topology* [Bau] that comprises the following types and attributes of and relations among objects (Fig. 0.6):

- type: *body*(*b*), *face*(*f*), *edge*(*e*), *vertex*(*v*)
- relations between objects and their boundary objects: an edge and its forward or backward vertex (*evf* or *evb*) an edge and its forward-left, forward-right, backward-left, or backward-right edge (*eefl*, *eefr*, *eebl* or *eebr*) an edge and its left- or right-hand face (*efl* or *efr*)
- attribute: object names and coordinate values of vertices.

The technique employed is structuring by directed graph, which is a natural selection since all the relations between objects are binary. The tool is ASP (Associative Structure Package, [Lan]) with which (a part of) the model is graphically represented as in Fig. 0.7.

This is the conceptual schema of which each tetrahedron with particular object names and coordinate values is an instance. What we term a framework, metamodel or schema description language is nothing but the set of those symbols (□, ○, △, ⊣) which occur in the schema and the rules for using them. □ and ○ denote an object and a relation respectively. Or, a node and an arc in the directed graph terminology. A line ending with ⊣ is a *ring* or circular list of which there exist three kinds. An *associative ring* starting with △ collects instances of a particular binary relation. Each *lower ring* headed by an object lists those associative *ringstarts* followed by relations in which the header object participates as the first argument. The other ring (*upper*

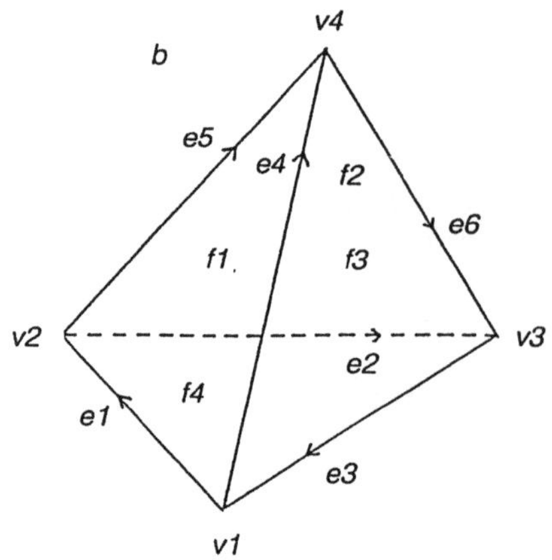

Fig. 0.5. A tetrahedral part

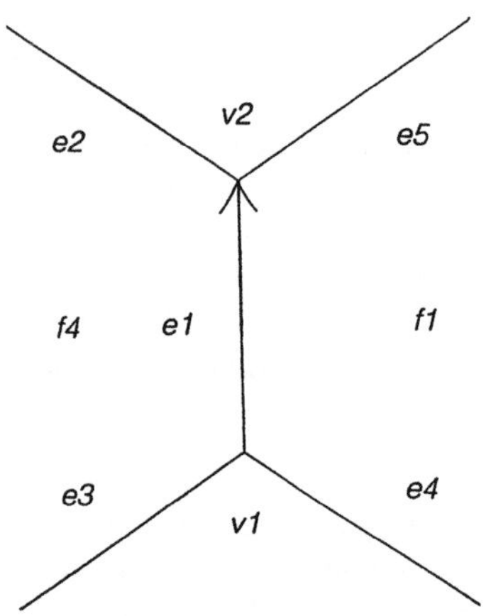

Fig. 0.6. The winged-edge topology

ring) starting with an object contains relations involving that object as the common second argument. The manipulation facilities provide the means for traversal on rings, retrieval and alteration of attribute values.

Let us now turn to the dynamic aspect and consider the problem of finding the solution to meet the design requirements. Suppose the requirements include that:

(1) one of the pairs of adjacent faces makes a right angle
(2) the faces in that pair are preferably acute triangles
(3) they have preferably the largest areas of the four faces.

'Preferability' means that pair $\{a, b\}$ is preferred to $\{c, d\}$ if and only if the number of acute triangles in the former is greater than or equal to that in the latter, and a or b has a larger area than c and d. It is not ambiguous but indeterminate as a criterion in that we do not generally have the unique

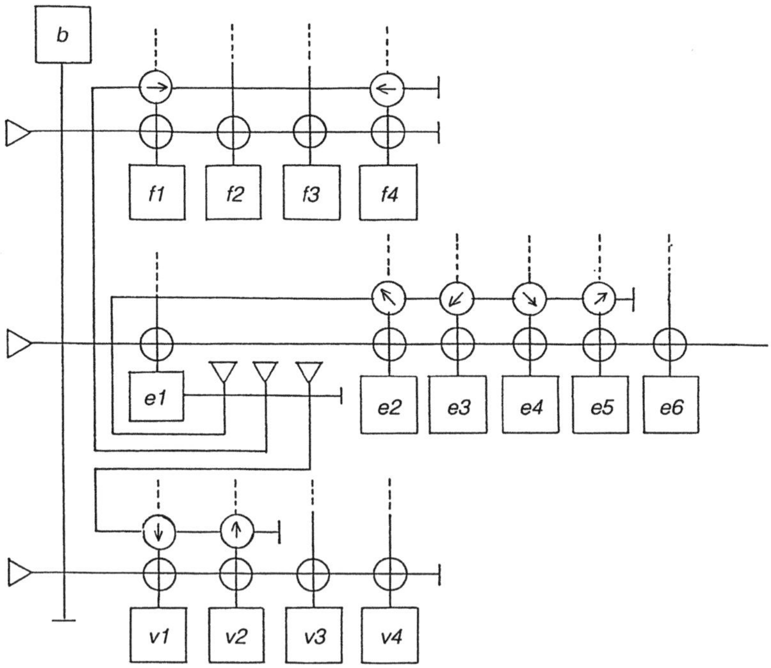

Fig. 0.7. The schema for a winged-edge structured tetrahedron

most preferred pair. For example, if the largest two faces (denoted by *t1* and *t2*) are both acute, they give the unique pair. If, however, the smallest two (*t3* and *t4*) are the sole acute triangles, {*t1*, *t2*}, {*t1*, *t3*}, {*t3*, *t4*} are all maximally preferred pairs. With further knowledge such as:

(4) at least one member of the pair must be acute, or
(5) at least one area in the pair must exceed all the others

we can narrow the extent of alternatives to {{*t1*, *t3*}, {*t3*, *t4*}} or {{*t1*, *t2*}, {*t1*, *t3*}} respectively. Which should be used may be decided by the help of still further knowledge or by the designer's direct instruction. The system shall perform the rotation to a right angle for each of the remaining alternatives and present the result to the designer for the final evaluation and approval (Fig. 0.8).

A lot more can be learned from this simple example by analyzing typical inquiry programs, trying to expand the schema for a more general class of objects and comparing the technique with other techniques. But these go beyond the scope of this introductory chapter, so we stop here and leave them for later discussions.

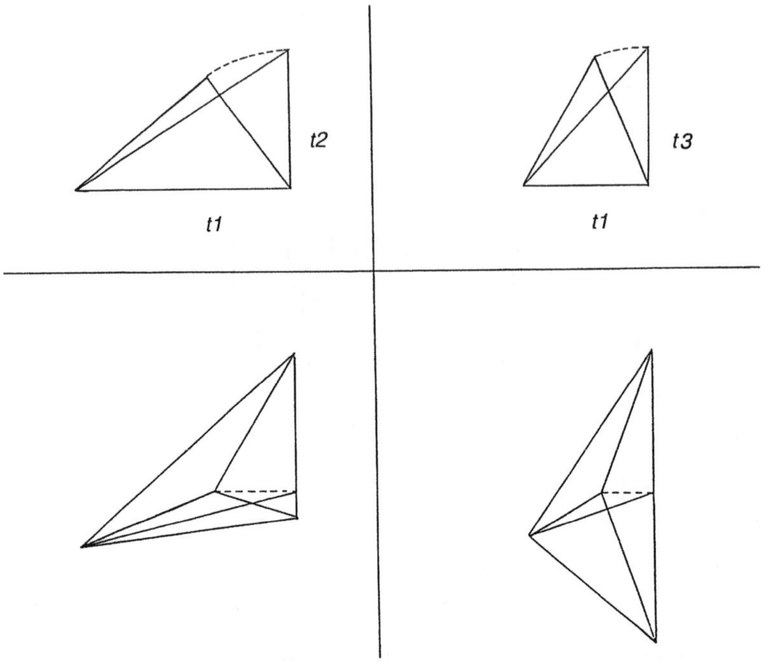

Fig. 0.8. Changing the shape to meet the requirements

0.7 On the Formalness and Abstractness of Language

One may object to our thinking of ASP's symbolism as a conceptual framework or schema language on the presumable ground that it is crudely intuitional, lacks rigorous formalism and does not dissociate itself from the aspects that are not at all conceptual. While we hold back our apology for the time being, we admit and accept the trend in recent years toward a thorough formalness and abstractness of linguistic framework. We indeed have singled out as the basic technique the use of interpreted predicate logic (Sect. 0.4 (4.3)) which is one of the typically formal and abstract approaches. But what is formalness, what is abstractness? At the risk of inviting more controversy, let us have them explicated as follows:

- Formalness is the state of being self-contained of a system which comprises a fixed set of primitive symbols, grammatical rules to combine them into higher level constructs and these constructs themselves. 'Self-contained' means that:
  - symbols are nothing but pure symbols,
  - the application of constructive rules is strictly mechanical,

  - the system recognizes as legitimate entities all those thus constructed, and
  - discards everything else. It is, so to speak, a syntactic stoicism.
- Abstractness is then a semantic stoicism in that the system denies itself the associations with all aspects other than the absolute core of meaning of the subject matter. 'Meaning core' is equivalent to what we call 'concept'.

Interestingly, these stoicisms contribute to the fertility of the system by making it more generally applicable and naturally expandable. In the opposite direction, as we have previously noted (Sect. 0.1, last paragraph), genericity implies the commitment to a particular position at an abstract or conceptual level.

The same spirit of abstractionism prevails in several branches of computer science, notably programming language, artificial intelligence and database management. The phrase 'conceptual modeling' is being used in recognition of the intrinsic commonality of problems and approaches in treating them ([Bro(1)], [Bro(2)]). Data/procedure abstraction, knowledge representation and data modeling are dialects in respective branches of this commonality. The degrees of formalness and formalities themselves differ from one to another, however. The differences may not be superficial as they possibly reflect or entail conceptual differences. These endeavors have been exerting profound influences on researches in product modeling and intelligent CAD, and are therefore worth close study. At the same time however, we must carefully examine the question of whether the proper and uncompromisable conception of our universe of discourse, that is, the world of design objects and processes, may not get lost in the deceptive similarities of forms. But on the more positive side, how should we expect a particular form to fit proper concepts? The ultimate ground will have to be given by the establishment of a self-sustaining discipline armed with methods and language that are clear and precise enough to be interpreted by computers. (See Sect. 0.1, last paragraph.) It will be the discipline of a general theory of design ([Tom], [Yos(3)]).

0.8 On the General Theory of Design

A general theory of design aims to elucidate and express in a rigorous form whatever facts, logic, methods and techniques may be presumed, conceived or employed by designers consciously or unconsciously, explicitly or implicitly during design processes. It is 'general' in that it is detached from the design of any specific kind of artifact. The immediate purpose or motivation is twofold, that is first, to obtain objective and scientific knowledge on design activities and second, to 'rationalize' or make them more efficient and economical [Hub]. From early in this century, long before the advent of automatic computing machinery, the task of exploring design activities has attracted researchers, but the achievements remained unremarkable until the 1970s, when we witnessed a rapid increase of research papers and international conferences in related areas. To mention a few, the first of the biennial ICEDs (International Conference on Engineering Design) was convened in 1981, then in 1984 the ISDS (International Symposium on Design and Synthesis) offered several sessions for design method, and in 1985 the whole working conference of IFIP WG 5.2 was dedicated to design theory for CAD. Interestingly, at the workshops for intelligent CAD in 1987, 1988 and 1989 sponsored by the same WG, discussions were centered around how we should and could understand design objects and processes, and envisage models thereof. It is obvious that researches in the theory of design and CAD are entering under strong, mutual influences. From this standpoint, the expectation is that a general theory of design will provide a theoretical or methodological foundation for CAD.

While the author shares that expectation with many CAD researchers, the theory of design per se will not be discussed in detail in the present book because first, the author feels it has not yet reached a state of maturity and hence is not an appropriate subject for an elementary book such as this, and second, he is not equipped to expound on various schools or to survey the state of the art in the field. It will be appropriate, however, if a few remarks are given here on the fundamental nature, problems and prospects of the theory.

First, the theory of design concerns itself not with any specific aspects but with the entirety of a range of highly intellectual and creative behaviors and activities of organized bodies of human beings. As such, it is more akin to the humanistic sciences such as sociology, psychology or cultural anthropology than design and engineering themselves. Second, this target domain of research is notoriously complicated and elusive for scientific approaches. The more holistic we try to be, the less substantial our discourse seems to get. We are not in the enviable situation that physical science enjoys where mathematics provides such a powerful and reliable framework, and where both disciplines have profoundly influenced and contributed to each other's development through presenting ever challenging and stimulating issues and ideas. Third, we may as well commit ourselves to a particular framework before starting to observe and examine factual details inevitably and legitimately, since what is captured within a framework must anyway be interpretable to computers and this fact substantially reduces the acceptable class of frameworks. This approach certainly limits our view but exactly because of that it will hopefully enable us to see and state the key issues lucidly and precisely. If we are allowed to draw another analogy between the theory of design and other disciplines, it is of the same methodological genus as structuralism in anthropology in that it transcends primary and individual observations, and concentrates on whatever can be talked about intelligently.

0.9 Program for Succeeding Chapters

We start by reviewing and summarizing the requirements of CAD databases which many authors have talked much about (Chap. 1). A special emphasis is placed on the contrast with the case of business data processing, which has virtually dominated the concerns of database researchers and hence is fairly well established. Consideration of intelligent CAD and knowledge bases is postponed to later chapters since they raise issues substantially different in their nature than ordinary databases.

We then select sample data models, each representing one of the widely known, well-founded and mutually distinct techniques, and analyze and evaluate them in the light of the requirements (Chap. 2). The samples include a few classical achievements in the 1960s, the CODASYL dbcs (database control system) and the relational model and its extensions. For readers who are not familiar with any of these, a concise description of the functions that determine modeling capabilities precedes each main part of the argument. The finding will be that none of these is sufficiently capable of modeling design objects.

Chapter 3 presents that part of first order predicate logic and model theory which is required for understanding the logical approaches to data modeling discussed in the next chapter. While informed readers may skip it, the author invites them nevertheless to notice the inclusion of several particular non-trivial notions and theorems such as Skolemization, Herbrand interpretations, completeness and incompleteness theorems. The reason is not the theoretical values themselves but the significance they carry for our conception of the world of design which is brought up in Chap. 5 and thereafter.

In Chap. 4, we continue to do the same as in Chap. 2 on the subjects that are among the most distinctive members of the logical school (in a broad sense), namely, data abstraction, logic programming, the deductive data model and semantic networks. They are attracting serious attention particularly in conjunction with intelligent or knowledge-based CAD. The primary issues are first, which we are talking about, a formal theory or an interpretation, second, what the domain of interpretation will be, third, what the meaning of a formula will be, and fourth, what kind of logic it is, or whether it is logic at all that we employ. Again, we will find that none of the approaches discussed here qualifies for our purpose.

It is one of the author's conclusions formed through the above criticism that what has caused the existing approaches to fail is the failure of their adherents to recognize the inherent externality and open-endedness of the universe to be modeled and/or to liberate themselves from the spell of the traditional subject-predicate world view. Another conclusion is a reinforcement of the previously stated position that a good framework must be formally and abstractly constituted with as few concepts and as limited a vocabulary as possible. (Ockham's principle! See Sects. 0.4 (4.3) and 0.7.) These conclusions suggest that we should turn to and study the (meta)theories concerning the structure of natural-scientific theories which deal with the same external and physical kind of entities as design objects. Particularly noteworthy are those of the logical positivist tradition with their strong tendency to formalness. Thus, after short philosophical discussions, we will come to choose D. Hilbert's formal-axiomatic method of constructing mathematical disciplines, specifically *Grundlagen der Geometrie* as the paradigm in accordance with which the framework is to be formulated. These are the contents of Chap. 5.

Next, Chap. 6 defines the fundamental structure of the model that can be described within the framework. Roughly and informally, the model looks as follows:

(1) The requirements that the objects are designed to meet are specified as formulas (axioms) of first order predicate logic. The collection of these formulas can be considered as constituting a formal theory.

(2) The database (object model) is an interpretation of this first order theory and will be its model (in the sense of predicate logic) when all the axioms become valid.

(3) Thus design process can be modeled as the process of enlarging and enriching the database from the empty state to the final state with all the requirements eventually being met.

(4) Within the database, each recognizable piece of design object is represented by an identifier which is a pure symbol and meaningless by itself. Functions and predicates can be defined over the union of this set of identifiers and whatever data types specific host languages permit. The meaning of a piece is nothing but the collection of all the functions and predicates in which it occurs as an argument. An identifier continues to identify the same piece of object through whatever changes the values of these functions and predicates undergo.

(5) The language provides application programs with all the means that predicate logic and set theory permit, including recursive functions and formulas for describing and manipulating databases.

Chapter 7 presents the syntax and semantics of the language formally and precisely, accompanied by some illustrative examples. The static functionality corresponds to the conceptual schema language of an ordinary data model while the manipulative part serves to describe the dynamism at the local, primitive level.

Modeling the design process at the more global and macroscopic level is the subject of Chap. 8. As previously stated (Sect. 0.5), each design step advances the state of the database one step further, which means that one or more axioms currently not satisfied are to be satisfied. But which axioms should be made true, and how? As also indicated, these questions cannot generally be answered definitely, which raises the possibility of exploiting intelligence for steering the course. But again, how? This is the issue the chapter addresses.

Because the full and efficient implementation of the above-explained framework is admittedly not easy, a subset was implemented and has been in practical use since 1979 for various CAD applications at various enterprises. So, we will give a brief overview of this subset in Chap. 9.

The author has made a particular effort to explicate the basic presumptions and world view on which the theory and the more concrete formulation of the linguistic framework have been developed. All of these are of course subject to criticism and change. Even if they are accepted as they presently exist, there remain many problems to be resolved before we can be confident of an efficient implementation and a broad range of applicability. These issues relate to the internal and external realms of modeling which are not the primary concerns of the book. We therefore conclude by just pointing out the problems and not going into the discussion of how to resolve them.

References

[Bau] Baumgart, B.G.: A Polyhedron Representation for Computer Visions. Proc. NCC, 1975

[Bro(1)] Brodie, M.L., Zilles, S.N. (eds.): Proc. of the Workshop on Data Abstraction, Databases and Conceptual Modelling. ACM, 1980

[Bro(2)] Brodie, M.L., Mylopoulos, J., Schmidt, J.W. (eds.): On Conceptual Modelling. Springer, 1984

[Gal] Gallaire, H., Minker, J., Nicolas, J.M.: Logic and Databases: A Deductive Approach. Computing Surveys, Vol. 16, No. 2, June 1984

[Hub] Hubka, V.: Attempts and Possibilities for Rationalisation of Engineering Design. Proc. Int'l Symp. on Design and Synthesis, The Japan Soc. Prec. Eng., 1984

[ISO] ISO TC184/SC4/WG1: The Working Draft of STEP 1.0. (To appear)

[Jac] Jacks, E.L.: A Laboratory for the Study on Man-machine Communication. Proc. FJCC 26 (1964)

[Lan] Lang, C.A., Gray, J.C.: ASP – A Ring Implemented Associative Structure Package. CACM, Vol. 11, No. 8, Aug. 1968

[Lil] Lillehagen, F.M., Dokken, T.: Towards a Methodology for Constructing Product Modelling Databases in CAD. In: Encarnação, J., Krause, F.-L. (eds.): File Structures and Data Bases for CAD. North-Holland, 1982

[Nic] Nicolas, J.M., Gallaire, H.: Data Base: Theory vs. Interpretation. In: Gallaire, H., Minker, J. (eds.): Logic and Data Bases. Plenum Press, 1978

[Ran] The Random House Dictionary. Random House, 1967

[Rei] Reiter, R.: Towards a Logical Reconstruction of a Relational Database Theory. In: [Bro(2)]

[Sho] The Shorter Oxford English Dictionary. Oxford Press, 1967

[Sut] Sutherland, I.E.: SKETCHPAD: A Man-machine Graphical Communication System. Proc. SJCC 23 (1963)

[Tom] Tomiyama, T., Yoshikawa, H.: Extended General Theory of Design. In: [Yos(1)]

[Web] Webster's Third New International Dictionary. G. & C. Merriam Company, 1969

[Yos(1)] Yoshikawa, H., Warman, E.A. (eds.): Design Theory for CAD. North-Holland, 1987

[Yos(2)] Yoshikawa, H., Gossard, D. (eds.): Intelligent CAD, I. North-Holland, 1987

[Yos(3)] Yoshikawa, H.: General Design Theory as a Formal Theory of Design. In: [Yos(2)]

Chapter 1
Requirements of CAD Databases

There are generally three approaches to the task of giving a systematic presentation of the requirements of CAD databases. First, we may survey as wide a range as possible of individual implementations and uses of database control software, collect the features and functions that are proper to CAD, extract from them application independent elements and state them in more abstract terms. This is the inductive approach. Second, we may take the deductive approach in which the database requirements derive from a still more abstract and comprehensive conception of the world of design objects and processes. Such a premise must be given by a general theory of design. Since both approaches raise obvious difficulties, we will have to rely on the third one which may imprecisely but conveniently be called abductive. That is, we assume a firm though not quite comprehensive position concerning our universe of discourse, examine the requirements contained in a selection of concrete examples and proposals, and refine and extrapolate them in accordance with this position. To use a popular metaphor, it is like looking closely at sampled specimens through a magnifying glass and portraying a complete picture constituted of those directly observed as well as unobserved portions.

What can and should we assume, then? It will at least be safe and appropriate to say this; what most essentially characterizes a CAD database lies in the fact that it is an informational model of a part of the real, physical world containing those entities (design objects) that we human beings create, sense, manipulate and think about. This seemingly plain and almost self-evident fact is the fundamental cause for many requirements unique to CAD (and often to CAM as well) which do not apply to business data processing where 'reality' independent of information thereof need not be presupposed. We expressed this view briefly in Sects. 0.3 and 0.9, and will elaborate further in Chap. 5; meanwhile let us proceed with this temporary sketch and summarize the result of examination. The specimens include the author's experiences in supporting various users in developing their applications, explicit statements of requirements and/or proposals for CAD data models, and reports on the implementations of databases for specific CAD applications (see the list of references for this chapter). (The author would like to remind readers that the term 'data model' implies genericity while the database is usually at instance level – see Sect. 0.4. Also that the considerations common to all database applications and implementation or administration issues are excluded from the following discussions.) We classify approximately two dozen requirement items into three categories as follows:

(1) *Structures and Semantics*

(1.1) Shapes, motions, tolerances, form features, etc.
(1.2) Long and variable-length attribute fields
(1.3) Abstract typing of attribute domains
(1.4) Organization by hierarchy
(1.5) Organization by classification
(1.6) More complex structures
(1.7) Manipulation of complex objects
(1.8) Support of multiple views
(1.9) Support of multimedia representations
(1.10) Semantics of objects, attributes and functions

(2) *Dynamic Aspect*

(2.1) Liberal (unrestricted) choice of access methods
(2.2) Dynamic extension and modification of schema
(2.3) Specification and validation of integrity constraints
(2.4) Security and protection
(2.5) Backtracking on the design path
(2.6) Arbitrarily long transactions
(2.7) Replication of arbitrary portions of database
(2.8) Private workspaces
(2.9) Support of distributed databases

(3) *Version Control*

(3.1) Support of alternate designs
(3.2) Control of versions and releases
(3.3) Maintenance of the history of changes
(3.4) Archive for past designs

1.1 Structures and Semantics

(1) *Shapes, Motions, Tolerances, Form Features, etc.*
While it is only too clear that these properties and functions must in one form or another be represented in the database (or maybe imbedded in programs), it is just as unclear how to represent them properly. Take shapes, for example. Any single representation scheme currently available is far from achieving the theoretical and practical ideals. The wireframe scheme is probably the simplest, most efficient and easiest for solid bodies, but not unambiguous, nor unique. BREP (boundary representation) and CSG (constructive solid geometry) are functionally far richer and unambiguous, but still not unique. Non-uniqueness often implies the non-existence of any particular reason for

choosing a particular representation over others, and even the non-existence of a 'natural' representation at all. For example, BREP representations of a torus are simply artificial, though a portion of the torus might be approximated by a patch completely determined by the corner points, tangent and twist vectors, or might as well be precisely represented by a set of parameters together with the boundary curves (Fig. 1.1). Using multiple representation schemes is in a sense quite natural since we see the patch either as a toric segment or as defined by boundaries depending on the context in which it is dealt with. Such a versatility is attained, however, only by virtue of the more abstract entity conceived by human beings and cohering all the varied forms of representations, and at the risk of inconsistencies and combinatorial explosions on the side of computers.

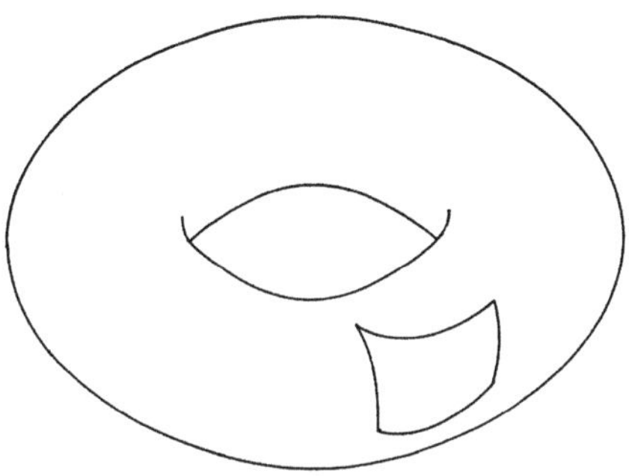

Fig. 1.1. A torus and a patch lying in it

Since the functionality of a design object, particularly of a mechanical product, is largely carried by the shapes, relative positions and motions of its constituent parts, the suitability of a specific geometric representation of a part depends on its functions and relations with others. Thus, a form feature (e.g., a through hole, groove, fillet, etc.) is to be described in terms not only of dimensions but of adjacent forms (e.g., top, bottom, side, etc.). Likewise, tolerances may be considered more appropriately as constraints among related pieces than as separate attribute values.

These issues are all well recognized and studied, so we will not elaborate any further on them but refer the reader to pertinent works ([Bae], [Brai(2)], [Eas(5)], [Lee(2)], [Req(1)], [Req(2)] and references therein). In view of the ever-increasing multitude of competing and complementary representation schemes, one of the most important points for the data model builder to heed is that of facilitating the inclusion of any combination of different schemes and changes of them without seriously impacting application programs.

(2) *Long and Variable-Length Attribute Fields*
Except for the extreme case where vertices are the only entities that are given attribute (that is, coordinate) values from which the remaining geometric data is constructed bottom-up, any scheme for representing shapes must accommodate variable-length attribute fields which tend to be very

long, much longer than those of usual business data. (Consider representing a cuboidal object some faces of which are free surfaces each of which in turn comprises several simple patches of third or higher degree.) Another typical example is an object being decomposed into a finite collection of elements for structural analysis, the number of which is not only large but subject to dynamic change.

(3) *Abstract Typing of Attribute Domains*
The implications of the need to allow variable-length fields are twofold. First, what is truly attributed to an entity is something that exists beyond the varying representations. Second, the separation of functional essence and implementation details, and the hiding of the latter from application programs, becomes more important, as while such attributes are more applications dependent, the handling of their values is more complicated and cumbersome. These clearly indicate that we should be able to specify some of the attribute domains as abstract data types (ADTs). The use of ADTs has already been proposed in the context of data models in general ([Row], [Schm], [Sto(1)]). On the other hand, several authors have tried to reformulate in more abstract terms such familiar data types in CAD as graphical and geometric entities ([Bat], [Buc(1)], [Mal], [Sto(2)]). The full integration of both lines of efforts is yet to be seen, however.

(4) *Organization by Hierarchy*
Hierarchy is one of the most basic principles for organizing data. The whole set of entities is divided into mutually disjoint clusters the entirety of which constitutes a partially ordered set. Further explanation is useful here. First, each cluster as a whole ranks higher or lower than one or more of the others (Fig. 1.2(a)). Second, each entity is related by the partial order to some (but not necessarily all) entities in higher or lower ranking clusters (Fig. 1.2(b)). Third, there may exist more than one hierarchical order in the same entity set. For example, a house is decomposed into roofs, outer walls, rooms and foundations. A room comprises ceiling, inner walls, and floor. On a wall there may be doorways, niches, panels, etc. This is so to speak a structural hierarchy for organizing the data in a house model. On the other hand, we may imagine slicing the whole construction horizontally at several heights and collecting those pieces and items contained wholly or partly in each slice into a cluster. Clusters are ordered according to their heights and are exposed separately to planar views when specified.

On the dynamic side, facilities must be provided to enable us to traverse up and down the hierarchical paths, look around and inspect within the same cluster, retrieve all the entities related to a particular one, or repeating such retrievals, take the transitive closure.

(5) *Organization by Classification*
The concept of class connotes the existence of a generic entity and the set of its instances for each class. In business applications, a generic entity is usually defined as a rigid record (or synonymously, a relation) type containing a list of

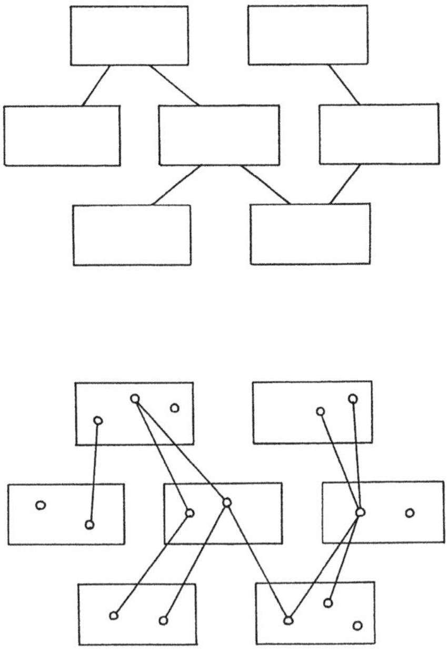

(a) Relations between clusters

(b) Relations between entities

Fig. 1.2. Hierarchical organization of the entity set

attribute names, keys and types of domains. Relations between entities such as hierarchies and accompanying constraints are defined in terms of generic entities. The case of CAD, however, requires greater flexibility and a higher degree of sophistication. For example, in the house model, the class 'rooms' can be introduced and is useful, but the description of a room's attributes does not simply fit into a fixed record structure. In fact, as mentioned above, a room contains many and varied elements the number and kinds of which differ from one room to another. There are components belonging to particular rooms such as ceilings, walls and floors, those shared by more than one room such as doorways, and standard parts or subassemblies such as plug receptacles, staircases, window frames, etc. Each of these elements may be a member of a respective class with its own assortment of attributes. Accessing their values is not so simple as by means of record type, key values, attribute name and/or indices, but may require structural and spatial pieces of information, and executions of procedures involving them.

(6) *More Complex Structures*
It is often contended and seems to be generally agreed upon by the CAD community that in CAD, databases need to implement more complex relations among entities than in business applications. At least $m : n$ binary relations and probably relations involving arbitrarily many terms must be available. This requirement is apparent even in such a simple object as a polyhedron (Sect. 0.6). Suppose we want to introduce a fixed ordering (e.g., the clockwise

direction) in each set of vertices belonging to a face. A natural way will be to have the ternary predicate

$$order(face,\ preceding\ vertex,\ succeeding\ vertex) \tag{1.1}$$

where the occurrence of a face is indispensable, as the same pair of vertices are ordered in the opposite direction on the other surface to which they belong. If we stick to the exclusive use of binary relations, we may as well define it as follows;

$$\begin{aligned} order(f, v1, v2) \equiv \exists e1 \exists e2 (&boundary(f, e1) \wedge boundary(f, e2) \wedge \\ &boundary(e1, v1) \wedge boundary(e1, v2) \wedge \\ &boundary(e2, v2) \wedge order(e1, e2)) \end{aligned} \tag{1.1$'$}$$

which essentially depends on the facts that boundary relations are binary, two adjacent edges determine a face and each edge has precisely two vertices. However, such conditions as these that can be conveniently exploited to reduce a predicate to more elementary ones do not always exist. A simple counterexample is the relation between two points and their center point. Another is the relation between the viewpoint (vp) and two spatial objects such that one is closer to vp than the other. Yet another, and well known, is from constructive solid geometry (CSG) in which a set operation (op) is applied to two objects (A and B) producing the third one (C), which may be written as

$$C = op(A, B) \tag{1.2}$$

or

$$op(A, B, C) \tag{1.2$'$}$$

The former might be taken as an instance of the binary predicate '='. This is a misconception, however, since it will only split the relation into two parts of which the more significant is the relation between $op(A, B)$, A and B which remains ternary.

In the house model, we might be interested in the loop formed by a set of adjacent segments of roofs or outer walls, and want to ensure that the opening is closed by another set of intersecting segments without causing undesirable concavities to appear (Fig. 1.3). To identify the participating segments and retrieve their geometric data, we may profitably exploit such relations as $loop(op,\ s1,\ s2,\ s3)$, $set(cs,\ s4,\ s5)$ and $intersect(op,\ cs)$ where op, si and cs stand for opening, segment and complex segment respectively.

As entities are often created or positioned relative to other entities, they are mutually related in an implicit but rather complicated way. For example, suppose that two circles $c1$ and $c2$ happen to share the same center, which may be accidental or must hold of necessity. If the latter, it must be possible to write an explicit condition ($center(c1) = center(c2)$) and when one of the terms changes, to change the other accordingly.

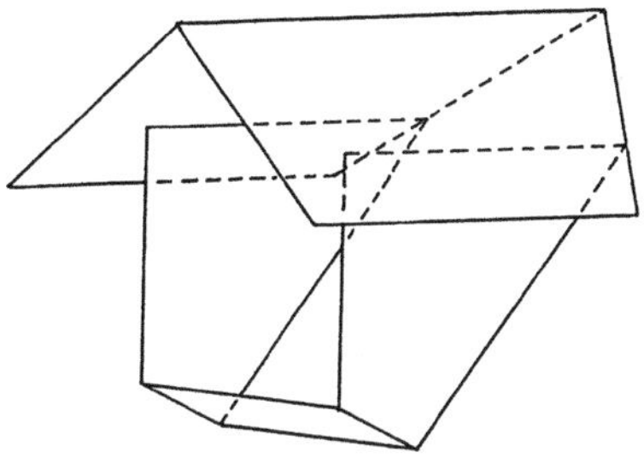

Fig. 1.3. Loops formed by planar segments in the house model

(7) *Manipulation of Complex Objects*
This is another focal point of requirements which is popularly advocated among CAD as well as database researchers. A close look at the related literature ([Adi], [Ath], [Bat], [Buc(1)], [Dem], [Has], [Lor(1)], [Lor(2)], [Rie], [Wei]), however, reveals that it actually involves several requirements of different natures. That is, first, a block of data large enough to correspond to an entire part, assembly or a set of drawings will have to be retrieved, transferred and locked with a single unit of manipulation. Second, provided record-based data modeling is applied, handling as a unity of pieces of data which are coherent but represented in a heterogeneous and variable set of records (or tuples) must be possible. Third, it must also be possible to regard a set of mutually related entities as a single entity, if this is desired. This third point differs significantly from the second despite the apparent similarity between them in that each constituent is in its essence an independent entity and does not represent a mere collection of attributes of some 'root' entity. (See also the discussion in the following paragraph.) This leads to the fourth requirement, that multiple views be supported to provide an external schema description capability, since it is up to each user or application to decide which way the related entities should be treated, as a set of individuals or as a single consolidated entity.

Often, entities constituting a complex object form a tree-like hierarchy at the top of which lies the root entity. Typical examples include:

- an assembly, sub-assemblies and parts,
- a room with its elements in the house model,
- an IC cell, sub-cells, pins, wires and primitive components,
- a polyhedron with the winged-edge structure, etc.

In case the hierarchy is indeed a tree into which references from outside are made only through the root, it is questionable whether subordinate entities should really be treated as independent entities. It might be more appropriate to consider them as representing an access cluster [Lor(2)]. One of the techniques for forging the structure of a complex object into a tree which is certainly much easier to manipulate than a general many-to-many relation is the use of 'link' entities containing all the necessary pieces of information to relate the 'parent' and 'child' entities. Only the link entities but not the (independent) child entities are included in the complex object. Another is actually copying referenced entities. Either technique has its own advantages and disadvantages, which we will not at this point discuss further. There is a particular reason, however, why the author brought up this issue here. That is, while it certainly involves serious storage and efficiency trade-offs, it goes beyond implementation concerns to the conceptual level, raising the problem of what we should consider as legitimate entities and how we should represent them. We will see later that our ultimate resort is the way we human beings recognize, conceive of and talk (in natural languages) about things and facts in the world of reality.

(8) *Support of Multiple Views*
We already noted above the possible choice between two views of collected pieces of data, that is, one as a single entity and the other as a structured set of multiple entities. Another familiar example is that between CSG and BREP representations of the same spatial object. The requirement for supporting multiple views is of course common to all the database applications and is often satisfied by providing a sub-schema description capability. Each application is given only those pieces of data that concern it and that have possibly undergone type and format conversion with unnecessary pieces suppressed. The case of CAD, however, is too subtle and complicated to accept such a solution. First, as we have seen of shapes, we seem to lack a single canonical schema from which all the views can derive. Second, objects may have to be viewed differently not only by different applications but by the same application. The issue here may rather be that of multiple representations. Third, different representations are guaranteed to be consistent if each of them is a correct representation, but they do not necessarily have to be equivalent or even comparable (that is, any one of them may not be able to be faithfully translated into any another). This reflects the fundamental truth that there can never arise logical contradictions so long we (correctly) talk about the the real, physical world, yet at the same time no finite combination of representation frameworks can ever fully describe it. This may as well be taken as a warning against an excessively radical semantic relativism [Bro].

(9) *Support of Multimedia Representations*
Multimedia databases are intended to represent pieces of information carried by diverse kinds of medium such as text, figure, visual image and voice in addition to the more usual data types (number, character string, record, etc.) already handled by most of the traditional database management systems and programming languages. These pieces of information may concern different aspects of a single object, or different objects which are independent but mutually related or constitute a more complex object as parts thereof. It is apparent that CAD databases are inherently multimedia oriented; explicit recognition of this fact and conscious efforts to establish data models to meet increasingly strong demand as well as to more systematically incorporate available technologies are the trends in recent years. In fact, it is the

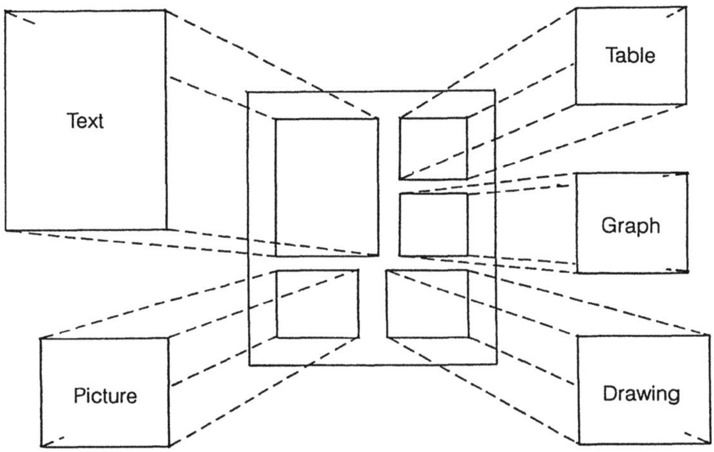

Fig. 1.4. Multimedia structure of engineering document

absolutely minimum capability of all the CAD databases that they support the system in dialoguing with users in graphical languages and in producing drawings with dimensions and annotations. In addition, many of them have functionally rich and multifaceted bodies of information constructed so that they can be used in carrying out various tasks involving diverse and unconventional media such as the following:

- presentation of moving images to simulate dynamic behavior of products being designed,
- rendering of spatial objects with colors, shades, reflections and other optical effects,
- creation of sequences of signals to cause the motions of cutters and workpiece holders of NC machine tools,
- measurement of fabricated parts and evaluation of the results by matching them against the data stored in the database.

There are other kinds of application environment than CAD, notably office information systems, for which multimedia databases have been under intensive study and are expected to play significant roles. Interesting methods and techniques are being developed that can be profitably utilized also for design and engineering applications. Suppose, for example, we want to produce an engineering document comprising pages, texts, figures and pictures. The same (object oriented) conceptual schema, physical presentation structure and user interface for interactive editing as used for the office or commercial type of document will apply (Fig. 1.4).

The only possible difference may come from the fact that in the case of engineering applications, there exist product models in the database containing all the necessary information (and much more than that) to create figures and drawings. Moreover, editing of drawings is not the primary task any more for most of the modern CAD systems that aim to support users in truly creative aspects of designing. Drawings per se tend to have significance only at the final presentation stage, and in some cases will be produced automatically from the product models with little or no human intervention. On the contrary, what are presented to and handled by people in the form of documents make the very substance of office information systems. We must therefore be careful not to miss the difference between CAD and engineering information (documents, more precisely) systems as to the proper level of abstractness of objects to be processed.

(10) *Semantics of Objects, Attributes and Functions*
Capturing the semantics of objects within a database is done by explicating human comprehension of them in a form interpretable to computers, specifically as statements in the conceptual schema or data dictionary/directory. It is one of the most important but also the most difficult and controversial issues in database theories. When trying to express the meaning of a thing or fact in a natural language, the usual procedure is:

- the word denoting the thing or fact in question is explicated in terms of other words and phrases which in turn are explicated in terms of yet other words and phrases,
- the process is repeated, spawning a multitude of words, some of which recur possibly with restricted meanings, and
- others are eventually transformed into primitive words that have intuitively such obvious meanings as to require no more explanation, or
- yet others are understood by appealing to examples, analogies, metaphors or metonymies.

There is of course nothing unusual or wrong about using definition by transformations and recursions in data modeling or any other fields of computer science provided the existence of fixed points is guaranteed. The problem is where to leave the transformative cycles and what to do with the residue. For example, the function of a cam mechanism is stated to be the transformation of a rotational movement into a linear movement, which may be quite sufficient for human programmers as well as mechanical engineers, and permits various terms, parameters and attribute values to be correctly interpreted for realizing the intended objectives. It is a long way from the semantics of these data being 'known' to the computer, however, since if we want to make a change to the kinematic characteristics of the tool end's movement, for example, we will have to have the program change depend entirely upon the programmer's knowledge of logical relationships among data items. Let us then state that a cam mechanism comprises an eccentric cam rotating at a constant speed around the center and a follower with the roller keeping contact with the cam contour, the center and the tool end (Fig. 1.5).

Furthermore, the relationship between the center and the two end points can be more explicitly defined so that the movement of one end may be

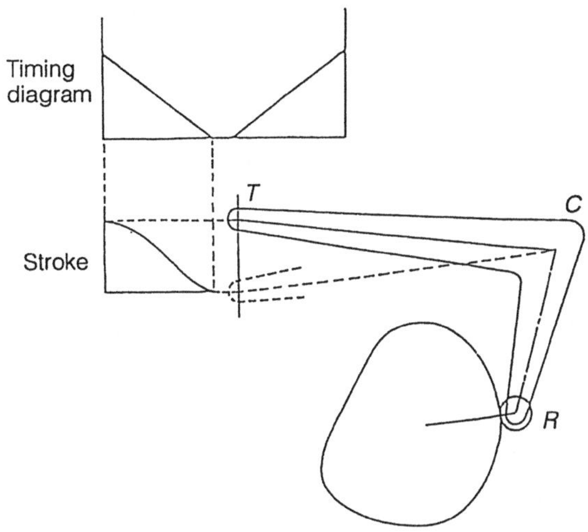

Fig. 1.5. A cam mechanism

described with respect to that of the other. That is,

$$lever(C, T, R) \tag{1.3}$$

$$\forall x(lever(C, T, x) \rightarrow (movmnt(x) = \frac{\overline{xC}}{\overline{TC}} movmnt(T))) \tag{1.4}$$

where $movmnt(x)$ denotes the function with the time or rotational angle as the independent variable and the displacement of point x from its original position as the value. We may go further and externalize the structure of this function or encapsulate it in the body of a procedure or an abstract data type. At any rate, while we may claim the meaning of a cam mechanism is now better conveyed to the computer than before, we must recognize that we stop explaining at some point where the meanings of things and facts are presumed to have been well established.

A great number of proposals have been made for 'semantic data modeling'. (See, e.g., [Bro], [Hul] for concise surveys.) One of the motivations behind the efforts has been to complement the syntax orientation of classical models, notably the relational model. We will not discuss them here but want only to indicate how the requirement to capture the semantics is understood by various authors in various ways. First, each object is decomposed into simple, elementary objects and reconstructed by joining them. This is considered necessary to identify and separate out each of the meaning elements which otherwise tend to be obscured by being mixed with each other in a large complex object. Better known techniques include the decompositions of relations into successively higher normal forms (3rd, 4th and projection/join). The most extreme case is the exclusive use of binary relations [Kow]. Second, new concepts and constructs are introduced with which more sophisticated structuring of databases is made possible. Typical examples are the distinction of entity and relationship [Che], hybridization of the CODASYL and relational models [Hay] and the special binary relations (*instance-of*, *is-a*, *part-of*, *member-of*) borrowed from work on semantic networks ([Brac], [Hul]). Structures play a fundamental role in restoring the semantics that has been lost in the 'flatness' of the relational model. Third, existing conceptual frameworks are selected and exploited as the bases for data modeling so that the syntactic formalities as well as semantic power of expression may be guaranteed to be attained to certain well-established extents. Limitations must of course be expected at the same time. These are mostly mathematical frameworks in a broad sense with logic, set theory and directed graphs being among the most widely used.

A final remark on the semantics is that in the case of CAD for discretely manufactured products, it seems to be intimately connected to the notion of functions which are often regarded as contrasting with attributes, hence consideration must be given to enabling databases to represent the functionality of design objects as well as their attribute values. We already touched on this issue in the above example of a cam mechanism.

1.2 Dynamic Aspect

(1) *Liberal (Unrestricted) Choice of Access Methods*
Objects which are given multiple representations, placed under multiple views and related to other objects through complex relationships permit a choice from a variety of methods to access them. Conversely, it must be possible for the application program to plan an access strategy that is the most natural and efficient for responding to the question at hand and to devise whatever set of attributes and relations may be necessary to carry it out. Of particular relevance to CAD is the problem of spatial search, that is, the retrieval of objects that reside in a specified region of space and/or satisfy a certain geometric condition. The difficulty inherent in this kind of problem is that the determination of precisely which objects are to be retrieved requires complicated calculations which can be executed only after the geometric attribute values of these objects are obtained. One way to avoid falling into this vicious circle is to plan for a two-step solution in which, first, a slightly larger set of objects than required is retrieved according to a simpler and coarser condition, and then, they are examined more closely with respect to the original condition. Let us see an example. Suppose we want to calculate and display the intersection of a plane with a free-form surface comprising, say, hundreds of patches. Suppose moreover that we want to do this for a number of planes that form, for example, a mutually parallel or coaxial, fan-like constellation (Fig. 1.6).

In the first step, to identify for each plane the patches that possibly intersect it, the use of a minimal quantity of information to indicate the approximate positions and extents of patches (e.g., the set of cubical or spherical

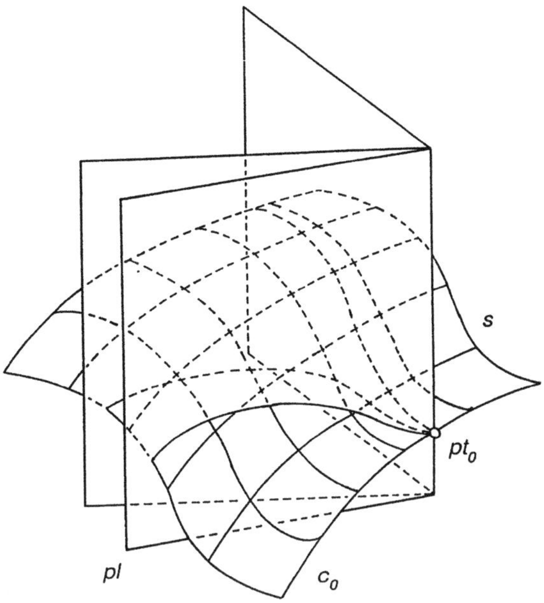

Fig. 1.6. Intersections of planes and surface patches

envelopes) will be appropriate. In the second step, we may exploit precise topological relationships such as adjacency and boundary among patches, curves and curve segments in order to obtain and process the detailed geometric attributes of them in the proper order. We should note that the operator has to give only a small number of parameters (e.g., the starting plane, linear or angular displacement between consecutive planes and number of planes) as compared to the complexity of the calculations involved, and tends to expect that the result will start to appear within a few seconds. We should also note that if he/she takes on the first step of the solution process manually, it will be a relatively easy task for him/her, as he/she is looking at the displayed pictures and the relative positions of pertinent geometric objects are immediately and intuitively apparent to his/her eyes. If the immediacy of human judgement is implanted in the retrieval condition, the efficiency of the retrieval process is expected to follow.

(2) *Dynamic Extension and Modification of Schema*
Unlike business databases for which schemas are relatively stable and if changes are needed, recompilations will usually suffice, CAD databases are exposed to possibilities of changes of views which may entail dynamic definitions of new attributes, relations, integrity constraints, retrieval conditions, etc. or modifications of existing ones. Even within the same application, we may find ourselves in a situation where an extension of the schema will meet additional requirements or contribute to improving efficiency. For example, in the above problem of taking intersections, experience might tell us that

designers are mostly interested in planes that are perpendicular to a particular plane. It will then be beneficial to augment the schema with the relations that facilitate 2-dimensional searches. Another motivation for extending the schema could come from the requirement for strengthening the constraints on the database so that discerning among design alternatives may become possible. This is one of the typical cases where updating the schema can be done without invalidating the current database contents or unduly affecting application programs.

(3) *Specification and Validation of Integrity Constraints*
An integrity constraint generally means a condition that the database must observe in order to be in a valid state. In the case of a typical business application, it is specified in terms of entity (record), attribute and relationship types. The database is normally in a valid state with all the constraints holding true, but may possibly be invalidated by executing a transaction. A transaction is a minimal sequence of primitive manipulations to cause a state transition, which usually lasts for a few seconds or less and will be made to back out if it actually falsifies any of the constraints, with all the effects of its updates being erased from the database.

CAD databases, however, demand entirely different interpretations and treatments of integrity constraints. First, they get satisfied piece by piece as the design proceeds and will become wholly true only at the end of the process. Second, it is possible and might be permissible that a designer's action falsifies some constraints while establishing some others. Third, a transaction tends to have an incomparably longer duration than that of a business application and cannot simply allow accomplishments to be nullified on which hours or days of creative and laborious efforts have been spent. Fourth, assertions have often to be made not only of type variables but of particular individuals. An example is the specification of an approximate size of the disc in a cam mechanism.

To examine whether the database state is kept valid is an expensive task even for business applications where the type of each transaction helps in distinguishing the points it may possibly fail to clear. For CAD databases, the normal state of which is 'invalid', it is already a serious problem to decide which constraints should be tested and when.

A probably more profound difference exists in the way integrity constraints are interpreted. First, those imposed on CAD databases more or less carry functional requirements or reflect designers' intentions. Under this interpretation, selecting the constraint to be established is an important decision in advancing the design process a step further. Giving an approximate size of a cam and demanding the precise dimensions to fit into it is an example. As another, let us consider the case where we stipulate that two circles *c1* and *c2* be coaxial. If the radius vectors $\vec{r}$ and the coordinate values (x, y, z) of the centers are predicated of them, these must be consistent with the coaxiality condition. Hence we have:

$$coaxial(c1, c2) \tag{1.5}$$

$$\begin{aligned}&\forall u \forall v (circle(u) \wedge circle(v) \wedge coaxial(u, v)\\ &\qquad \rightarrow (\vec{r}(u) \perp (x(u) - x(v), y(u) - y(v), z(u) - z(v))) \wedge\\ &\qquad\quad (\vec{r}(v) \perp (x(u) - x(v), y(u) - y(v), z(u) - z(v))))\end{aligned} \tag{1.6}$$

The latter constraint is due to the redundancy of relating two circles explicitly by means of the predicate *coaxial* as well as implicitly through a set of geometric attributes. This example thus shows the second of the characteristic interpretations of design integrity constraints, that is, their role in removing possible contradictions among multiple representations, the necessity of which has already been pointed out. Third, a more restricted form of constraint will permit some of the attributes and/or relations to be uniquely determined from others. For example, formulas (1.3) and (1.4) together with the movement of the tool end (T) allow the movement of the roller (R) to be automatically calculated. This leads to reinterpreting the constraints rather as deductive laws or axioms and integrity maintenance as logical inference.

(4) *Security and Protection*
These serve to guarantee that every part of the database be accessed by the right users (application programs) in the right ways at the right times. Two different issues are involved here. One is to limit access to particular parts to authorized users for various reasons including among others confidentiality, incompleteness and possible conflicts with the data privately owned by each user. The other is to prevent the database contents from being changed in such a way as to cause unexpected loss of information to any of its potential or actual users. While these may concern database applications of any sort, they are of particular relevance to CAD in view of such facts as the following:

- that the database undergoes an extraordinarily rapid and extensive growth during a relatively short period in the design process,
- that it is exposed to frequent changes and updates even after the design is completed and released to subsequent stages of development and manufacturing,
- that many and unspecified users engaged in the same project perform write and read operations against the same database, some of which must be in strictly ordered sequences while others may be done concurrently, and
- that the above takes place in an environment where the schema may dynamically change, where each transaction lasts for a very long time and where there may be users working on private spaces as well as the public area.

Maintaining the database in a secured and protected state with respect to the second issue is tantamount to enforcing a set of conditions that remove potential inconsistencies caused by illegal updates during concurrent transactions. These conditions also are often called integrity constraints, but are of an entirely different sort than those discussed in the preceding paragraph. Let us therefore call them 'protocols' for concurrency control [Banc] to maintain 'consistency' or a 'consistent state' of the database as opposed to 'integrity constraints' and 'validity' or a 'valid state'. Concurrency protocols require the 'serializability' of transactions, that is, that the results of their execution be the same as those obtained by a serial execution. This means that the following implication holds:

If (*integrity constraints observed by each serially executed transaction*) *and*
(*concurrency protocols observed*) *then*
(*integrity constraints observed by concurrently executed transactions*)

but not vice versa. In fact, the set of integrity constraints could be null in which case the first and third condition are trivially satisfied though concurrency protocols could still be violated. We might be able to say that integrity constraints concern the logical and static validity of the operand (database) while concurrency protocols enforce the executional and dynamic orderliness of the operators (transactions).

(5) *Backtracking on the Design Path*
The inherent indeterminacy of the design process and the resultant possibility of backtracking has already been indicated (Sect. 0.5). Because design errors may be found long after the moments when they are actually made, it must be possible to backtrack arbitrarily far from any point to any previous point in the design process. Precautions must be taken in addition not to have the results of other transactions that have been running concurrently, if any exist, overly or unnecessarily affected. It should be noted that the falsification of integrity constraints as such does not necessarily signify an error, as a designer's actions may quite normally trade in certain previously established constraints for certain others. Backtrackings may either be required by the system when, for example, it encounters unrecoverable failures or recognizes the existence of logical contradictions among integrity constraints, or be voluntarily initiated by designers when they find anything undesirable with the work so far completed and decide they have opted for wrong alternatives.

(6) *Arbitrarily Long Transactions*
Let us first restate the definition and fundamental properties of the notion of transaction in a more formal manner which we touched on in the above discussions of integrity and consistency maintenance. That is:

- A transaction is a partially ordered set of primitive database operations (that is, read, write, lock, unlock, etc.) with unique minimum (initial) and maximum (terminal) elements (which could be null operations) where the order indicates the sequence of executions (Fig. 1.7). Operations neither preceded nor succeeded by each other can be executed in parallel.

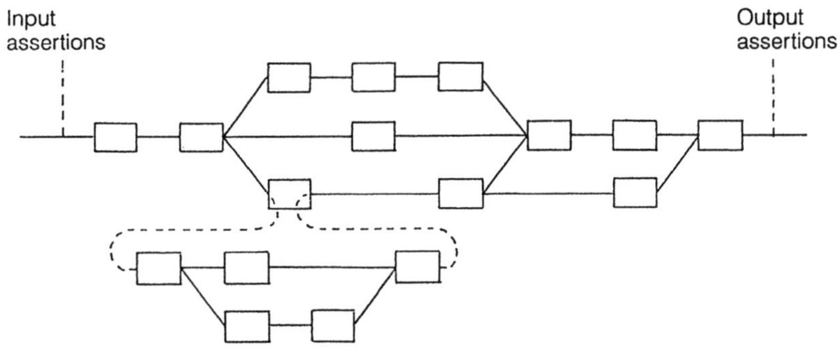

Fig. 1.7. A partially ordered transaction

- A set of assertions on the input and another on the output data are assigned to the entry and exit points respectively of each transaction. The input assertions must have been established prior to the intial operation, otherwise the transaction will not start. Its success is defined as the output assertions' holding true at the end of execution of the terminal operation.
- Once started, a transaction must either succeed or fail inevitably as a whole. This means, first, that the transaction is judged to have failed if the exit is not reached or the output assertions do not hold, and second, that a failure forces it to back out and the database to return to the state at the entry erasing every bit of the effect of its execution. This characteristic of the transaction concept is referred to as 'atomicity' [Gray].
- A set of protocols is stipulated for all concurrent transactions to observe so that potential database inconsistencies resulting from illegal interactions and interleaving among them may be avoided. One of the popular techniques used in devising such protocols is locking and unlocking portions of the database to the effect that crucial competitions among concurrent transactions are resolved in forced (and hopefully correct) serialization of primitive operations.

The above applies to any sort of database applications. What distinguishes CAD from others, especially business data processing, is the extraordinarily long duration of a transaction and the resultant particularities of more qualitative nature such as the following:

- A transaction can neither simply back out abandoning the work completed up to the point of failure, nor arbitrarily lock the part of database it may possibly access, the size of which could be very large in proportion to the duration of lockup.
- A rather straightforward solution is to take a copy of the pertinent part of the database and place it in a private space on which one transaction can work in an exclusive mode while the common space remains accessible to all transactions.
- The computational complexity and sophistication of a transaction substantially increases the need for subcontracts with others to execute more

or less specialized and self-contained functions (e.g., structural analysis, calculation of mass properties, rendering of surfaces on displays, etc.). These transactions will often be organized into a hierarchical structure (nested transactions), run in parallel to one another and share workspaces or semi-public spaces for exchange of data.

- The totality of data to be processed by a transaction is considered to form a complex object (Sect. 1.1 (7)) and should be able to be treated as such. The determination of its extension, however, is not as straightforward as by means of a root entity and repeated applications of hierarchical references. A designer might want to examine, for example, a selected portion of a drawing and make additions and modifications in it, which requires a spatial search of objects together with accessing through explicit references.

(7) *Replication of Arbitrary Portions of Database*
It is one of the common practices of designers working with CAD to make replications of portions of the database and put them to various tentative or permanent uses. The case of moving a complex object to a private workspace was mentioned above. The result of the work may be one of the alternate solutions for a single design task or serve as a new version of a previously released part. Or perhaps designing an entirely new part or product exploiting an existing one as a prototype is intended, which will require the restoration of pertinent data in the archival store back into the active database. Yet another case is the placement of instances of a generic part at multiple locations possibly with different transformations and parameter values.

It may be questioned whether and why the above is unique to CAD and does not apply equally to other kinds of applications. Is it not in fact quite usual that when updating a master file, many records and data items are simply copied from the old to the new version? The answer will be that the business data are copied for the purposes of safety, convenience and efficiency, with copies and the originals always denoting the same entities, whereas in CAD they could represent different and independent objects. Copied invoices which may not be arbitrarily altered can be compared to the different occurences of a numeral on a sheet of paper which carry the unchangeable meaning of a number as an abstract entity. Physical separation alone decisively separates the meaning of design objects, however, which are intrinsically physical existences.

(8) *Private Workspaces*
The use of private spaces was pointed out in connection with long, concurrent transactions. Some of these spaces are not purely private but rather semi-public, shared between 'clients' and 'subcontractors' ([Banc], [Kim]). Transactions and workspaces thus interrelated will form a hierarchy-like structure as depicted in Fig. 1.8.

The need for private spaces is not limited to a multiple user or parallel processing environment, however. A more basic use will be for working on alternate plans for the product currently being designed or versions for one

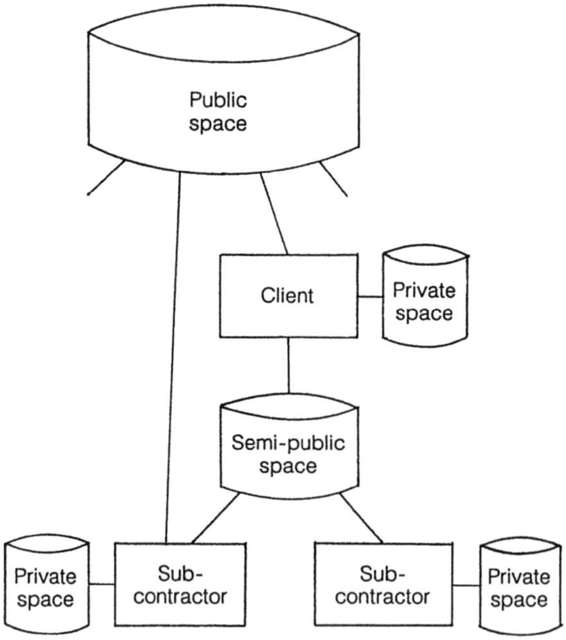

Fig. 1.8. Public, semi-public and private spaces

previously completed and released. It is obvious that the database must facilitate the coexistence of alternatives and their comparative evaluation before a permanent change is made to the contents of the public space by incorporating the final selection from them. It must also facilitate the dynamic allocation, expansion and disallocation of private spaces as the needs arise and diminish.

(9) *Support of Distributed Databases*
Distributing data to multiple stations and putting them under more or less independent managements is common to many applications. While it appears to be primarily a matter of physical economy and efficiency, serious conceptual issues will be raised concerning identity and consistency if data may overlap and creation and updating are allowed to take place at different stations. A frequent and characteristic example in CAD is for each local (graphics) computer to retain a separate database during the active period of interaction and to up-load it to the host computer at appropriate timings to be integrated into the central database. The purpose is of course to shorten the response time by reducing the host-local communications and accesses to the central database. It will then be desirable to allow the local database to define a certain limited class of objects, e.g., simple geometric objects, as well as to handle attentions and retrieve stored data. Even when the capability is restricted to the latter, correlation of objects which are separately stored but identical must be devised and is not a trivial task. This is mainly

because the local database is structured in line with the graphic representations of objects and their manipulations in an interactive environment, hence its schema is more in the specialized, external realm than in the abstract, conceptual realm. What is needed here is a bidirectional mapping between two schemas.

1.3 Version Control

(1) *Support of Alternate Designs*
It will be the case with almost every design activity that alternate solutions for one task are searched, proposed and evaluated before the final decision is reached. To facilitate simulating this practice on a CAD system will require several capabilities of the database manager, some of which have already been discussed in connection with other facilities. First, the portion of a design for which alternate solutions are to be created must be clearly identified and distinguished from remaining portions. A natural and convenient way to make such an identification is to find the minimum or maximum collection of complex objects containing or contained in the designated spatial region respectively. Second, that portion will be checked out (copied) from the public or semi-public to a private workspace for exclusive use during the redesigning. Third, its interface with other portions must be kept intact and a warning must be issued when violations are about to be made. This is necessary so that changes may not propagate beyond the boundaries. Fourth, by the same token, alternatives for the same design objects will be mutually related, say by the 'equivalence' relation, enabling designers to switch back and forth easily between them. Finally, the selected alternative is checked in to the public or semi-public workspace and integrated into the entire design object after it goes through validation of integrity constraints. It shoud be noted that the validation can be restricted to the internal structure of the current target of redesigning and is made much simpler than otherwise by virtue of the above-mentioned stipulation on the interface with the outside.

Usually, all or most of the alternatives except the final choice will be discarded at the end of design activity, which is typically marked by the release of the result to succeeding stages. This could be accomplished by the disallocation of the associated semi-public and/or private workspaces.

(2) *Control of Versions and Releases*
The creation of technical contents of a version is done in much the same way as that of alternate plans in the original design. Additional features unique to versioning must be taken into consideration, however, in order to properly control the release of each version and the series of releases. First, unlike design alternatives, any portions of versions once released are not allowed to be deleted or modified, as they are usually already in the hands of downstream processes of design or manufacturing, or even of customers. Second,

a single release may contain versions for more than one part or portion of a product, hence coherence or consistency among them must be checked out before it is finally approved. This set of versions constituting an authorized release is sometimes called a 'configuration' ([Kat(1)], [Kat(2)], [McL]). Third, the enforcement of much more formal and strict regulations is mandatory for authorizing a release than for creating a design alternative since related work in more numerous and diverse organizations could be affected. Notifications must be sent, possibly with different levels of detail depending on the roles and responsibilities of the recipients. For example, designers of related parts or products should be informed of the changes of interface, manufacturers should have the complete and most detailed set of data on the parts in question, while field organizations will need to know which versions are to be replaced by new ones as well as the updated maintenance procedures. Fourth, it is important to permanently record the genealogy of versions and configurations, that is, which have derived from which, in order to manage development, manufacturing and maintenance of products over the entire life cycle during which different versions of the same parts and products coexist in plants, warehouses and marketplace. Last, the availability of archival storage together with pertinent utilities for storing and restoring the complete design data to and from it is crucial. When restored, the data must appear precisely the same to designers as when it was being actively worked on. Though seemingly trivial, this is really not so since both the logical and physical environments in which the data is placed may be quite different, entailing its transformation and adjustment to keep its essential invariants. (An example of such adjustable factors is the system-generated identifier or 'surrogate' of an entity, the sole purpose of which is to distinguish an entity from others in any given environment.)

(3) *Maintenance of the History of Changes*
The need to keep a chronological record of design changes is twofold. One is to provide a common and absolute measure of ordering for set of versions of different parts which are created and maintained more or less independently of each other. Versions may be partitioned into 'layers' according to their date of creation, which serves as the basis for forming configurations [Kat(1)], [Kat(2)]. The other is to implement a simplified genealogy of versions, the true structure of which is quite complicated as can be seen from Fig. 1.9. Each part has its own subset of versions which is structured along the horizontal (alternatives) and vertical (derivatives) axes, and there exists a multitude of such substructures.

(4) *Archive for Past Designs*
Storage of finished designs in a permanent form is increasingly important, as CAD databases tend to contain far richer semantics than mere computerized drawings. In fact, in the extreme case where the complete set of drawings is produced automatically without any human intervention, the database is virtually the sole representation of the design. The use of an archival store is multifold. First, it will contain all the successive revisions as well as the

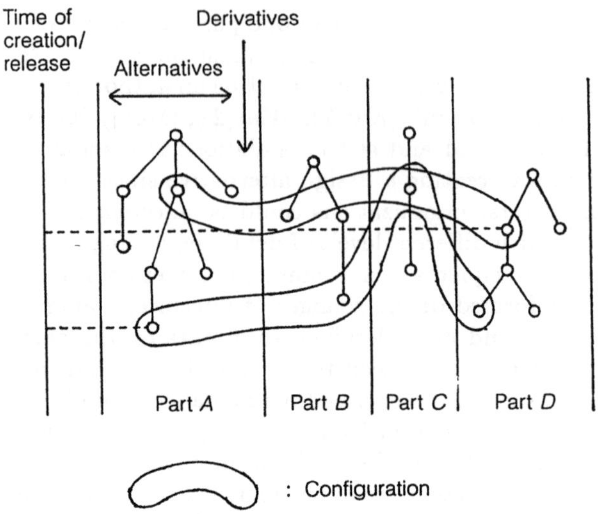

Fig. 1.9. The genealogy of versions with associated chronology

original release and serve as the ultimate recourse for design specifications and implementation details throughout the life of the product. This point was already discussed above. Second, past designs will often give good first approximations to new ones of similar kinds. To turn potential benefits into reality in practical environments, however, some keys or clues to effectively narrow down the search space must be devised. Here, another possibility exists to exploit experts' empirical knowledge to make the work of designing more rational. Third, the same pieces of information as in the archival storage which result from a stage of design will be transferred to downstream stages for further detailing and augmentation. For this purpose, the establishment of protocols and adherence to them at related stages are mandatory. Fourth, it is often desired to save the database contents in an off-line file even in the midst of a design process, usually as a precaution against possible accidental loss of vital information. There is no reason why the procedures for handling the save file and its format should differ from those for a permanent archive.

This last point indicates the requirement that we should provide tools and utilities with functions that are independent of specific purposes for which the information in the archive is used. It further implies that the fundamental structures to be kept invariant through transfers and transformations should be self-described (in a schema description level language) and visible from outside and also that room should be left for application-specific aspects to be appropriately treated.

References

[Adi] Adiba, M., Nguyen, G.T.: Information Processing for CAD/VLSI on a Generalized Data Management System. Proc. 10th VLDB, 1984

[Arm] Armitage, S., Hall, P.A.V.: Conceptual Schema for CAD. Computer Aided Design, Vol. 9, No. 3, July 1977

[Ath] Athay, R.J.: Object Models for Computer Aided Design: An Overview. SIGGRAPH Computer Graphics, Aug. 1978

[Atk] Atkinson, M., Wiseman, N.: Data Management Requirements for Large Scale Design and Production. SIGDA Newsletter, ACM, March 1977

[Bae] Baer, A. et al.: Geometric Modelling: A Survey. Computer Aided Design, Vol. 11, No. 5, Sept. 1979

[Banc] Bancilhon, F. et al.: A Model of CAD Transactions. Proc. 11th VLDB, 1985

[Band(1)] Bandurski, A.F., Jefferson, D.K.: Data Description for Computer Aided Design, Proc. ACM SIGMOD Workshop, 1975

[Band(2)] Bandurski, A.F., Jefferson, D.K.: Enhancements to the DBTG Model for Computer Aided Ship Design. Proc. Workshop on Databases for Interactive Design, 1975

[Bar] Baron, N. et al.: An Approach to the Integration of Geometrical Capabilitities into a Data Base for CAD Applications. In: [Enc(2)]

[Bat] Batory, D.S., Buchmann, A.P.: Molecular Objects, Abstract Data Types, and Data Models: A Framework. Proc. 10th VLDB, 1984

[Bau] Baumgart, B.G.: A Polyhedron Representation for Computer Vision. Proc. NCC, 1975

[Bel] Bell, J.: Data Modelling of Scientific Simulation Programs. Proc. ACM SIGMOD Conf., June 1982

[Bena] Benayoune, M., Preece, P.E.: Methodology for the Design of Databases for Engineering Applications. Computer Aided Design, Vol. 18, No. 5, June 1986

[Benn] Bennett, J.: A Database Management System for Design Engineers. Proc. 19th Design Automation Conf., 1982

[Bha] Bhateja, R., Katz, R.H.: VALKYRIE: A Validation Subsystem of a Version Server for Computer-Aided Design Data. Proc. 24th Design Automation Conf., 1987

[Bo] Bo, K.: Data Base Design. Lecture Notes in Computer Science 89, Springer-Verlag, 1980

[Brac] Brachman, R.J., Schmolze, J.G.: An Overview of the KL-ONE Knowledge Representation System. Cognitive Science 9, 1985

[Brai(1)] Braid, I.C., Hillyard, R.C.: Geometric Modelling in ALGOL 68. ACM SIGPLAN Notices, Vol. 12, No. 6, May 1977

[Brai(2)] Braid, I.C.: On Storing and Changing Shape Information. SIGGRAPH Computer Graphics, Aug. 1978

[Bro] Brodie, M.J.: On the Development of Data Models. In: Brodie, M.J. et al. (eds.): On Conceptual Modelling. Springer-Verlag, 1984

[Buc(1)] Buchman, A.P.: Current Trend in CAD Databases. Computer Aided Design, Vol. 16, No. 3, May 1984

[Buc(2)] Buchmann, A.P.: An Architecture and Data Model for CAD Databases. Proc. 12th VLDB, 1986

[Cav] Cavagna, C., Cugini, U.: Data-Structure for the Description and Handling of Engineering Drawing. Computer Aided Design, Vol. 9, No. 1, Jan. 1977

[Cha] Challis, M.F.: Typing in Data Base Models. In: [Enc(2)].

[Che] Chen, P.P.: The Entity-Relationship Model – Toward a Unified View of Data. ACM TODS, Vol. 1, No. 1, March 1976

[Cho] Chou, H.-T., Kim, W.: A Unifying Framework for Version Control in a CAD Environment. Proc. 12th VLDB, 1986

[Dai] Daini, O.A.: Numerical Database Management System: A Model. Proc. Int'l Conf. on Management of Data, June 1982

[Daß] Dassler, R. et al.: Databases for Geometric Modelling and Their Application. In: [Enc(2)]

[Dem] Demolombe, R.: STREL: An Extended Relational Model for Structured Objects. Information Processing, 1986

[Eas(1)] Eastman, C.M., et al.: A Database for Designing Large Physical Systems. Proc. NCC, 1975

[Eas(2)] Eastman, C.M.: The Concise Structuring of Geometric Data for Computer Aided Design. In: Data Structures, Computer Graphics and Pattern Recognition. Academic Press, 1977

[Eas(3)] Eastman, C.M.: Systems Facilities for CAD Databases. Proc. 17th Design Automation Conf., 1980

[Eas(4)] Eastman, C.M.: Recent Developments in Representation in the Science of Design. Proc. 18th Design Automation Conf., 1981

[Eas(5)] Eastman, C.M., Preiss, K.: A Review of Solid Shape Modelling Based on Integrity Verification. Computer Aided Design, Vol. 16, No. 2, March 1984

[Enc(1)] Encarnação, J., Neumann, T.: A Survey of DB Requirements for Graphical Applications in Engineering. Lecture Notes in Computer Science 81, Springer-Verlag, 1980

[Enc(2)] Encarnação, J., Krause, F.-L. (eds.): File Structures and Databases for CAD. North-Holland, 1982

[Foi(1)] Foisseau, J. et al.: On CAD Data Modelling Through Mechanical Engineering Design. Proc. Int'l Conf. Interactive Techniques in CAD, Bologna, 1978

[Foi(2)] Foisseau, J., Vallette, F.R.: A Computer Aided Design Data Model: FLOREAL. In: [Enc(2)]

[Grab(1)] Grabowski, H., Eigner, M.: Employing A Relational Data Structure in a CAD System. Proc. Int'l Conf. on Interactive Techniques in Computer Aided Design, Bologna, 1978

[Grab(2)] Grabowski, H., Eigner, M.: Semantic Datamodel Requirements and Realization with a Relational Datastructure. Computer Aided Design, Vol. 11, No. 3, May 1979

[Grab(3)] Grabowski, H., Eigner, M.: A Data Model for a Design Data Base. In: [Enc(2)]

[Gray] Gray, J.: The Transaction Concept: Virtues and Limitations. Proc. 7th VLDB, 1981

[Hal] Hall, P.A.V.: Requirements of Databases for Design. Proc. Conf. on Database Technology, 1976

[Har] Hardwick, M.: Extending the Relational Database Data Model for Design Applications. Proc. 21st Design Automation Conf., 1984

[Has] Haskin, R.L., Lorie, R.A.: On Extending the Functions of a Relational Database System. Proc. ACM SIGMOD Conf., 1982

[Hay] Haynie, M.N.: The Relational/Network Hybrid Data Model for Design Automation Databases. Proc. 18th Design Automation Conf., 1981

[Hei] Heiler, S. et al.: An Object-oriented Approach to Data Management: Why Design Databases Need It. Proc. 24th Design Automation Conf., 1987

[Hol] Hollaar, L. et al.: The Structure and Operation of a Relational Database System in a Cell-oriented Integrated Circuit Design System. Proc. 21st Design Automation Conf., 1984

[Hul] Hull, R., King, R.: Semantic Database Modeling: Survey, Applicatinos, and Research Issues. ACM Computing Surveys, Vol. 19, No. 3, Sept. 1987

[IEEE] Proc. Data Base Week: Engineering Design Applications. IEEE, 1983

[Joh] Johnson, H.R. et al.: A DBMS Facility for Handling Structured Engineering Entities. In: [IEEE]

[Jul] Jullien, C., Leblond, A.: A Database Interface for an Integrated CAD System. Proc. 23rd Design Automation Conf., 1986

[Kal(1)] Kalay, Y.E.: A Relational Database for Non-manipulative Representation of Solid Objects. Computer Aided Design, Vol. 15, No. 3, Sept. 1983

[Kal(2)] Kalay, Y.E.: A Database Management Approach to CAD/CAM Systems Integration. Proc. 22nd Design Automation Conf., 1985

[Katz(1)] Katz, R.H.: A Database Approach for Managing VLSI Design Data. Proc. 19th Design Automation Conf., 1982

[Katz(2)] Katz, R.H. et al.: A Version Server for Computer-Aided Design Data. Proc. 23rd Design Automation Conf., 1986

[Kim] Kim, W. et al.: A Transaction Mechanism for Engineering Design Databases. Proc. 10th VLDB, 1984

[Kore] Korenjak, A.J., Teger, A.H.: An Integrated CAD Data Base System. Proc. 12th Design Automation Conf., 1975

[Kori] Koriba, M.: Database Systems: Their Applications to CAD Software Design. Computer Aided Design, Vol. 15, No. 5, Sept. 1983

[Kow] Kowalski, R.: Logic for Data Description. In: Gallaire, H., Minker, J. (eds.): Logic and Data Bases. Plenum Press, 1976

[Kut] Kutay, A.R., Eastman, C.M.: Transaction Management in Engineering Databases. In: [IEEE]

[Lac] Lacroix, M., Pirotte, A.: Data Structures for CAD Object Descriptions. Proc., 18th Design Automation Conf., 1981

[Laf] Lafue, G.M.E.: Integrating Language and Database for CAD Applications. Computer Aided Design, Vol. 11, No. 3, May 1979

[Lee(1)] Lee, Y.C., Fu, K.S.: Integration of Solid Modeling and Database Management. Proc. 20th Design Automation Conf., 1983

[Lee(2)] Lee, Y.C., Fu, K.S.: A CSG Based DBMS for CAD/CAM and Its Supporting Query Language. In: [IEEE]

[Lin] Linden, C.A.: Grammars Which Describe Large Bodies of Data. Computer Aided Design, Vol. 10, No. 1, Jan. 1978

[Lor(1)] Lorie, R., Plouffe, W.: Complex Objects and Their Use in Design Transactions. In: [IEEE]

[Lor(2)] Lorie, R.A., Daudenarde, J.-J.P.: On Extending the Realm of Application of Relational Systems. Information Processing 86, 1986

[Mal] Mallgren, W.R.: Formal Specification of Interactive Graphics Programming Languages. MIT Press, 1982

[McL] McLeod, D. et al.: An Approach to Information Management for CAD/VLSI Applications. In: [IEEE]

[Mei(1)] Meier, A., Lorie, R.A.: A Surrogate Concept for Engineering Databases. Proc. 9th VLDB Conf., 1983

[Mei(2)] Meier, A.: Applying Relational Database Techniques to Solid Modelling. Computer Aided Design, Vol. 18, No. 6, July 1986

[Neu(1)] Neumann, T.: CAD Data Base Requirements and Architectures. Lecture Notes in Computer Science 89, Springer-Verlag, 1980

[Neu(2)] Neumann, T., Hornung, C.: Consistency and Transaction in CAD Database. Proc. 8th VLDB, 1982

[Neu(3)] Neumann, T.: On Representing the Design Information in a Common Database. In: [IEEE]

[New] Newell, M.E., Evans, D.C.: Modelling by Computer. In: Allan, J.J. (ed.): CAD Systems. North-Holland, 1977

[Phi] Phillips, R.J. et al.: AESOP: An Architectural Relational Database. Computer Aided Design, Vol. 11, No. 4, July 1979

[Req(1)] Requicha, A.A.G.: Representations of Rigid Solid Objects. Lecture Notes in Computer Science 89, Springer-Verlag, 1980

[Req(2)] Requicha, A.A.G.: Representations for Rigid Solids: Theory, Methods, and Systems. ACM Computing Surveys, Vol. 12, No. 4, Dec. 1980

[Rie] Rieu, D., Nguyen, G.T.: Semantics of CAD Objects for Generalized Databases. Proc. 23rd Design Automation Conf., 1986

[Rob] Roberts, K.A. et al.: A Vertically Organized Computer-Aided Design Data Base. Proc., 18th Design Automation Conf., 1981

[Row] Rowe, L.A., Schoens, K.A.: Data Abstraction, Views and Updates in RIGEL. Proc. ACM SIGMOD Conf., 1979

[San] Sanborn, J.L.: Evolution of the Engineering Design System Data Base. Proc. 19th Design Automation Conf., 1982

[Schm] Schmidt, J.W.: Type Concepts for Database Definition. Proc. Int'l Conf. on Data Bases, Haifa, Israel, Aug. 1978

[Schu] Schutzman, H.B.: ICHABOD A Data Base Manager for Design Automation Applications. Proc. 22nd Design Automation Conf., 1985

[She] Shenoy, R.S., Patnaik, L.M.: Data Definition and Manipulation Languages for a CAD Database. Computer Aided Design, Vol. 15, No.3, May 1983

[Sid] Sidle, T.W.: Weakness of Commercial Data Base Management Systems in Engineering. Proc. 17th Design Automation Workshop, 1980

[Sim] Simon, R.L.: CAD/CAM – The Foundation for Computer Integrated Manufacturing. Proc. 20th Design Automation Conf., 1983

[Spa] Sparr, T.M.: A Language for a Scientific and Engineering Data-Base System. Proc. 19th Design Automation Conf., 1982

[Sta] Staley, S.M., Anderson, D.C.: Functional Specification for CAD Databases. Computer Aided Design, Vol. 18, No. 3, April 1986

[Sto(1)] Stonebraker, M.: Adding Semantic Knowledge to a Relational Database System. In: Brodie, M.J. et al. (eds.): On Conceptual Modelling. Springer-Verlag, 1984

[Sto(2)] Stonebraker, M. et al.: Application of Abstract Data Types and Abstract Indices to CAD Databases. In: [IEEE]

[Ulf] Ulfsby, S. et al.: TORNADO: A DBMS for CAD/CAM Systems. Computer Aided Design, Vol. 13, No. 4, July 1981

[Val] Valle, G.: Relational Data Handling Techniques in Computer Aided Design Procedures, Allan, J.J. (ed.): CAD Systems. North-Holland, 1977

[War] Warn, D.R.: VDAM – A Virtual Data Access Manager for Computer Aided Design. Proc. Workshop on Databases for Interactive Design, ACM, 1975

[Wei] Weiss, S. et al.: DOSS: A Storage System for Design Data. Proc. 23rd Design Automation Conf., 1986

[Wel] Weller, D., Williams, R.: Graphic and Relational Data Base Support for Problem Solving. Computer Graphics, Vol. 10, 1976

[Wil(1)] Williams, R.: On the Application of Relational Data Structures in Computer Graphics. Information Processing 74, North-Holland, 1974

[Wil(2)] Williams, R. et al.: Data Structures in Computer Graphics. Proc. Workshop on Database for Interactive Design, 1975

[Wil(3)] Williams, R., Giddings, G.: A Picture-Building System. IEEE Trans. Software Eng., Vol. SE-2, No. 1, March 1976

[Zar] Zara, R.V., Henke, D.R.: Building A Layered Database For Design Automation. Proc. 22nd Design Automation Conf., 1985

[Zha] Zhang, J., Wang, R.: Some Consideration on the Data Model of Geometric Databases. Proc. 21st Design Automation Conf., 1984

[Zin] Zintl, G.: A CODASYL CAD Data Base System. Proc. 18th Design Automation Conf., 1981

Chapter 2
Criticism of Past and Current Data Models

The purpose of this chapter is to evaluate selected representatives from a variety of past and current methods for constructing CAD databases. The evaluation will be made in the light of requirements presented in the preceding chapter, particularly those which concern modeling capabilities (Sects. 1.1 and 1.2). If it is permissible to state the conclusion beforehand, we will find that each of these methods has one or more serious shortcomings which causes it to fail to meet some of the requirements.

It is well known that one of the sources of CAD databases is in the pioneering efforts to develop graphics packages in the late 1960s. This is evidenced, for example, by the surveys by J. C. Gray [Gray] and R. Williams [Wil(1)] with their titles referring to CAD and graphics data structures respectively, yet with largely overlapping subject matter. We tend to look upon what these efforts produced as mostly outdated, and revisit them as having sense, if any, only from a historical viewpoint. This is not really so, however. For one thing, they have produced viable technologies on which many of the later developments are essentially based. For another, through critical analyses of them, we may recognize more clearly the comparative merits of modern approaches.

Another source is the application of data modeling techniques intended primarily for business transaction processing. There are three classical achievements in this field, namely, the hierarchical, network (CODASYL) and relational data models. (See [ACM] for a comprehensive survey.) As the hierarchical model is considered to be a subset of the CODASYL model, we may reasonably concentrate on the remaining two. There have been a good many undertakings to expand the original models so that more of the meaning of the real world entities may be captured, resulting in what are grossly called 'semantic models'. Our analyses will cover a few typical examples of these.

We will leave several important topics over for later discussions, including the concept of semantic network and its integration with database technologies, deductive data models and object-oriented approaches. These, together with related methods in a logical vein such as logic programming and abstract data typing, will be dealt with in Chap. 4.

There have been countlessly many ad hoc developments of databases for individual CAD applications. The sheer volume and diversity of them allows us to take only a cursory glance at these efforts. This is justifiable, however, because first, our primary interest is in abstract and generic frameworks rather than specific instances, and second, a close examination of any given ad hoc database usually reveals that what remains after depriving its schema of domain specificities fits well into one or more of the general purpose data models.

2.1 Graphics Packages in the 1960s

The traits of early CAD databases are, among others, that

- graphic (geometric) entities are given the utmost importance,
- these entities are almost identified with design objects themselves, and that
- in many cases, they are organized into a set of vertically (hierarchically) and horizontally interrelated rings.

While it might well be considered as a vintage technology today, the ring structure was once esteemed so highly that it was expected to be the 'canonical' data structure for CAD [Gray] and provide the basis for most databases in the not too distant future [CAD]. This was probably because the 'topology' of graphic entities, such as their boundary and adjacency relations seemed to be naturally manifested in a set of rings.

We select ASP (Associative Structure Package) as the representative of ring-based packages because it is one of the most general and widely known of its class. That it is of relatively low level and offers only primitive operations contributes to revealing the peculiarity as well as the versatility of the ring structure.

Of methods other than ring structure, LEAP (an associative language, [Fel]) and the set-theoretic data structure [Chi] are noteworthy. The value of these as well as of ASP lies in their methodological distinction, general applicability and considerable degree of completion.

2.1.1 Ring Structure, Typically ASP

(1) *Structure*
The vocabulary and grammar of ASP [Lan] has been outlined (Sect. 0.6). There are three primary categories of entities, that is, ***element***, ***associator*** and ***ringstart***. Each entity is given a unique identifier when created. In addition, an element may contain an arbitrary number of arbitrary kinds of data items, an associator may contain one data item called ***association***, and a ringstart contains its *name* and *type*.

The lower ring of ringstarts of an element together with all the associative rings headed by these ringstarts is called the dependent structure of that element. We may well consider the set of elements as hierarchically organized through dependency relations. Starting at an element or a ringstart, all the lower level entities can be reached by following rings and moving down hierarchies. Conversely, every entity except the topmost one must be able to be reached in this way. An entire ASP structure is placed under one of two exceptional entities, the *universe element* and the *universe ringstart*. Isolated fragments of structure which are inaccessible are not allowed to exist. Consequently, the deletion of an element entails the deletion of its dependent structure. Also, an element will be deleted if all the associators on its upper ring are deleted.

(2) *Basic Functions*
Basic ASP functions fall into four categories as follows:

- creation of elements, associators and ringstarts, and placing these on appropriate rings,
- deletion of elements, associators and ringstarts,
- retrieval of the next entity on a ring, and
- storage and retrieval of data, associators, names and types into and from entities.

These are provided in the form of functions and procedures callable from application programs written in a suitable host language (e.g., FORTRAN). Let us list only those functions as examples which will be used in a sample program shown below:

lrf($s|e$) : get the ringstart following s (ringstart) or e (element) on a lower ring,
arf($a|s$) : get the associator following a (associator) or s on an associative ring,
ael(a) : get the element on whose upper ring a lies,
snam(s) : get the name of s.

(3) *Sample Program*
Suppose the structure representing the winged-edge topology of a tetrahedron has been already created (as in Fig. 0.6) and the problem is to find the face (?f) standing opposite to a given face ($f1$) with respect to a given common boundary ($e1$). A program for this will be as follows:

```
begin
  rs := lrf(e1); rsname := snam(rs);
  ⟨The identifier and name of the first ringstart on the lower ring
  of e1 are obtained and stored in rs and rsname respectively.⟩
  while rsname ≠ ef do
  begin
    rs := lrf(rs); rsname := snam(rs)
  end;
  ⟨The lower ring of e1 is searched for the ringstart with the name ef.⟩
  as := arf(rs); ?f := ael(as);
```

⟨The identifier of the first associator on the associative ring of ef and that of the element (face) whose upper ring contains this associator are obtained.⟩
if $f = f1$ **then** $f :=$ **ael**(**arf**(as))
⟨If the face just obtained is the initially given $f1$, then the face in question can be reached via the next (and the other) associator on the same ring.⟩
end.

(4) *Criticism*
(4.1) *Functional Limitations*
First and foremost, the undue complication of the above solution for a rather simple problem should be noted. This is usually inevitable with ASP or any ring-based package in which only binary predicates exist to relate elements and in which moreover, elements related to each other can be accessed only through the dependent structure of either one or the other. If we are allowed to exploit predicates to access any of the arguments directly, we may be able to write

$$\textbf{evaluate}\ ef(e1, ?f) \wedge (?f \neq f1) \tag{2.1}$$

where the prefix '?' indicates a free variable, or even more simply

$$\textbf{evaluate}\ fef(f1, e1, ?f) \tag{2.2}$$

assuming further that ternary predicates can be defined.

(4.2) *An Inverse Problem*
Suppose the problem is to find the edge ($?e$) which is the common boundary of two given faces ($f1$ and $f4$). The symmetry of arguments for predicates allows us to write

$$\textbf{evaluate}\ ef(?e, f1) \wedge ef(?e, f4) \tag{2.3}$$

or

$$\textbf{evaluate}\ fef(f1, ?e, f4) \tag{2.4}$$

which is in apparent harmony with (2.1) and (2.2) above. In contrast with this, programming in ASP will result in a substantial modification of the original code.

(4.3) *Lack of Predicates of Higher Degrees*
As already argued (Sect. 1.1 (6)), we often find more profound reasons than the simplicity of programs for requiring predicates of arities greater than two. It is difficult for ASP to accommodate this requirement. One popular method which is applicable to any framework based on binary predicates is to decompose each tuple of related elements for each predicate (say, $(a_1, a_2, \ldots, a_n)$ for which $p(a_1, \ldots, a_n)$ is true) into two parts (say, $(a_1, (a_2, \ldots, a_n))$) and consider each inner tuple as an independent element. Another is to create an element representing the whole tuple to which each argument is related

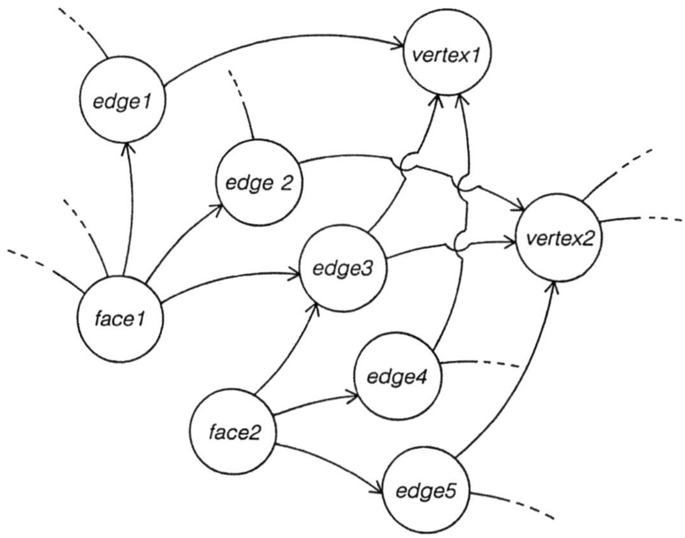

Fig. 2.1. A part of the directed graph for the topology of a polyhedron

through a binary predicate. (See Sect. 4.6.1 for a proposal by R.A. Kowalski [Kow] along the same lines.)

The difficulty is the system must see to it that a tuple considered as an element will automatically cease to exist when any (in case of the first method) or all (in case of the second) of its constituents are deleted. This is a meta-rule which cannot be stated within ASP or any language in question and is known only to the system. The introduction of such a meta-rule should be avoided as far as possible.

(4.4) *Lack of Symmetry in Structure*
A structure based on binary predicates is illustrated naturally as a directed graph, as in Fig. 2.1. Looking at the graph and trying to decide how to represent it within the computer memory, one might be tempted to store the identifier (name, index, address or whatever) of the node at the head of an arc in the node at its tail.

This has two implications, one technical and the other philosophical. First, it is necessary to accommodate at each node the origination of a variable number of different kinds of arcs. The dependent structure of a node comprising two levels of rings is the device for this, certainly a very clever device. Second, one tends to see arcs as belonging to the starting nodes, that is, facts (expressed by relations) as belonging to elements (entities or subjects), which leads to (or perhaps reflects) what one may call an 'entity-centered' or a 'subject-predicate' world view. The phrase 'dependent structure' is indeed very indicative. A thorough analysis of this view and its relevance to data modeling will be given in Chap. 5. Meanwhile, we note that a lack of symmetry is implicit in the restriction of organizational axes to binary predicates, as it derives from the dichotomy of interrelated concepts which will be yet more unwarranted with relations among three or more terms.

(4.5) *Lack of Formalities*
Lack of formalities in defining structures often leads to confusion and misinterpretations as to the real semantics of syntactic constituents and their combinations. In the above, we focused our attention on the relations between elements and treated other constituents (ringstart and associator) as if they were of lower level (that is, internal schema level rather than conceptual). Alternatively, we may well consider the latter as standing on equal terms with elements as conceptual entities. Either interpretation is possible and may produce a formally reconstructed ASP schema description language. It will be quite possible to precisely state constraints, such as the prohibition of isolated fragments of structure, in such a language. (That is why we referred to ASP symbolism as a conceptual framework in Sect. 0.6.) The result will be essentially different depending on which interpretation is given, however.

(4.6) *Irrelevance of Attributes to Structure*
Optional pieces of information (data, name, type and association) have little relevance to the structure. We might want, for example, to associate a procedure to calculate the length from the coordinate values of vertices with the class of edges, yet have no way to do this. Nor can an application program simply request the system to give it the edge with the greatest length. In the context of solid geometric modeling, this is somewhat like making too much of topology at the cost of geometry. Furthermore, functions of two or more variables, like predicates of higher degrees, cannot be defined. If we need to define the distance between two faces, one way to do this will be to store the value in the associator that relates these faces. The question is, however, how this associator stands to faces. Is it an independent entity? If so, the distance is an attribute of this single entity rather than a function of two entities. (Another manifestation of the entity-centered view of the world!) Otherwise, maintaining the dependency lies with application programs, which is not stated explicitly anywhere in the schema. Defining functions of three or more variables will be hopeless.

2.1.2 LEAP, an Associative Language

(1) *Structure*
The world of LEAP [Fel] consists of *items*, which form much the same concept as the elements of ASP. There are two organizational axes in LEAP, one for associating items and the other for constructing *sets* of items. An association is an ordered triple of items (a, o, v) for which the relation

$$a \cdot o \equiv v \qquad (2.5)$$

holds true. a, o and v stand for 'attribute', 'object' and 'value' respectively. Hence, the above relation is interpreted as stating that the value of attribute

a of the object o is v. Sets of items can be defined either by enumeration or by set-theoretic operations on existing sets.

There are three subclasses of items: *declared*, *created* and *association* items. Declared items are constants and exist permanently. Those created by an application program during its execution can also be deleted. An association item is nothing but an association (a, o, v) considered as an item, and is created or deleted by asserting or negating relation (2.5). Declared and created items may have data associated with them.

(2) *Basic Functions*
LEAP, as an extension of ALGOL 60, offers all the constructs available in ALGOL. Additional functions which are proper to LEAP include, among others, the following:

- declaration of items and item variables,
- creation of items denoted by **new** or bracketed triple $[a{\cdot}o \equiv v]$, and deletion of these,
- built-in functions **first**, **second** and **third** to be applied to an association item to yield the corresponding component,
- declaration of set variables,
- construction of sets by means of:
  - enumeration of items,
  - evaluation of
    $\{x : \exists y \exists z((y{\cdot}z \equiv x) \wedge (y \in P) \wedge (z \in Q))\}$ denoted by $P{\cdot}Q$
    or
    $\{x : \exists y \exists z((y{\cdot}x \equiv z) \wedge (y \in P) \wedge (z \in Q))\}$ denoted by $P'Q$
    where P and Q are sets of items, and
  - set-theoretic operations (union, intersection and difference) on existing sets,
- use of the reserved word **any** as an item expression to denote an arbitrary item,
- assignment of item and set expressions to variables of corresponding types,
- inserting or removing an item into or from a set respectively.

It is said that the concept of loop statement is the raison d'etre for the entire system. It takes the form

foreach ⟨associative context⟩ **do** ⟨statement⟩

where the ⟨associative context⟩ is a triple, an elementary boolean expression or a logical conjunction of these, which contains one or more loop (free) variables and stipulates the condition for these variables to satisfy. The ⟨statement⟩ will be executed for each combination of instances of the loop variables.

(3) *Sample Programs*
(3.1) Let us first sketch a program for the same problem as in the preceding section.
(3.1.1) The items (declared or created) are the geometric entities (*body*, *faces*,

edges, *vertices*) and the symbols to denote topological relations between geometric entities. The latter include:

> *bf*(*body-face*),
> *efl*(*edge-face-left*), *efr*(*edge-face-right*),
> *eefl*(*edge-edge-forward-left*), *eefr*(*edge-edge-forward-right*),
> *eebl*(*edge-edge-backward-left*), *eebr*(*edge-edge-backward-right*),
> *evf*(*edge-vertex-forward*), *evb*(*edge-vertex-backward*)

and possibly others.

Each *vertex* has its coordinate values associated with it.

(3.1.2) Associations are created to represent instances of topological relations such as $[bf{\cdot}b \equiv f1]$, $[efl{\cdot}e1 \equiv f4]$, $[eebr{\cdot}e1 \equiv e4]$, etc.

(3.1.3) The identifiers *faces*, *edges* and *vertices* are declared as being of type **set**.

(3.1.4) A program to find the face in question is

> **begin**
> **foreach** $efl{\cdot}e1 \equiv ?f$ **do** $faces := \{?f\}$;
> **foreach** $efr{\cdot}e1 \equiv ?f$ **do** $faces := faces \cup \{?f\}$;
> $faces := faces - \{f1\}$
> **end.**

(3.2) The inverse problem may be programmed similarly

> **begin**
> **foreach** $(efl{\cdot}?e \equiv f1) \wedge (efr{\cdot}?e \equiv f4)$ **do** $edges := \{?e\}$;
> **foreach** $(efr{\cdot}?e \equiv f1) \wedge (efl{\cdot}?e \equiv f4)$ **do** $edges := edges \cup \{?e\}$
> **end.**

(3.3) LEAP is almost ideally suited for the CSG scheme for representing solid bodies. Each triple

$$(op, a, b),$$

where *op* is one of the pre-defined set-theoretic operators, and *a* and *b* are the operands, plays the double role of denoting the operation and its result. Either operand may in turn be another triple, and the nesting may be arbitrarily deep.

(3.4) Let us present another very small program just to show the power and flexibility of LEAP language. The problem is to construct a set of pairs (f, v) of face and vertex such that v does not lie on f. While LEAP does not accommodate constructing cartesian products of sets directly, the difficulty can be alleviated by declaring an item (say 'π') to denote the operation of taking the cartesian product and creating an association $[\pi{\cdot}a \equiv b]$ for each pair of elements (a, b) of component sets. A program is then

begin
faces := $bf{\cdot}b$; *vertices* := $(evf{\cdot}(efl'faces)) \cup (evb{\cdot}(efl'faces))$;
$S := \phi$;
foreach $(?f \in faces) \wedge (?v \in vertices - (evf{\cdot}((efl{\cdot}?f) \cup (efr{\cdot}?f))))$
do put $[\pi{\cdot}?f \equiv ?v]$ **in** S
end.

(4) *Criticism*
(4.1) *Advantages*
The LEAP language is strictly at a conceptual level and free of implementational details (such as rings), and facilitates non-procedural descriptions of tasks to be done. This is why programs in LEAP are simpler and more understandable than those in ASP. What is more, LEAP's organizational capability is substantially greater and comes close to that of FOPL. Let us elaborate on this.

First, an arbitrary binary predicate can be interpreted in I, the universe of items, by declaring the predicate symbol as an item and considering the set of triples whose first argument is this item as its extension.

Second, the names 'attribute', 'object' and 'value' do not carry any meaning but serve merely as placeholders, and an association item $[a{\cdot}o \equiv v]$ may as well be written as (a, o, v). Thus, the universe of associations which is the fundamental structure of the world as seen by LEAP can be defined as the extension of a built-in ternary predicate 'association' in the set I^3.

Third, taking subsets of I^3, associating association items with other items and applying these operations in turn enables the interpretation of arbitrary predicates of three, four and more arguments in I.

(4.2) *Functional Limitations*
Looking at the sample programs in Subsects. (3.1) and (3.2), one may wonder if it is simpler and better to write

begin
foreach $((efl{\cdot}e1 \equiv ?f) \vee (efr{\cdot}e1 \equiv ?f)) \wedge (?f \neq f1)$ **do** *faces* := $\{?f\}$
end

and, for the inverted problem

begin
foreach $((efl{\cdot}?e \equiv f1) \wedge (efr{\cdot}?e \equiv f4)) \vee ((efl{\cdot}?e \equiv f4) \wedge (efr{\cdot}?e \equiv f1))$
do *edges* := $\{?e\}$
end.

Unfortunately, neither is possible with LEAP, as conjunction is the only logical operator that is available. Strangely enough, set-theoretic versions of the associative contexts

foreach $?f \in ((efl{\cdot}e1) \cup (efr{\cdot}e1)) - \{f1\}$

and

foreach $?e \in ((efl'f1) \cap (efr'f4)) \cup ((efl'f4) \cap (efr'f1))$

are quite legitimate.

Next, suppose we want to assert $b{\cdot}y \equiv x$ for a given y and for each x in $a{\cdot}o$. The simplest and most natural form of statement will be

make $\forall x((a{\cdot}o \equiv x) \rightarrow (b{\cdot}y \equiv x))$

which is again impossible. Instead, either

foreach $a{\cdot}o \equiv ?x$ **do make** $b{\cdot}y \equiv ?x$

or

make $b{\cdot}y \equiv (a{\cdot}o)$

will do the job.

Conversely, let us consider retrieving such y as satisfy $b{\cdot}y \equiv x$ for every x in $a{\cdot}o$. The phrase

foreach $b{\cdot}?y \equiv (a{\cdot}o)$

will evaluate the formula

$$\exists x((a{\cdot}o \equiv x) \wedge (b{\cdot}y \equiv x))$$

instead of the intended formula

$$\forall x((a{\cdot}o \equiv x) \rightarrow (b{\cdot}y \equiv x)).$$

Generally, LEAP enables us to write non-procedural statements to evaluate formulas of the form

$$\exists x(p(x) \wedge q(x,y)) \tag{2.6}$$

but not

$$\forall x(p(x) \rightarrow q(x,y)), \tag{2.7}$$

although both kinds of formulas occur frequently as retrieval conditions.

(4.3) *Lack of Minimality, Orthogonality and Completeness*

As indicated above, triples may well be considered as ternary atomic formulas, and those containing just one free variable have extensions denoted by set expressions. Often, associative contexts may be defined in terms of either triples or set-theoretic boolean expressions to the same effect. The duality is not complete, however. On one hand, set expressions can be constructed which do not denote any extensions of triples or their combinations, as we have seen. On the other, combinations of triples containing two or more free variables have no corresponding set expressions.

The weakest spot of LEAP is probably not the existence of this or that shortcoming but the difficulty of perceiving immediately what is possible and what is not. The difficulty derives principally from the excessive abundance of concepts which interact with each other and yet are not organized into any self-contained framework. We have indicated that the world of LEAP is close to that of FOPL in its structural richness. The closeness falls short of equivalence, however.

(4.4) *Ambiguities*
Ambiguities about the roles of item and association are among the major causes for the lack of conceptual orthogonality. As has been pointed out, an item can participate in associations as either the predicate symbol or a component, and an association in turn can act as either an instance of relations (a fact) or an entity. An example given in Feldman/Rovner [Fel] is

$$\textbf{make}\ \mathit{number}\cdot[\mathit{part}\cdot\mathit{hand} \equiv \mathit{finger}] \equiv \textbf{new}(5)$$

where '5' is the datum associated with the newly created item. We suspect that what the bracketed triple really points to may not be the association itself but a particular component *finger* of it, hence that

$$\textbf{make}\ \mathit{number}\cdot\textbf{third}[\mathit{part}\cdot\mathit{hand} \equiv \mathit{finger}] \equiv \textbf{new}(5)$$

or

$$\textbf{make}(\mathit{number}\cdot\mathit{finger} \equiv \textbf{new}(5)) \wedge (\mathit{part}\cdot\mathit{hand} \equiv \mathit{finger})$$

might be more appropriate. (Incidentally, the selector **third** can only be applied to an identifier and the verb **make** to a single triple, but we took the liberty of slightly expanding the language.)

The example in Subsect. (3.3) takes advantage of the ambiguity of association items. There, the second and third components stand on equal terms with respect to the first, and the whole triple as an item denotes none of its components but something new, the result of the operation. It will hence be more agreeable to use this triple in further associations than in the above case. The semantic difference between the two cases is known only to application programs (or rather programmers), and no syntactic distinction is possible.

Another subtlety about the concept of association item is that every bracketed triple, and only bracketed triples, but no **make** statement, will create an item to represent the association. We question why. Perhaps the reason is not particularly profound. If we modify the language so that every association may have its representing item, however, the consequence will be that the universe of associations is a proper subset of that of items.

(4.5) *Underlying World View*
The preceding arguments, particularly the last one above, indicates that what underlies LEAP is after all the 'entity-centered' world view. The extra sophistication of the language could be considered as the cost for reconciling the required symmetry and multidimensionality of relations with this view.

(4.6) *Irrelevance of Attributes to Structure*
Much the same can be said about LEAP as about ASP as to the irrelevance of attributes (data) to structure. That data cannot be associated with association items makes it yet more difficult than in ASP to define functions of

two or more arguments. A possible means, though probably not very recommendable, is to create an item to represent each tuple of arguments in the domain of the function.

2.1.3 Set-Theoretic Data Structure (STDS)

Childs [Chi] attempts to formulate a framework in which to organize data purely and exclusively in terms of sets. This is a very singular approach despite the fact that many data models incorporate some of the set-theoretic concepts and manipulations in one way or another. It was almost disregarded as inapplicable to CAD and graphics [Wil(1)], probably because of its purity and lack of such devices as rings and pointers, which are so obvious to our intuition and seem indispensable for describing geometric entities. As we will see, STDS is in fact more powerful than most of the contemporary packages in defining complex structures.

(1) *Structure*
There is the base set β of elements called *datum names*, which are similar to declared items of LEAP and have data associated with them. A potentially infinite number of sets can be constructed on the basis of β, each identified by a unique name given by an application program. The set of these names is denoted by μ. The methods of constructing sets include taking:

- the subset of an existing set by enumerating its elements or stipulating the condition for them to satisfy,
- the union and intersection of sets, the difference between two sets,
- the cartesian product of sets, and its projection on a component set,
- the converse of two components of a cartesion product,
- the join of two cartesian products on the common set of components,
- the power set of an existing set,
- selecting the following member sets from a set a of sets:
  - $\mathcal{G}(x : a) \equiv \{b : (b \in a) \wedge (x \subset b)\}$,
  - $\mathcal{D}(x : a) \equiv \{b : (b \in a) \wedge (projection_1(x) \subset b)\}$,
  - $\mathcal{R}(x : a) \equiv \{b : (b \in a) \wedge (projection_2(x) \subset b)\}$ where b is a cartesian product, and
- any finite combination of these.

Two binary predicates '$\in$' and '$\subset$' are the only organizational axes in this set of sets. As each set or datum name could occur as element in multiple super-sets, a sophisticated structure must be maintained to represent sets economically and guarantee consistency of their relations. This structure is strictly internal to the system and transparent to application programs.

(2) *Basic Functions*
Functions are provided to implement the methods listed above. In addition, we have functions

- to evaluate boolean expressions combining atomic expressions '$b \in a$' and/or '$b \subset a$',
- to count the numbers of elements in sets, and
- to retrieve specified portions of data associated with datum names.

(3) *Sample Program*
Let us consider again the case of the winged-edge tetrahedron. The set of datum names (β) comprises the *body*, *faces*, *edges* and *vertices*. The coordinate values are stored as data associated with the latter. We define the set

$$a \equiv \{(o, v) : a{\cdot}o \equiv v \text{ in LEAP notation}\}$$

for each topological relation and the sets *faces*, *edges* and *vertices* which are precisely the same as with LEAP. The program will then be:

begin
  $faces := (efl \cup efr)[\{e1\}] - \{f1\}$
end.

where the square brackets mean taking a projection from a cartesian product:

$$a[b] \equiv \{y : \exists x((x \in b) \wedge ((x, y) \in a))\}.$$

That for the inverse problem will be:

begin
  $edges := [f1](efl \cup efr) \cap [f4](efl \cup efr)$
end.

(4) *Criticism*
(4.1) *Advantages*
Interestingly, the conceptual restraint of STDS has brought it expressive power as well as methodological clarity. It is more capable than LEAP and is regarded as virtually equivalent to FOPL. For example, the extensions of dual formulas (2.6) and (2.7) are expressed as

$$Q[P] \quad \text{and} \quad \sim(\sim Q)[P]$$

respectively. Here, P and Q denote the extensions of p and q respectively, and '$\sim$' denotes taking the complement in a suitable super-set.

(4.2) *Obscurity of Semantics*
Restraining the concepts and terminology to those of set theory tends to obscure the inherent meaning of a logical formula, which is manifested only in its formation but not by its extension. For example, the formula

$$p(\tilde{x}) \wedge q(\tilde{y})$$

will have its extension expressed as

$$P \cap Q,\ P{\times}Q,\ Q|P \quad \text{or} \quad P[2 = 1]Q$$

depending on which of the following forms the formula takes,

$$p(x) \wedge q(x),\ p(x) \wedge q(y),\ p(x) \wedge q(x,y) \quad \text{or} \quad p(x,y) \wedge q(y,z).$$

(The last one is in the notation not of [Chi] but of [Codd(3)].) It will not be easy to perceive immediately from the diversity of these expressions the shared meaning embodied in the original formula, that is, that the properties p and q are predicated of certain elements $(\tilde{x}, \tilde{y})$.

Another example is the extreme case of an arbitrary collection of closed formulas which are true in the current state of the database. All of them have the same extension TRUE, yet their meanings may be entirely different from each other. While each set has a unique name in STDS, different sets may have the same construction, hence the same extension. We take this as indicating that extensions alone do not decide semantics, and further that a set may change its extension yet maintains its meaning. A crucial question is whether such is the case with sets constructed from more basic sets. If so, a set could be severed from the method of construction used at its original definition. Otherwise, we will have to make a distinction between primitive and constructed sets. The latter will cause STDS to be yet closer to FOPL in which we have two distinct levels of formulas, atomic and compound formulas, with the definitions of the latter fixed and their extensions completely determined by those of the former.

(4.3) *Complication of Notation*
It seems that the conceptual simplicity of STDS is attained at the cost of notational complexity. As indicated above, the impossibility of explicitly denoting components of a cartesian product could make it quite difficult to express the extension of a formula when the lists of free variables differ from one sub-formula to another.

(4.4) *Emphasis on Procedure*
It is likely that we tend to think of a set expression as the procedure to yield the elements. For example, we may think of $Q[P]$ as the result of accepting each tuple in Q the first component of which is in P and then disregarding that first component. Evidence for our conjecture is that a procedure is provided by STDS corresponding to each basic set operation. A compound set expression will be comprehended as the result of nested execution of these basic procedures. In this sense, STDS is to FOPL what relational algebra is to relational calculus [Codd(3)].

2.2 Ad hoc Developments of CAD Databases

In this section, we will select some examples from the works done for rather specific purposes, try to realize the conceptual basis and to evaluate the most noticeable features of them from a more general point of view.

(1) *Description of Shapes*
BREP and CSG are among the best known and most widely used schemes

for representing spatial forms. It has been indicated that if we try to abstract organizational principles from these and incorporate them into more general frameworks, we may well be led to a ring-structure based package and a LEAP-like associative language.

In BDS (Building Description System, [Eas(1)], [Eas(2)]), polyhedral objects are modeled as hierarchical collections of topology, geometric constraints in the form of equations, values of attributes in local coordinates and transformations into world coordinates. Starting as a topological pattern (e.g., a polyhedron), an object will get more detailed and specialized at each lower level until it is completely defined and positioned at a particular place in the world. The topology is implemented as a set of rings connecting constituent parts. This is an interesting integration of the schema, classes and instances of objects, and could fit nicely into certain specific domains. It will nevertheless be a vulnerable approach for a general purpose data model to take, as it lacks the flexibility to have schemas separated from instances and multiple views associated with a single database.

BDS was later incorporated into a system called GLIDE (Graphical Language for Interactive Design, [Eas(3)]) which supported among others the following functions:

- extending the schema dynamically by means of 'modules' which encapsulate new and extended data type definitions, and the operators allowed on these,
- defining the tree of 'frames' which qualify the identifiers of permanent items to enable naming in local contexts,
- converting data types by 'typecase' statements to enable the application program to access new types without requiring any change to it,
- guaranteeing the database integrity by executing SIRs (System Invoked Routines) just prior to accessing the objects they are assigned to, and
- creating 'alternatives' in the form of checkpoint files which contain sets of database update operations.

These all concern important and difficult issues of design databases. The extension of the schema should not be limited to extending data types, while doing it dynamically is seldom supported even by modern data models. Integrity management raises not only technical but also conceptual questions. We should ask particularly whether and how integrity constraints can be distinguished from derivation rules, which has become one of the key questions in the recent trend toward deductive databases. Taking checkpoints is certainly one of the possible approaches to implementing design alternatives. The problem is too complex, however, to enable us to find a satisfactory solution in a naive answer such as this. At least, we must be able to replace an arbitrary part of a design without affecting the interface between it and the remaining parts. The final point we want to pay attention to is the embodiment in modules of the concept of data abstraction. While this is quite agreeable, the application of this important concept should be explored on a wider scale than only for schema extension.

In Braid [Bra], a shape called 'assembly' is described in two levels, first as a combination of constituent assemblies (instances), each of which is described in terms of yet other assemblies and so on, and second as a body in the winged-edge topology. Methods for combining assemblies include union, intersection, difference and juxtaposition. A prototype assembly may occur as different instances with different transformations. A shape thus described is, in short, the integration of a CSG-like semi-tree and BREP structures. An interesting feature is the conversion of a sub-tree into a single BREPed body by the *evaluate* command. It is within the discretion of the application program to decide which sub-tree should be evaluated and when.

Probably the most remarkable thing about Braid's work is the use of ALGOL 68 to specify the data structures and define the operators to manipulate them. It has been shown that with a sufficiently rich organizational capability, a high-level programming language can be effectively used for constructing CAD systems.

(2) *VDAM (Virtual Data Access Manager)*
VDAM [War] is an extension of APL (Associative Programming Language, [Dod]) which was developed to provide data management functions for a CAD system called CADANCE (Computer Aided Design And Numerical Control Effort) at General Motors. (Incidentally, APL is another ring structure package from the 1960s and is known to have provided a basis for the CODASYL effort, which has resulted in the network data model.)

A VDAM structure comprises two kinds of constituents, *item* and *set*. Each item represents an object, and a set relates a root item to member items. A set can be compared to an associative ring in ASP, the ringstart together with all associators on it. Access to member items is managed by an array of pointers introducing a linear order among members of the set. A *file* is a hierarchical network in which items and sets alternate with each other, and a collection of files makes up a database.

VDAM provides functions to create or delete items and sets, and to insert or remove items into or from sets. It also provides functions to access items and sets either by moving up and down the hierarchy or directly by *names*. The organizational principle of VDAM is essentially the same directed graph as that of ring-based packages. As such, VDAM shares many of the functional and conceptual limitations of ASP, though the access mechanisms are made simpler and more direct thanks to the augmented functionality of relational terms (that is, sets as compared with ringstarts or associators).

(3) *FLOREAL*
FLOREAL (Foisseau/Valette [Foi]) is a development based on the experience with a system called GERMINAL which comprises an ALGOL 68-like language for defining and manipulating data types, and a problem-oriented language L-MECAO for mechanical design. There are four basic concepts in the FLOREAL schema description language, that is, *type*, *object*, *relation* and *function*. A type is just like a static data type in usual programming languages. The method of defining an object is by far the most characteristic

feature of FLOREAL. For each object, there may be a set of *views* which are defined in terms of types, other objects, functions and relations among them. Each *view* is equivalent to another in the sense that it fully defines the object and that *coherence* must be stipulated between different views. For example, the object *square* may be defined by either the view *points* or the view *sides*, and by the condition that the *distance* between two end *points* be equal to the *length* of a *side*. A part of the definition may thus be:

```
type point(abs, ord: real), side(ori, ext: point);
function dist(p1, p2: point), length(s: side): real;
relation right-angle(s1, s2: side);
object c
  constant l: real;
  view points
    point p1, p2, p3, p4;
    dist(p1,p2) = l;
    ...
  end view;
  view sides
    side s1, s2, s3, s4;
    right-angle(s1,s2);
    length(s1) = l;
    ...
  end view;
  coherence points.p1 = sides.s1.ori;
  ...
  end coherence
end object.
```

Clearly, FLOREAL's approach is in the same spirit as that of abstract data typing.

(4) *COMPAC*

COMPAC (Dassler et al. [Dass]) deals with different levels of objects, from assemblies, sub-assemblies, and parts to geometric elements, and offers a wide range of application capabilities including design, drawing, kinematic simulation, finite elements analysis, process planning and NC data preparation. Objects are related to each other through binary predicates in one-to-many correspondences, except for those between geometric elements and their boundaries, which are many-to-many. Thus they form a hierarchical network structure. In the input stage, workpieces are defined as boolean combinations of primitive parts which are separately described in either volume- or face-oriented representation. These descriptions are converted by the COMPAC processor into the internal representations which are based on the ASP structures.

While few things are new in COMPAC from the viewpoint of data model theory, it is worth noticing that several proven techniques are incorporated in it to suitably describe different types of objects for different purposes.

(5) *ADDL (A Data Definition Language)*
Lacroix/Pirotte [Lac(2)] begins by criticizing available data models, namely, the CODASYL (network), IMS (hierarchical) and relational models and semantic extensions of the latter. It is contended that one of the serious disadvantages shared by these is the lack of capability to model complex structures inherent in design. ADDL's approach to meet the requirement is to provide a set of rules for constructing new domains (types) of objects from other domains starting from basic domains of atomic (unstructured) objects. These rules include:

- a set constructor to define the powerset of the original domain,
- aggregation to define the cartesian product of component domains,
- forming the union of two or more domains,
- mapping from one domain to another, and
- nested applications of the above, including recursive definitions.

In the model of a circuit, for example, the nets can be defined as a mapping from the set of net names to the power set of terminals. A terminal is either a circuit terminal name or an ordered pair ⟨*element occurrence name*, *element terminal name*⟩. Each of these names is an atomic object.

There is a striking parallelism between ADDL and STDS, and as such, ADDL shares most advantages and disadvantages of STDS as previously discussed. (One may well look upon ADDL as an application of STDS, which seems to be an interesting coincidence of theoretical positions, however, as [Lac(2)] does not make any explicit or implicit reference to Childs' papers.)

2.3 The Network (CODASYL) Model and Its Extensions

It seems ironical that the CODASYL model is usually considered as supporting principally business data processing, while APL, one of its immediate predecessors, was particularly intended for CAD applications. This may not be inexplicable. For one thing, a great emphasis is placed in CODASYL on the facilities to describe business database schemas, specifically strict and static typing of objects and relations. For another, little attention has been paid to augmenting its organizational power to meet the CAD requirements for more complex structures. The fundamental structure remains the same hierarchical graph as with the ring-based packages.

Two things other than the organizational axes must be pointed out in the CODASYL model. First, the separation of the conceptual realm (schema description and data manipulation) from implementation concerns (storage and efficiency considerations) has been pursued. The Journal of Development

[CODA(2)] purged from the conceptual schema many features which are not really conceptual and proposed providing the data storage description language (DSDL) to describe these. Second, an extensive set of functions and devices have been incorporated to facilitate access control, integrity management, concurrency control, recovery from system failures, data reorganizations, etc., which are a must in any data-intensive application environment. It is certainly a great benefit of relying on the CODASYL or any general data management system that these facilities are readily available. Yet we will not go any further into discussing these, as they are of a primarily administrative nature and have little relevance to the realm of conceptual modeling.

The CODASYL languages (DDL and DML) are extensive languages with a rich repertoire of data types and manipulation commands. It is neither possible nor necessary to present them here in full detail. Therefore, for fairness, we must have it clearly stated that what follows is only the small part that is relevant to our limited purposes and may not necessarily reflect their overall capacity. Besides, it is expressed in an informal style and accompanied by rather liberal (yet hopefully correct) interpretations.

2.3.1 Structure and Basic Functions

(1) *Structure*

(1.1) *Constituents and Constraints*

The database consists of *records*, each of which is an aggregate of data items or attribute values and identified by a system-generated database key. Each data item is of some type available in the host language. An arbitrary number of binary predicates called *sets* can be defined between records. For each set, one of the arguments is the *owner* and the other the *member*. A set whose owner is the system-provided imaginary record *system* is actually a unary predicate. The other kind of unary predicate is *area*, which divides the whole database into mutually disjoint areas.

If we let records correspond to elements and sets to dependent structures, we can easily see the similarity of the above structure to that of ASP. The record *system* corresponds to the *universe element*. The correspondence to VDAM constituents is even more direct. The similarity to a ring structure does not extend any further, however. The following set of rules and constraints strongly characterizes the CODASYL model:

- Every record must be of a unique *type*. A record type defines the template of *names* and *types* of data each record of that type may have.
- A record type may define one or more combinations of data names as *keys*. Each key identifies records of this type and can be used for direct retrieval of records from the database. It may also be specified as ascending or descending to introduce a linear order in the set of records of this type.
- Every set must also have a *type*. A set type defines the one owner record type and one or more permissible member record types.

- For each set type, the collections of members belonging to different owners must be disjoint with each other. A set will then relate records in a one-to-many correspondence. In other words, a set can be defined as a mapping from a collection of member records to a collection of owner records.
- For each member record type of each set type, a sequence of data names is designated as the sort control key which defines a local order within that record type. Multiple local sequences may be optionally collated into a single sequence by specifying the order of member record types.
- For each record type, the whole set of records of that type must be in the same area.

Clearly, the concept of record type plays the most fundamental role. It is not that extant records are classified into types, but that they come to exist only as instances of generic objects called record types. That records of the same type are homogeneous with respect to the list of attributes they may have carries the utmost significance for database theories as well as practice. It signifies that the universe of discourse comprises a fixed number of templates and things which are bundles of attributes in conformity with the templates. Such is typified by an assortment of vouchers, lines of tables and items of ledgers. Thus, business transaction processing provides the CODASYL model with a paradigm.

(1.2) *The Data Description Language (DDL)*
For each **schema**, one or more entries should be provided to describe each of the following major categories of constituents:

- **area** entry,
- **record** entry,
- **set** entry.

Once given, a schema is not susceptible to change and remains fixed. We leave further explanation of the DDL to the sample program given later.

(2) *Basic Functions (The Data Manipulation Language DML)*
The DML comprises approximately twenty kinds of statements, of which some are for red tape operations and the following are more important for our purposes:

- **store** to create a new record and store it in the database,
- **find** to establish the specified record as the **current** record,
- **erase** to erase the specified record from the database,
- **connect** to make the specified record a member of the specified set,
- **disconnect** to remove the specified record from the membership of the specified set,
- **modify** to change the values of one or more data items of the specified record,
- **get** to retrieve the specified record (values of data items) from the database.

The specification of a record can be done explicitly (e.g., by means of a key value, an absolute position – **first** or **last** – in a linear order, etc.) or implicitly (e.g., by means of a position relative – **next** or **prior** – to the current position, an owner-member relation with the current record in a set, etc.). The concept of **currency** is intimately related to the access path dependence or navigational peculiarity of the network structure. That is, the methods of access to (implicitly specified) records are based on three principles:

- just one record is accessed at one time,
- an access is made from the previously secured (current) record or position, and
- it is made to a record in the structural proximity of the current record or position.

It should be noted that these principles equally underlie the access mechanisms of ring-based packages. Particularly in ASP, the program variable holding the identifier of a ringstart or an associator controls the currency.

(3) *Sample Program*
A part of the schema for the winged-edge structure of a polyhedron looks as follows:

```
schema name is polyhedron;
area name is geometric-entities;
area name is other-entities;
record name is body within geometric-entities;
record name is face within geometric-entities;
...
record name is vertex within geometric-entities
   key vertex-name is name duplicates are not allowed
   02 name type is character 2
   02 coordinate
      03 x type is real
      03 y type is real
      03 z type is real;
record name is edge-face-link within other-entities
   02 bounding-edge-name source is name of owner of bounding
   02 bounded-face-name source is name of owner of bounded;
set name is bounding
   owner is edge
   order is permanent insertion is sorted within record-type
   member is edge-face-link
      insertion is automatic retention is fixed
      key is ascending bounded-face-name
      set selection is through bounding owner identified by edge-name;
```

set name is *bounded*
owner is *face*
order is permanent insertion is sorted within record-type
member is *edge-face-link*
insertion is automatic retention is fixed
key is ascending *bounding-edge-name*
set selection is through *bounded* **owner identified by** *face-name*;

(3.1) The problem of finding the face, given a face and an edge (Sect. 2.1.1 (3)), is programmed as follows:

⟨The variables *edge-name* and *f* are assumed to hold the names of given edge and face respectively. The name and attribute values of the face being sought will be assigned to the variables *bounded-face-name* and *face*.⟩
begin
find *edge-face-link* **via** *bounding*;
⟨Select the set whose owner is identified by *edge-name* following the set selection method as specified in the definition of *bounding* and find its first member.⟩
get *bounded-face-name*;
⟨Get the name of the bounded face stored as a data item of the *edge-face-link*.⟩
if *bounded-face-name* = *f* **then**
begin
find next *edge-face-link* **record of** *bounding* **set**;
get *bounded-face-name*
end;
⟨If the bounded face retrieved is the face given initially, find the next *edge-face-link* in the same set *bounding* and get the *bounded-face-name*.⟩
find owner record of *bounded* **set**;
get *face*
⟨Find the face that owns the *edge-face-link* retrieved in the set *bounded* and get the associated data.⟩
end.

(3.2) The program for the inverse problem ('find the edge on the common boundary of two given faces') is:

⟨The variables *face-name* and *bounded-face-name* are assumed to hold the names of two faces. The name and attribute values of the edge being sought will be assigned to the variables *edge-name* and *edge*.⟩

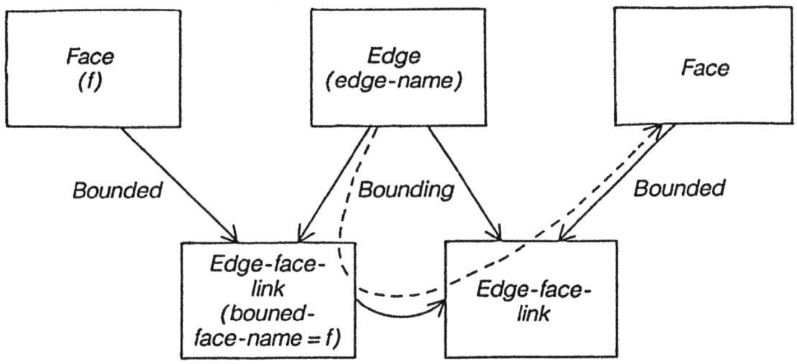

a *Finding the face, given a face and an edge*

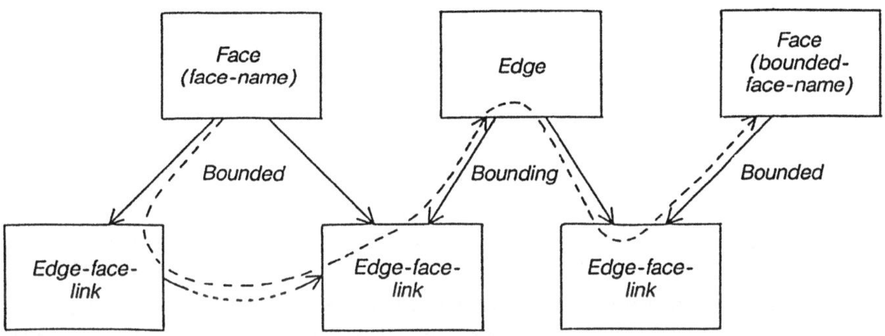

b *Finding the edge on the common boundary of two given faces*

Fig. 2.2. Finding an object via set

begin
  find any *face* **using** *face-name*;
  repeat
    find next *edge-face-link* **record of** *bounded* **set**;
    find owner record of *bounding* **set**;
    find *edge-face-link* **via current of** *bounding* **using** *bounded-face-name*;
  until successful
end.

2.3.2 Criticism

As we have noted, the common underlying structure (that is, directed graph) makes the CODASYL model similar to ring-based packages, while the strict and static typing of data distinguishes it from them. This is manifested even in such a simple problem as conditional search for a tetrahedral edge or face. By comparing two programs (Sects. 2.1.1 (3) and 2.3.1 (3.1)), we find that these run largely parallel to each other, yet differ in one small but nonneglible point; the search for a ringstart on the lower ring of an object is missing in the CODASYL DML's case. What causes this is obviously the inclusion of the specification for set type ***bounding***, particulary the method for selecting its instance. It also enables immediate accesses from owner to member records via set without the intervention of associators. Generally, the strict typing of records and sets, together with the requirement that every instance be of one and only one type, will more or less take the load off application programs. This is not without cost, however. The schema may not be altered dynamically. The structure tends to be inflexible. Evidence of the latter weakness is the failure to admit $m : n$ correspondence within a set type, of which the boundary relation between faces and edges offers an example. The usual way to alleviate the difficulty is to introduce so called 'link records' by which to decompose a relation into two $1 : n$ relations. The ***face-edge*** relation is thus replaced by ***bounding*** and ***bounded***, which are different set types and hence are allowed to share the same records (***edge-face-link***). Link records are nothing but associators in disguise.

In view of its strong business orientation, we will not be surprised if there are rather few applications of the standard CODASYL model to CAD. Korenjak/Teger [Kor] is one of the few examples, and deals with the hierarchical structure of integrated circuits, the same subject as that of Lacroix/Pirotte's ADDL (Sect. 2.2 (5)). It is interesting to note the difference between the two approaches. For example, the relation between ***nets*** and ***elements*** they connect is modeled in ADDL as a function over the set of ***net names*** whose value is a set of ***terminals***, and by Korenjak/Teger as a set whose owner is a ***net*** and whose member is a ***pin*** (terminal). In the former, a terminal is actually a pair ⟨***element name***, ***terminal name***⟩ while in the latter, a pin is also owned by a ***part*** (element) in a separate set. Pins will be needed to link both sets even if we are interested only in nets and elements. ADDL can accommodate this case more flexibly. A yet simpler and more direct representation will be to define a binary predicate on the cartesion product ***nets*** × ***elements*** or a function from pins to that set.

Proposals have been made to make CODASYL model more suitable for CAD, of which Bandurski/ Jefferson [Ban] is a typical one. The proposed enhancements include among others:

- the direct representation of many-to-many relations,
- the removal of application-oriented capabilities from the schema DDL to the sub-schema DDL, such as ***set selection***, ***virtual result*** and ***source***, ***encoding/decoding***, etc.,
- the facility for defining additional sets in a sub-schema as parts of the user's view,

- the facility for recursively including in a view all the records below a given owner,
- the provision of procedures in a sub-schema to enable application programs to follow a single linear structure instead of traverse a tree,
- the mechanism to physically copy a selected part of the database into a working file and to produce reports of conflicts between the master and working files.

TORNADO (Technical Oriented Network Data Organization, Ulfsby et al. [Ulf]) which was originally developed for ship design, the same application domain as that of Bandurski/Jefferson, also incorporates several enhancements such as:

- the availability of variable length records,
- the availability of dynamic length *table* records,
- the mechanism to access a record by (user given) name,
- the unrestricted combination of record types and set types,
- the direct manipulation of many-to-many relations by hiding link records from application programs, and
- the avoidance of currency and other measures to simplify application programming.

The intended application of Zintle [Zin] is again the field of electronic CAD for which it is acknowledged that the structures offered by the CODASYL model are sufficient. It is contended that the real disadvantages are:

- the lack of flexibility of application programs with regard to changes of data structures to improve efficiencies, and
- the access to one record at one time.

The remedy is to define a high-level DML (HLDML) which allows application programs:

- to formulate database orders independently from record types and sets,
- to handle multiple logically connected records at one time, and
- to be free from the influence of changes in the data structures.

It appears from the syntax of HLDML that the language is strongly problem-oriented. The question is, then, why users prefer to rely on the CODASYL model, aside from their good experiences with it and the availability of a ready-to-use management system. Could it be that the CODASYL data management system is expected to serve as an internal schema description and structure manipulation language?

In summary, the CODASYL model and its implementation offers a framework which is approximately equivalent to ring-based packages in the light of its capability for conceptual modeling. Despite various efforts to enhance it, the fundamental limitations caused by the underlying structure and strict typing are not overcome.

2.4 Relational Model and Its Extensions

Among the more remarkable facts about the relational model are that it was originally proposed as an antithesis to preceding achievements and that theoretical foundations were much emphasized. It is no exaggeration to say that the history of data model theories commenced with the advent of the relational model. As with any theory-laden approach, we must direct our attention to the aspects of reality which have been discarded by the relational approach as well as those which have been abstracted and conceptualized within the theory. It was intended to provide a means of describing the 'natural' or simplest possible structures of information only, and to make application programs free from dependencies on such auxiliaries as ordering, indexing and access path. From the beginning, however, there was a question whether several conceptual devices which were indispensable for capturing the semantics of real-world entities were thrown out together with the garbage. A multitude of efforts have been made to salvage the lost treasures. As we will see gradually, the results are far less than satisfactory.

2.4.1 Structure and Basic Functions

(1) *Structure*
(1.1) *Constituents*
According to [Codd(1)] and [Codd(3)], a 'relation' R is a subset of the cartesian product

$$R_1 \times \cdots \times R_n \tag{2.8}$$

of 'domains' R_i and the database is a collection of relations. In a 'normal' form of relation, every domain is simple (nondecomposable) and actually a sub-type of some data type available in the host language. N-tuples in a relation are also called its 'elements'. Different domains may be of the same type. To distinguish the 'role' of a domain or its occurrence at a particular position (i) within (2.8) above, the term 'attribute' is used to denote the pair (i, R_i). It is in fact a function (projection)

$$\pi_i : R_1 \times \cdots \times R_n \to R_i \tag{2.9}$$

or its restriction on R.

An attribute or combination of attributes (π_I) is called a 'primary key' or simply a key of R if and only if all attribute values of an element of R are uniquely determined by those of π_I and I is minimal of such sets of indices. When the list of attributes of R contains a primary key of another relation S (which may be the same relation), that list is called a 'foreign key'. It is construed as a pointer by which each element of R cross-references an element of S.

(1.2) *Schema Description*
A schema description language will allow:

- the declaration of relations,
- the declaration of domains and attributes for each relation, and
- the identification of one or more primary keys for each relation.

(2) *Basic Functions*
(2.1) *The Relational Algebra*
A collection of operations on relations, called a relational algebra, embodies the data retrieval capacity of the relational model and includes:

- traditional set operations (union, intersection, difference and cartesian product) as applied to relations as sets of elements (tuples),
- projection (π_I) on a partial product ($\Pi_{i\in I}R_i$),
- θ-join of R on a list of attributes (indices) I with S on J defined by

$$R[I\theta J]S \equiv \{(r,s) : (r \in R) \wedge (s \in S) \wedge (\pi_I(r)\theta\pi_J(s))\} \tag{2.10}$$

  where θ is a binary predicate (e.g. $=, <, >$, etc.) defined over the common partial product $\Pi_{i\in I}R_i(= \Pi_{j\in J}S_j)$,
- division of R on a list of attributes I by S on J defined by

$$R[I \div J]S \equiv \{\pi_K(r) : (r \in R) \wedge (\pi_J(S) \subset \pi_I(R \cap \pi_K^{-1}(\pi_K(r))))\} \tag{2.11}$$

  where

$$I + K = \{1, \ldots, n\} \quad \text{(the list of indices of } R\text{)}$$
$$\Pi_{i\in I}R_i = \Pi_{j\in J}S_j \quad \text{and}$$

- θ-restriction of R on lists of attributes I and J defined by

$$R[I\theta J] \equiv \{r : (r \in R) \wedge (\pi_I(r)\theta\pi_J(r))\} \tag{2.12}$$

  where θ is a binary predicate (e.g. $=, <, >$, etc.) defined over the common partial product $\Pi_{i\in I}R_i(= \Pi_{j\in J}R_j)$.

When redundancies exist among relations, constraint statements of the form

$$\alpha(R, S, \ldots, T) = \beta(R, S, \ldots, T)$$

can and must be specified to maintain the consistency of the database, where α and β are formulas composed of the above operators and possibly functions and predicates defined over the component domains.

(2.2) *An Example*
A schema for the tetrahedron with winged-edge topology and some additional properties may be specified as follows:

relation *body*($B, F1, F2, F3, F4, E1, E2, E3, E4, E5, E6,$
$V1, V2, V3, V4,$ *density*), **key** B,
face(F, *orientation*, *finish*), **key** F,
edge($E, FL, FR, EFL, EFR, EBL, EBR, VF, VB$), **key** E,
vertex(V, *position*), **key** V;

domain $B, F1, \ldots, F4, E1, \ldots, E6, V1, \ldots, V4, F, E, V, FL, FR,$
$EFL, EFR, EBL, EBR, VF, VB$: **char**(5),
position, *orientation* : **record**(x, y, z : **real**),
density : **real**,
finish : **integer**.

B, $F1$, ..., $V4$ denotes the *body*, *face1*, ..., *vertex4* respectively. The roles of foreign keys FL, FR, ..., VB are the same as in Sect. 2.1.2 (3.1.1). As the underlying plane of each face is determined by the vertices on its boundary, redundancy exists among the face and vertex relations, and entails the stipulation of an integrity constraint such as

$$perpendicular(\pi_2(\pi_2(edge)),\ vector(\pi_2(\pi_8(edge)),\ \pi_2(\pi_9(edge)))).$$

The face lying opposite to a given face ($f1$) across a given edge ($e1$) can be obtained by

$$face := (\pi_{FL}(edge[E = e1]) \cup \pi_{FR}(edge[E = E1]))[F{\neq}f1] \tag{2.13}$$

which is the same as the case of STDS (Sect. 2.1.3 (3)) except for notational differences.

(2.3) *The Relational Calculus*
As characterized in [Codd(3)], the approach of relational algebra is 'constructive' and of an intermediate level. A 'calculus-oriented' language called ALPHA was proposed as one which was 'descriptive' and of a higher level [Codd(2)], [Codd(3)]. The term 'calculus' was used in an apparent analogy with the calculus of logic.

- There are three categories of terms in ALPHA:
  - *constants* $a_1, a_2, \ldots$ to denote attribute values or elements in component domains,
  - *variables* $r_1, r_2, \ldots$ to denote tuples in relations, and
  - $r.A$ to stand for $\pi_A(r)$ where A is an attribute of the relation to which r belongs.
- Predicates are either unary $P_1, P_2, \ldots$ to indicate membership in relations $R_1, R_2, \ldots$ respectively, or binary predicates ($=, <$, etc.) defined over the component domains. Atomic formulas will hence take either one of the forms:
  - $P(r)$, called a 'range term',
  - $r_1.A_1\theta r_2.A_2$ and $r.A\theta a$, called 'join terms'.
- A well-formed formula (WFF) is generally a boolean combination of subformulas each of which has the form

  $$U_1(r_1) \wedge \cdots \wedge U_n(r_n) \wedge V(r_1, \ldots, r_n) \tag{2.14}$$

  where U_i and V must satisfy the following set of conditions, called 'range separability':
  - for each i, U_i does not contain any quantifier or any variable other than r_i,

- atomic formulas in U_i are all range terms which are 'union compatible' (defined over the same cartesian product of domains),
- the logical operators permitted in U_i are $\wedge$, $\vee$ and $-$,
- all and only variables r_i are free variables in V,
- atomic formulas in V are all join terms, except in Γ explained below,
- the logical operators permitted in V are $\wedge$, $\vee$ and $-$ except for the case below, and
- the quantifiers may occur only in V in such forms as

$$\forall s(\Gamma(s) \rightarrow \Delta(s, r_1, \ldots, r_n)) \quad \text{and} \quad \exists s(\Gamma(s) \wedge \Delta(s, r_1, \ldots, r_n))$$

where Γ and Δ must satisfy the same conditions as U_i and V do respectively.

– Data manipulation statements of major importance include:
- **get** ⟨work space⟩ $(r_1.A_1^{(1)}, r_1.A_2^{(1)}, \ldots, r_n.A_1^{(n)}, r_n.A_2^{(n)}, \ldots) : \langle \text{WFF}(r_1, \ldots, r_n) \rangle$ to evaluate (determine the values of free variables that satisfy) $\text{WFF}(r_1, \ldots, r_n)$, apply the projection on $A_1^{(1)} \times \cdots \times A_1^{(n)} \times \cdots$ and deliver the result in the work space,
- **hold** ⟨work space⟩ $(r_1.A_1^{(1)}, r_1.A_2^{(1)}, \ldots, r_n.A_1^{(n)}, r_n.A_2^{(n)}, \ldots) : \langle \text{WFF}(r_1, \ldots, r_n) \rangle$; ⟨host language statements to possibly alter the contents of work space⟩;
- **update** ⟨work space⟩ where the **hold** functions to the same effect as a **get** except that it is used in conjunction with the **update** to change the database contents,
- **put** ⟨work space⟩ ⟨relation name⟩ to create tuples with the attribute values as stored in the work space and put them in the specified relation, and
- **delete** $r : \langle \text{WFF}(r) \rangle$ to delete from the database those tuples (denoted by r) that satisfy $\text{WFF}(r)$.

(2.4) *An Example*
The relational calculus counterpart of statement (2.13) will be

$$\textbf{get } ws\ s : face(s) \wedge (s.F \neq f1) \wedge ((s.F = e1.FL) \vee (s.F = e1.FR)). \quad (2.15)$$

The inverse problem will be answered by

$$\begin{aligned} \textbf{get } ws\ t : edge(t) \wedge (((t.FL = f1) \wedge (t.FR = f4)) \vee \\ ((t.FR = f1) \wedge (t.FL = F4))). \end{aligned} \quad (2.16)$$

(2.5) *Completeness of the Relational Algebra*
It has been proved [Codd(3)] that given a relational database, the extension of an arbitrary WFF (2.14) can be expressed as a sequence of operators of relational algebra applied to base relations. This fact, called the completeness of relational algebra, is understood as providing a constructive proof of the

completeness of the relational calculus as a special case of first order formal theory (see Theorem 3.2).

(3) *Sample Programs*
Let us present two more small examples to be examined later for evaluating the power of the ALPHA language.

(3.1) The database comprising relations *supplier*, *part*, *project* and *supply*:

relation *supplier*(*sname*, *loc*), **key** *sname*,
part(*pname*, *price*), **key** *pname*,
project(*jname*, *mgr*), **key** *jname*,
supply(*sname*, *pname*, *jname*, *qty*), **key** (*sname*, *pname*, *jname*);
domain *sname*, *loc*, *pname*, *jname*, *mgr* : **char**(10),
price : **fixed**(6.2),
qty : **integer**.

(3.1.1) To obtain the names and locations of the suppliers that supply some parts to all projects together with the names of parts and projects being supplied:

$$\textbf{get}\ ws\ (s, t.pname, t.jname) : supplier(s) \land supply(t) \land (s.sname = t.sname) \land \forall j(project(j) \rightarrow \exists u(supply(u) \land (s.sname = u.sname) \land (j.jname = u.jname))).$$

(3.1.2) To find names of such parts that are supplied by sole suppliers:

$$\textbf{get}\ ws\ t.pname : supply(t) \land (\textbf{icount}(t, pname, sname) = 1)$$

where **icount** is a library function to count the number of *snames* occurring in those tuples of relation *supply* that share a certain *pname*.

(3.1.3) *Propagation of Deletion of Tuples*
When a *supplier* (say *a*) is deleted from the database, all tuples of relation *supply* with this *sname* should be automatically deleted. If furthermore it is the user's intent to have all parts deleted for which that *a* is the sole supplier, the following should be executed:

$$\textbf{delete}\ s, t, p : supplier(s) \land supply(t) \land part(p) \land (s.sname = t.sname = a) \land \forall u(supply(u) \rightarrow ((u.pname \neq p.pname) \lor (u.sname = a))).$$

(3.2) The database comprising relations to represent circuits in terms of the types of elements, their positions and the nets connecting them:

relation *element*(*id*, *input*, *type*), **key** *id*,
pin(*element.id*, *no*, *position*, *type*), **key** (*element.id*, *no*),
use(*super-element*, *no*, *sub-element*, *position*), **key** (*super-element*, *no*),
net(*super-element*, *no*, *sub-element*, *pin.no*), **key** (*super-element*, *no*, *sub-element*, *pin.no*),
segment(*net.element.id*, *net.no*, *no*, *orig*, *dest*), **key** (*net.element.id*, *net.no*, *no*);

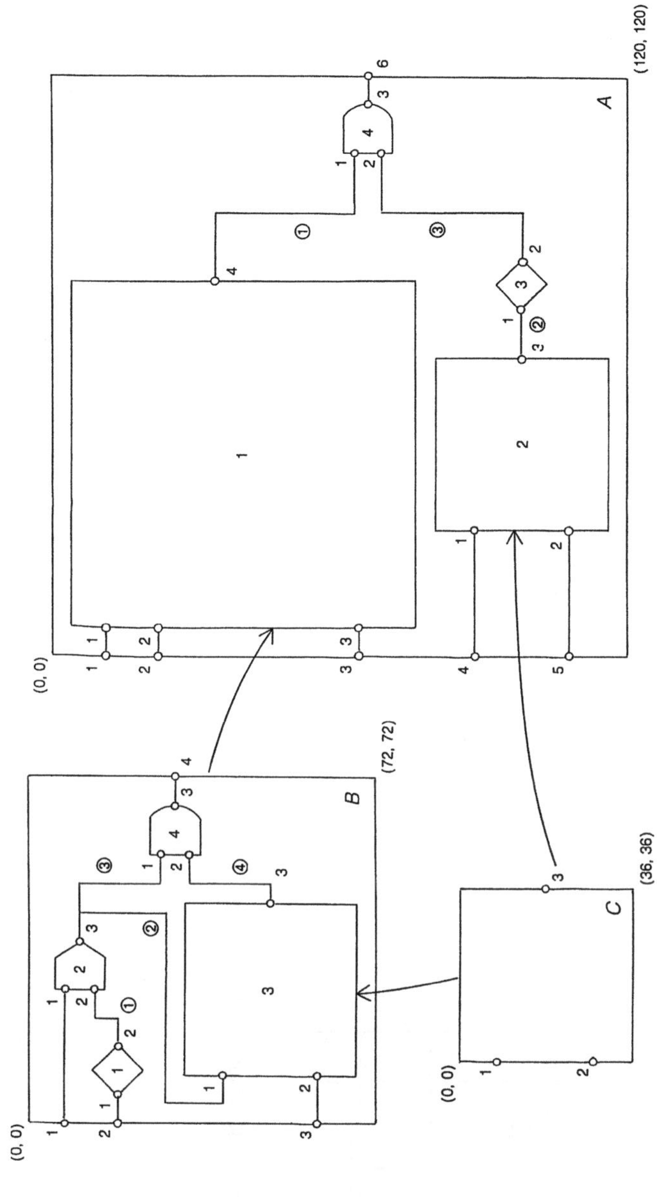

Fig. 2.3. **A circuit and its relational representation**

domain *element.id* : **char**(5), *super-element*, *sub-element* : *element.id*,
input, *pin.no*, *use.no*, *net.no*, *segment.no* : **integer**,
element.type : {'AND', 'OR', 'NOT', 'COMPLEX'},
pin.type : {'IN', 'OUT'},
position, *orig*, *dest* : **record**(x, y : **real**).

(3.2.1) To obtain all elements that are connected to input pins of a given element (e):

element	id.	input	type
	A	5	'Complex'
	A3	1	'Not'
	A4	2	'Or'
	B	3	'Complex'
	B1	1	'Not'
	B2	2	'And'
	B4	2	'Or'
	C	2	'Complex'

use	super-element	no.	sub-element	position
	A	1	B	4, 6
	A	2	C	80, 26
	B	3	C	32, 10

segment	net. element id	net no.	no.	orig.	dest.
	A	①	1	34, 78	34, 92
	A	①	2	34, 92	63, 92
	A	①	3	63, 92	63, 104
	B	④	3	33, 46	33, 56

pin	element id	no.	position	type
	A	1	0, 11	'In'
	A	2	0, 22	'In'
	A	6	66, 120	'Out'
	A3	1	98, 72	'In'
	C	3	18, 36	'Out'

net	super-element	no.	sub-element	pin
	A	①	1	4
	A	①	4	1
	B	④	3	3
	B	④	4	2

Fig. 2.3. (Continued)

get $ws\ x : element(x) \wedge$
$\exists m \exists n (net(m) \wedge net(n) \wedge (m.(super\text{-}element, no) =$
$n.(super\text{-}element, no)) \wedge (m.sub\text{-}element = e) \wedge (n.sub\text{-}element = x) \wedge$
$\exists p (pin(p) \wedge (p.element.id = e) \wedge (p.no = m.pin.no) \wedge (p.type = \text{'IN'}) \wedge$
$\exists q (pin(q) \wedge (q.element.id = x) \wedge (q.no = n.pin.no) \wedge (q.type = \text{'OUT'}))).$

(3.2.2) To verify that every input pin is connected to some output pin, and vice versa. This will be accomplished by executing the following statement and seeing if the result is empty:

$$
\begin{aligned}
&\textbf{get } ws\ p : pin(p)- \\
&\quad \exists q(pin(q) \land \exists m \exists n(net(m) \land net(n) \land (m.(super\text{-}element, no) = \\
&\qquad\qquad\qquad\qquad\qquad\qquad n.(super\text{-}element, no)) \\
&\qquad\qquad\qquad \land(m.(sub\text{-}element, pin.no) = p.(element.id, no)) \\
&\qquad\qquad\qquad \land(n.(sub\text{-}element, pin.no) = q.(element.id, no))) \\
&\qquad\qquad\qquad \land((p.type = \text{'IN'}) \rightarrow (q.type = \text{'OUT'})) \\
&\qquad\qquad\qquad \land((p.type = \text{'OUT'}) \rightarrow (q.type = \text{'IN'}))).
\end{aligned}
$$

2.4.2 Criticism

(1) *Advantages*
The elimination of data dependencies and the successful formulation of high-level logical languages are counted as the major contributions of the relational model, both of which derive from the uniform representation of all information in terms of attributes. The advantage of ALPHA, or relational calculus language in general, is to allow concise and non-procedural specifications of conditions for elements to satisfy. We can readily see this if we compare statements (2.15) and (2.16) with those with other languages for the same problems. The symmetry between the two statements should also be noted. It is an instance of what Codd [Codd(1)] calls 'symmetric exploitation of relations', a distinguished feature of the relational model which many other data models lack.

(2) *Facts Versus Things, or the Limitations of Expressive Power*
The uniformity has not been attained without cost, however. One of the fundamental losses incurred is the confusion between facts and things, or the elimination of ways to explicitly express the relationships among things as such. This is an overkill of the capacity of traditional methods which is so vital for capturing the semantics of the real world.

By definition, relations represent relationships among domains. Tuples will then assert facts about attribute values. A close look at the languages, specifically ALPHA, however, will reveal that tuples are actually treated as things rather than facts. First, they are called 'elements' or 'entities', second, they maintain identities through arbitrary changes of (non-key) attribute values, and third, quantifications are applied not to attribute but to tuple variables. The concept of foreign key further complicates the issue, as tuples containing foreign keys represent some facts about the things pointed to. Surprisingly, the issue did not even bother the founder of the theory, who disposed of it in the following way:

'... to treat the data in a data bank as consisting of two parts, one part consisting of entity descriptions (for example, descriptions of suppliers) and the other part consisting of relations between the various entities or types of entities (for example, the *supply* relation).

This distinction is difficult to maintain when one may have foreign keys in any relation whatsoever. In the user's relational model there appears to be no advantage to making such a distinction ....' [Codd(1)]

Can such a position be justified in the light of modeling capability? The answer must be no, it cannot, because it will impose unnatural, undesirable and unnecessary restrictions on the language. Take for example the problem of Sect. 2.4.1 (3.1.1). With liberal use of quantifications, we may write a simpler and more natural version

$$\textbf{get } ws\ (s, t.pname, t.jname) : supplier(s) \wedge supply(t) \wedge \\ \forall j(project(j) \rightarrow \exists p \exists q(supply(s.sname, p, t.jname, q))) \qquad (2.17)$$

which is unfortunately not a legitimate statement of ALPHA. The simplest of the class of expressions in question is

$$s : \forall j(J(j) \rightarrow P(s, j)) \qquad (2.18)$$

which is legal in standard FOPL but not in ALPHA. By replacing (s, j) by a single variable p and $P(s, j)$ by a corresponding join term, we obtain

$$p[1] : P(p) \wedge \forall j(J(j) \rightarrow (p[2] = j[1])) \qquad (2.19)$$

which is permitted in ALPHA yet has a meaning quite different from (2.18). (Indeed, the answer set of the above will be either empty or the entire relation P, depending on whether J contains more than one element or not.) A correct formula with valid syntax will be

$$p[1] : P(p) \wedge \forall j(J(j) \rightarrow \exists q(P(q) \wedge (p[1] = q[1]) \wedge (q[2] = j[1]))). \qquad (2.20)$$

Comparing (2.18) with (2.20) and generalizing the case, we may well conclude that:

- a tuple is considered as representing an entity or thing even when it contains one or more foreign keys,
- the database is a collection of thus represented entities,
- a relation defines a unary predicate, and at the same time a bundle of attributes (functions) over the database, and
- such unary predicates and the binary predicates derived through attributes from those in underlying domains constitute the sole axes along which to organize the database and general multi-term predicates are not really available.

(3) *Facts Versus Things (continued)*
While it is nice to have library functions such as **icount** available, we may mold a retrieval condition for the problem in Sect. 2.4.1 (3.1.2) after a more general pattern such as

$$pname : \forall s1 \forall s2((\exists j \exists q(supply(s1, pname, j, q)) \wedge \\ \exists j \exists q(supply(s2, pname, j, q))) \rightarrow (s1 = s2)) \qquad (2.21)$$

if quantifying attribute variables is again permitted, or equivalently, if tuples are considered as representing facts about independent entities.

We have stated that all tuples of *supply* with a particular supplier name should be automatically deleted when that name is deleted from the *supplier* relation. While this might sound only too logical, it is based on the implicit distinction between two relations, one representing facts and the other entities, which is not supported by the relational model. Hence, the propagation of deletions must always be explicitly stipulated by application programs regardless whether it is application dependent or independent.

(4) *Things Versus Aspects*
The description of an object is often distributed over more than one tuple in different relations, which when properly joined will restore that object in its entirety. The case of circuits in the preceding section is an example. The question is whether each of these tuples represents a specific aspect of the same object or they are mutually related yet self-subsistent entities. Without knowing the answer, the proper execution of joins will be impossible. It is hard to tell, however, simply by looking at the schema or relations since the answer depends on how these relations are interpreted. The problem is that the interpretation is not made explicit but left to application programs, causing undesirable interactions between programs and the database.

In the case of circuits, the description of a single net is split into multiple tuples even within the same relation (*net* and *segments*). This is apparently due to the inability of a rigidly formatted relation to accommodate a natural description of the fact that nets may be connected to an indefinite number of pins and may have an indefinite number of segments. A tuple in *net* relates a net to a pin, while one in *segment* describes a part of a net. The join of two relations may produce many tuples which are not only redundant but almost nonsensical.

(5) *Ambiguity of the Concept of Key*
The concept of key, which plays the double role of identifying and describing entities, is based on two premises: first, an entity is identified by a selected set of attributes, and second, the selection can be made not for each individual entity but for each a priori type of entity. While these are true of vouchers and entries in account books, we may rightly question the existence of such a set of attributes in the case of physical design objects. It seems customary and quite normal that, on one hand, we keep viewing things as maintaining their identities through whatever changes of attribute values they may undergo, and on the other, we often regard things with the same attribute values as separate. To mention a very simple case, we make such statements as that two (different) points occupy the same position (the sole attribute of entities of type *point*) or this (same) point changes its position from here to there.

The distinction between identification and description is not wholly blurred in ALPHA, however, as modifying key values is prohibited. This could make the matter worse if the key is selected from the properties which may change without affecting the identity (such as the position of a moving point). If instead an artificial key is devised (such as ***net.no*** or ***pin.no***), this creates an additional responsibility for the application program to have under its control a kind of data which is of little significance for it, a very undesirable effect.

(6) *Type Versus Homogeneity of Attribute List*
A relation schema defines the type of its elements in much the same way as a record type does of record instances in the CODASYL model, where a 'type' is virtually identified with a homogeneous setting of attributes. This may again be acceptable for business applications, but it is not acceptable for CAD, where on one hand, no fixed and formatted collection of attributes can adequately describe objects, and on the other, objects may continue to be the same regardless of changes in attribute settings. For example, we will have to prepare a separate relation for each number of segments if we want to represent a net as a single tuple, and we will have to expect that it may leave one relation for another if we allow changes in the segment construction of a net. Yet, how could it be possible that an entity which may exist only as an instance of its type remains the same when it becomes of a different type? (See also the discussion in Sect. 2.4.4.)

2.4.3 Semantic Extensions of Relational Model

It would be unfair to ignore the multitude of efforts to extend the relational model so that the meaning of the universe of discourse may be better captured. In the following, we will limit ourselves to taking a look at some classical achievements, not only because there are simply too many proposals but also because the majority of them are technically very detailed and irrelevant to our direct concerns. A few more recent developments will be examined later (see Chap. 4). The topics we want to discuss here include:

- theory of normal forms and normalizations,
- classification of relations into different types,
- abstraction by aggregation and generalization, and
- introduction of domain relational calculus.

(1) *Normal Forms and Normalizations*
A relation schema R^* is said to be <u>not</u> in 3NF (the third normal form) [Codd(4)] if and only if there exists a division of attributes into mutually disjoint subsets X, Y and Z such that:

- X does not contain a key,
- XY contains a key, and
- any relation $R(= R[X, Y, Z])$ of type R^* can be expressed as

$$R[X, Y, Z] = R[X, Y] * R[X, Z] \tag{2.22}$$

where the right side is the join on the common set of attributes (X).

Note that X contains a key in $R^*[X, Z]$ as inferred from the second and third conditions above.

R^* is said to be not in 4NF (the fourth normal form) [Fag(1)] if and only if there exists a division of attributes as above except that XY need not contain a key. It is obvious that being in 4NF implies being in 3NF.

R^* is said to be not in PJNF (projection-join NF) [Fag(2)] if and only if there exists a division of attributes into (not necessarily disjoint) subsets $X_1, \ldots, X_n$ such that:

- any relation $R(= R[X_1, \ldots, X_n])$ of type R^* can be expressed as

$$R[X_1, \ldots, X_n] = R[X_1] * \cdots * R[X_n] \tag{2.23}$$

  which is not a logical consequence of the stipulation for certain sets of attributes to be keys of R.

It is again obvious that being in PJNF implies being in 4NF, hence in 3NF. (1NF and 2NF are also defined, but we may dismiss both as they are very weak and have little relevance to our analysis here.)

Given an arbitrary relation schema, we can always convert it by repeating decompositions into a set of projections which are in a normal form. This process of normalization is intended to make clear the semantics of a relation schema. The idea behind it is that:

- each of the projections which when joined will recover the original relation is considered as carrying its own meaning independently of others, hence
- a decomposable relation schema embodies a mixture of different meanings, and
- accepting such a mixture tends to not only complicate but also to blur the semantics of a relation schema.

Questions will be raised, however, as to whether the intended goal is really achieved:

- The way of decomposing a given relation schema might not be unique. Where can we find the meaning of a component of one decomposition in another in which it does not exist?
- A new local relationship, hence decomposability, could become extant as the result of taking a projection of a relation schema.
- Conversely, related attributes could be separated into different components causing their relationship to be nonextant within any one component schema. An interrelational relationship (to coin a term!) will have to be specified to salvage the loss.
- The larger the number of relations gets, the more difficult it will become to maintain consistency, as relations can in principle be updated independently of each other.
- If decompositions are carried out to the ultimate, we will obtain

$$R \equiv R[K_1, A_1] * \cdots * R[K_r, A_r] \tag{2.24}$$

where each K_i is a key and A_i is a single attribute not contained in K_i. While such a decomposition does not seem to buy anything for some authors (e.g., Fagin [Fag(2)]), others (e.g., Kowalski [Kow]) advocate it as being more advantageous than general n-ary relations. We seem to lack a common semantic ground on which to evaluate syntactic alternatives.

We might well view a tuple as a sentence in which the key acts as the subject of which other attributes are predicated. Normalization is then a process to convert a complex sentence schema into a conjunction of simple ones, with the simplest comprising the subject and only one attribute. It will also be understandable if an attempt is made to have the subject be devoid of any particular property and the elementary sentences be coherent with each other. A rather naive yet always dependable means to accomplish this is to provide an artificial key or 'surrogate', with which (2.24) will become

$$R \equiv R[K, A_1] * \cdots * R[K, A_r]. \tag{2.24$'$}$$

If furthermore we allow attribute values to be undefined for some entities, it will be more natural to define the database schema as the collection of separate binary relations

$$\Sigma = \{P_1(K, A_1), \ldots, P_r(K, A_r)\} \tag{2.24$''$}$$

than that of n-ary decomposable relations. This is what Kowalski [Kow] actually proposes. It should be noted, however, that much of the semantics will now have to be captured by interrelational constraints in which n-ary predicates could be resurrected.

(2) *Classification of Relations*
Schmidt/Swenson [Sch] classifies relations into five types:

- Type 1 represents objects that exist independently of any attributes or other objects together with non-repeating simple attributes and/or the keys of non-repeating complex attributes. ('non-repeating' means that the value is uniquely determined by the object (like the type of a circuit element) whereas a 'repeating' attribute may have multiple values for the same object (like the segment of a net). A 'simple' attribute is such that no other attributes may not be predicated of it (like the type of a circuit element) whereas a 'complex' attribute will be further characterized by other attributes (like the segment of a net with two end positions).
- Type 2 contains repeating simple attributes or the key of a repeating complex attribute.
- Type 3 represents non-repeating complex attributes together with non-repeating simple attributes and/or the keys of non-repeating complex attributes of their own.
- Type 4 represents repeating complex attributes and contains repeating simple attributes or the key of a repeating complex attribute.
- Type 5 represents relationships between independent objects.

Roughly speaking, relations of type 2 and 3 are the results of normalizations into 4NF and 3NF respectively, and those of type 4 are the combination of these. If all related tuples of types 2, 3 and 4 are joined together with a type 1 tuple, the original complex independent object will be restored. Relations of type 5 have no interaction with the interrelational bundling of other types. The underlying view for this separation is that there are two kinds of object, namely one existing on its own and the other used only for describing another object. It is questionable, however, whether such a distinction is really possible. For instance, pins, which may well be defined as a repeating complex attribute of an element, may equally well be related to a net, another independent object. Generally, a type 5 relation can describe as complex a situation as complex independent objects do, hence be subject to similar decompositions. More types will be spawned. A question then arises whether such a differentiation of relation types really reflects the semantic structure of the world, or whether it is motivated by the technical goal of attaining an optimal representation of information.

The approach taken by Chen [Che] ('Entity-Relationship' (ER) model) is much the same as the above. It starts with the commitment to

> 'the more natural view that the real world consists of entities and relationships'

which is precisely the opposite to the previously quoted proclamation by [Codd(1)] (Sect. 2.4.2 (2)). It is said that

> 'An *entity* is a "thing" which can be distinctly identified. ... A relationship is an *association* among entities.'

and also that the method proposed is to

> 'separate the information about entities from the information about relationships'.

There are four kinds of relation in the ER model, that is:

- *regular entity* relations to represent entities which can be identified without using any relationships with other entities,
- *weak entity* relations to represent entities which will be identified only by means of relationships with other entities,
- *regular relationship* relations to represent relationships among regular entities, and
- *weak relationship* relations to represent relationships among entities some of which are weak entities.

It will not be hard to observe the correspondence as follows between the above kinds and the five types of Schmidt/Swenson's model:

- Types 1,2 and 5 correspond to regular entity, weak entity and regular relationship relations respectively.
- Types 3 and 4 are promoted to types 1 and 2, hence correspond to regular entity and weak entity relations respectively.

The concept of a weak relationship relation, which has no counterpart in Schmidt/Swenson's model, seems useful for representing such a case as the relation between a net and a pin.

The ER model has had a considerable influence on practical as well as theoretical works on the management of databases, including those for CAD. This is partly (or perhaps primarily) due to the distinction made between entities and relationships, or things and facts, on which many researchers and practioners would agree. So does this author, yet not without reservations. Two questions still linger, namely:

- Will the notion of a 'weak' entity be a truly viable one? Should we rather postulate explicitly the conditions for the existence of certain entities to depend on that of others?
- Why should we stick to the concept of 'relation' to represent entities which by definition would exist independently of any relations with other entities? (Will the term 'relation' not signify also by definition the involvement of two or more things?)

(3) *Abstraction by Aggregation and Generalization*
In Smith/Smith's model [Smi], relations are organized into hierarchies along the *aggregation* and *generalization* axes. These will cause multi-level vertical and horizontal decompositions of relations. An aggregate will comprise the keys of one or more relations of lower level (in the sense of aggregation) together with its own attributes. A relation (say R) may define a 'cluster' of relations (say $S_1, \ldots, S_n$) of lower level (in the sense of generalization) in which each S_i is characterized by a specific value of a given attribute A of R. Each S_i may have additional attributes of its own. More than one cluster may be defined for the same relation.

Syntactically, decompositions along the aggregation axis may be realized by normalizations into 3NF or 4NF. These are not semantically equivalent, however, as every relation whether it results from aggregation or decomposition must represent a generic object (that is, an object type) which can be properly named by a natural language noun. A relationship between objects is itself defined as a named object. An implication of this is that the distinction between things and facts is again confused. (Or perhaps, the concept of relationship as such does not even exist.)

Also syntactically, the relationship between a parent relation and members of a cluster will be implemented by means of foreign keys. Semantically, one may question whether the member relations represent independent objects or mere aspects of the parent. The rules for integrity maintenance seem to indicate the latter. Then, will a split along generalization axis be anything more than a matter of technical convenience?

(4) *Introduction of Domain Relational Calculus*
Some authors (Ullman [Ull], Lacroix/Pirotte [Lac(1)], Chang [Chan]) propose a version of relational calculus in which quantifications are applied to attribute variables. It has been shown ([Ull]) that both versions are equiva-

lent in their expressive powers. We have indicated that the use of attribute variables would indeed be more natural when relations represent relationships between independent objects. There are cases, however, where relations describe single objects. What we need is a more flexible language in which variables can be used to denote both tuples and attributes, and the range of a variable is not restricted to a single relation or its projection on a domain.

It should be clear from the above analysis that each of the proposed semantic extensions solves one or more of the difficulties of the relational model, yet no one extension solves all the difficulties. This is not surprising, because the difficulties are inherently due to the record orientation of the relational model, that is, to rigid typing, homogeneity of the attribute list, the concept of key, etc., which remains unchanged for all extensions. In concluding the present section, the author would like to point out that the record orientation alone has been much criticized from the general semantic viewpoint of data models. In this regard, readers are especially referred to [Ken(1)] and [Ken(2)], which form one of the typical yet most radical instances of such criticisms. Another reference we want to mention in particular is the report by ISO ([ISO(1)] followed by [ISO(2)]). As its title indicates, the emphasis is on the conceptual realm of data schema definition which concentrates on the meaning of information. Although no specific language is proposed, the use of interpreted predicate logic for defining a conceptual schema is suggested.

2.4.4 Applications of the Relational Model to CAD

There have been numerous reports on applying the relational model to CAD. Some of the efforts, particularly those in early days, are based on almost the same model as originally proposed by [Codd(1)] and [Codd(2)] ([Wil(2)], [Wil(3)], [Wil(4)], [Wel], [Val], [Phi]). Others more or less include extensions and/or modifications to the original model to better suit their specific universes of discourse. Let us take a few exemplary cases of the latter and see how the extensions/modifications are actually done.

The approach of Grabowski/Eigner ([Grab(1)] and [Grab(2)]) is apparently under the strong influence of [Sch] and [Che], and is characterized by the following features:

- There are three basic types of constituent of schema, that is, *entity*, *association*, and *attribute*.
- For each entity type, a fixed set of attributes is defined, one of which is the system controlled key to uniquely identify each occurrence of that type. An entity type can be further split into not necessarily disjunct sub-types.
- An association type defines a predicate over two or more entity types. The identifier consists of the system keys for participating entities. It may have one or more attributes of its own.
- At any given time, the set of occurrences of an entity or association type is considered as a relation among the attributes. All entity or association types, as relation schemas, must be in 4NF or PJNF respectively.

- Attributes are differentiated as *simple*, *compound* (i.e., record), *simple repetitive*, *compound repetitive* and *dynamic* (i.e., procedural) types.
- There are two types of association, that is, *independent* and *characterizing* associations. In either case, an occurrence can exist only if all of the participating entity occurrences exist. What differentiates them is that:
  - an occurrence of an entity type can exist independently of any occurrence of any independent association type in which that entity type participates,
  - an entity type will participate in a characterizing association type either as a superior or as a characterizing (attributive) entity type,
  - an occurrence of an entity type will entail that of each characterizing association type in which that entity type participates as a superior entity, and
  - an occurrence of an entity type can exist only if there exist some occurrences of some characterizing association type in which that entity type participates as a characterizing entity.
- For each association type and for each entity type participating in it, the maximum and minimum numbers of occurrences of the former that contain any specific occurrences of the latter must be specified as a special kind of integrity constraint.

A characterizing entity is nothing but a bundle of attributes which is given the status of an entity, which allows alternate descriptions of a superior entity type. For example, the planar line segment may be defined either as

- a single entity with a quadruplet of real type attributes, or as
- a superior entity related through a characterizing association to two end points, each of which represents a couple of coordinate values.

This approach enjoys certain practical as well as theoretical advantages over the standard relational model. One advantage which is particularly relevant to CAD is the automatic maintenance of positional relationships between geometric elements which is enabled by the introduction of the system key and the distinction of association from entity (Fig. 2.4).

Conceptual ambiguity and structural rigidity, which tend to limit expressive power, have not been wholly eliminated, however. To point out some more serious flaws:

- As a repetitive attribute is described in its own relation schema, the same system key value will appear in more than one relation. Which will the system key serve to identify, a tuple or an abstract entity? With the understanding that the latter is the case, we question why repetitive attributes cannot be included in the same schema as other attributes.
- An entity may possibly contain as an attribute the key of a complex attribute, which will blur the distinction between entity and association.
- On the other hand, what is the ontological status of a characterizing entity? That it may be related as an attribute to multiple entities is syntactically the sole reason for it to be regarded as an entity. A superior entity may also

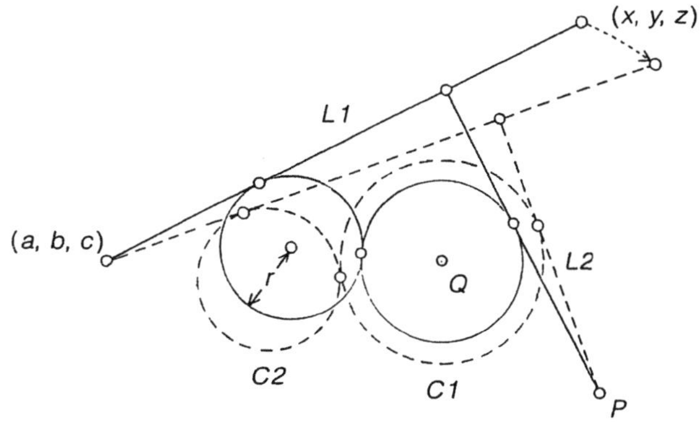

L1 **line**, *(a, b, c)*, *(x, y, z)*
L2 **line-thru**, *P*, **perp-to**, *L1*
C1 **circle**, *Q*, **tan-to**, *L2*
C2 **circle-tan-to**, *L1*, *C1*, *r*

Fig. 2.4. Maintaining positional relationships

be related to multiple subordinate entities, however. Let us consider, for example, the case of a polygon having its edges as characterizing entities. Intuitively, the polygon can be deleted leaving any part of the boundary intact, but not vice versa. On the contrary, the formal rule allows removing any number of edges so long as there remain at least the minimum number of edges after the removal. The legitimacy of the asymmetry between superior and subordinate entities is questionable.

- The above example indicates also the inadequacy of limiting the number of superior entity occurrences as an integrity constraint.
- Perhaps the introduction of characterizing entity and characterizing association is the way to implement the equality relationship holding between different attributes. If so, it should be treated as something that belongs to the internal schema level, while at the conceptual level a more abstract yet precise formulation of integrity constraints (such as by means of closed logical formulas) will be more appropriate.
- That the key of an association type consists of those of participating entity types is tantamount to defining a (single valued) function over the cartesian product of entity sets of these types. If we admit the repetitive type of attribute of a single entity type, why don't we do the same of multiple entity types? Or more generally, should we not be able to define associations in which all arguments, including attributes, participate on equal terms?
- Typing is as rigid as before with every entity permitted to exist only as an instance of a certain type which is characterized by a fixed setting of attributes. A simple case which needs a more flexible treatment is that of a polygonal object and its edges, which maintain their identities through arbitrary changes to the types of edges (Fig. 2.5) With regard to this point,

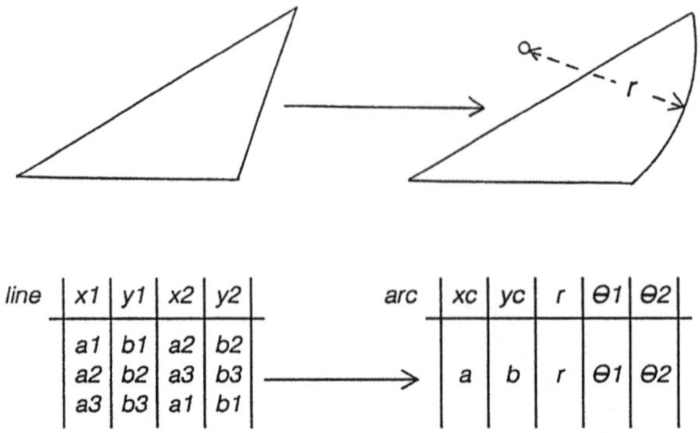

Fig. 2.5. Identities maintained through changes of types

readers are referred to Challis [Chal] which develops interesting arguments on alternate ways of typing: rigid and static versus flexible and dynamic.

Haskin/Lorie [Has] and Lorie et al. ([Lor(1)], [Lor(2)], [Lor(3)]) also adopt the means of a system generated key to identify objects and have them cross-reference each other. Otherwise, their approach differs noticeably from that of Grabowski/Eigner as seen below:

- There is no such thing as association, but cross-references are made within tuples that themselves represent entities, hence relationships are limited to binary type.
- Relations form a tree in which a tuple normally references another of superior type. Each root tuple heading a chain of successively referencing tuples defines a 'complex object' which can be retrieved as a whole by specifying that root. It is treated as a unit of locking, and checked in and out between private and public data spaces.
- A repetitive attribute is represented by a set of tuples which share the same superior entity. An example is the series of segments of a net on an IC chip (see Sect. 2.4.1 (3.2)). To facilitate processing such a set properly, the introduction of an order among the tuples within each set is proposed.
- A 'cursor' mechanism is provided to enable piecewise processing of extraordinarily long attribute fields, which will be useful particularly for such applications as finite element analysis and numerically controlled machining.

The importance of handling complex objects and long attribute fields cannot be overemphasized. It does not seem always easy, however, to fit these ideas naturally into the above framework. First, we might want to retrieve all superstructures that commonly interface a given substructure or the set of entities that satisfies a specified condition on their attributes. The provision of general n-ary associations which can be accessed via any of the arguments seems inevitable. Second, attribute fields which are not only long but themselves structured will facilitate representing repetitive attributes in a more straighforward manner. Handling of multiple tuples as if they formed an integral unit of information, not to speak of ordering among them, contradicts the original spirit of the relational approach. Moreover, these tuples do not themselves represent any real entities but are mere aspects of a superior entity. Separation of the relation to contain them, if it is technically preferable, had better be hidden from the conceptual schema and application programs.

Haynie [Hay] presents an interesting possibility to combine the desirable features of the relational and network data models. It aims at organizing data in a network which offers itself as a natural model for design objects, specifically a computer, without 'defiling' the simplicity and cleanliness of the relational approach. The network structure is also expected to attain higher performance and reduced data redundancy.

- Within a relation of an owner type, each tuple contains as the value of an attribute ('access attribute') the identifier of the relation of its member tuples, which may in turn be another owner relation.
- The net effect of the above from the network model's standpoint is that:
  - the collection of member records owned by a particular record in a set is given a name and can be referenced by that name,
  - within each set, the type of member records must be unique,
  - linear structures among member records by keys and sort key are missing, instead simultaneous retrievals of multiple records are enabled.
- The same can be seen from the relational viewpoint as:
  - the admission of multiple instances of a relational schema, and
  - the relaxation of the condition that all relations be in the first normal form (that is, no attribute may take relations as its values).
- Data definition and manipulation facilities are provided by a modified SQL including among others dynamic creation, deletion and expansion of relational schemas.

As pointed out in the beginning of the present section, the relational approach has been proposed as an antithesis to previous, particularly heavily structured models. Hence, hybridization with the network model does not sound very hopeful. While Haynie has successfuly devised a mechanism to incorporate a hierarchical structure into the pure relational model, it is essentially no more than a scheme to implement a repetitive attribute with recursive structures of its own and manipulate it as a whole.

We have taken a mere glimpse at the diverse attempts to describe design objects à la relational model. The very diversity tends to make us suspect that there may exist a fundamental difficulty that cannot be overcome but only alleviated in one way or another. As we have seen already and will see in more detail, what we need is liberation from the spell of rigid typing of data and the underlying entity-centered world view.

References

[ACM] ACM Computing Surveys: The Development of Data-Base Technology. Vol. 8, No. 1, March 1976

[Ban] Bandurski, A.E., Jefferson, D.K.: Enhancements to the DBTG Model for Computer-Aided Ship Design. Proc. Workshop on Data Bases for Interactive Design, 1975

[Bra] Braid, I.C.: On Storing and Changing Shape Information. SIGGRAPH Computer Graphics, Aug. 1978

[CAD] Computer Aided Design, Special Issue: Databases. Vol. 11, No. 3, May 1979

[Chal] Challis, M.F.: Typing in Data Base Models. In: [Enc]

[Chan] Chang, C.L.: DEDUCE 2: Further Investigations of Deduction in Relational Data Bases. In: [Gal]

[Che] Chen, P.P.-S.: The Entity-Relationship Model – Toward a Unified View of Data. ACM TODS, Vol. 1, No. 1, March 1976

[Chi] Childs, D.L.: Description of a Set-theoretic Data Structure. Proc. FJCC, 1968

[CODA(1)] CODASYL Data Base Task Group Report. ACM, 1971

[CODA(2)] CODASYL Data Description Language Journal of Development. 1978

[Codd(1)] Codd, E.F.: A Relational Model of Data for Large Shared Data Banks. CACM, Vol. 13, No. 6, June 1970

[Codd(2)] Codd, E.F.: A Data Base Sublanguage Founded on the Relational Calculus. Proc. ACM-SIGFIDET Workshop on Data Description, Access and Control, 1971

[Codd(3)] Codd, E.F.: Relational Completeness of Data Base Sublanguages. Courant Computer Science Symposia 6, Data Base Systems, Prentice-Hall, 1971

[Codd(4)] Codd, E.F.: Further Normalization of the Data Base Relational Model. Ibid.

[Dass] Dassler, R. et al.: Databases for Geometric Modelling and Their Application. In: [Enc]

[Dod] Dodd, G.G.: APL – A Language for Associative Data Handling in PL/1. Proc. FJCC, 1966

[Eas(1)] Eastman, C.M. et al.: A Database for Designing Large Physical Systems. Proc. NCC, 1975

[Eas(2)] Eastman, C.M.: The Concise Structuring of Geometric Data for Computer Aided Design. In: Data Structures, Computer Graphics and Pattern Recognition. Academic Press, 1977

[Eas(3)] Eastman, C.M.: System Facilities for CAD Databases. Proc. 17th Design Automation Conf., 1980

[Enc] Encarnação, J., Krause, F.-L. (eds.): File Structures and Databases for CAD. North-Holland, 1982

[Fag(1)] Fagin, R.: Multivalued Dependencies and a New Normal Form for Relational Databases. ACM TODS, Vol. 2, No. 3, Sept. 1977

[Fag(2)] Fagin, R.: Normal Forms and Relational Database Operators Proc. SIGMOD, 1979

[Fel] Feldman, J.A., Rovner, P.D.: An ALGOL-Based Associative Language. CACM, Vol. 12, No. 8, Aug. 1969

[Foi] Foisseau, J., Vallette, F.R.: A Computer Aided Design Data Model: FLOREAL. In: [Enc]

[Gal] Gallaire, H., Minker, J. (eds.): Logic and Databases. Plenum, 1978

[Grab(1)] Grabowski, H., Eigner, M.: Semantic Datamodel Requirements and Realization with a Relational Datastructure. In: [CAD]

[Grab(2)] Grabowski, H., Eigner, M.: A Data Model for a Design Data Base. In: [Enc]

[Gray] Gray, J.C.: Compound Data Structure for Computer Aided Design: A Survey. Proc. ACM National Meeting, 1967

[Has] Haskin, R.L., Lorie, R.A.: On Extending the Functions of a Relational Database System. Proc. ACM SIGMOD Conf., 1982

[Hay] Haynie, M.N.: The Relational, Network Hybrid Data Model for Design Automation Databases: Proc. 18th Design Automation Conf., 1981

[IEEE] Proc. Data Base Week: Engineering Design Applications. IEEE, 1983

[ISO(1)] ISO TC97/SC5/WG3 N695: Concepts and Terminology for the Conceptual Schema and the Information Base. April 1982

[Iso(2)] ISO TC97/SC21/WG5-3 N991: Assessment Guidelines for Conceptual Schema Language Proposals. Aug. 1985

[Kent(1)] Kent, W.: Data and Reality. North-Holland, 1978

[Kent(2)] Kent, W.: Limitations of Record-Based Information Models. ACM TODS, Vol. 4, No. 1, March 1979

[Kor] Korenjak, A.J., Teger, A.H.: An Integrated Database System. Proc. 12th Design Automation Conf., 1975

[Kow] Kowalski, R.A.: Logic for Data Description. In: [Gal]

[Lac(1)] Lacroix, M., Pirotte, A.: Domain-Oriented Relational Languages. Proc. VLDB, 1977

[Lac(2)] Lacroix, M., Pirotte, A.: Data Structures for CAD Object Description. Proc. 18th Design Automation Conf., 1981

[Lan] Lang, C.A., Gray, J.C.: ASP – A Ring Implemented Associative Structure Package. CACM, Vol. 11, No. 8, Aug. 1968

[Lor(1)] Lorie, R., Plouffe, W.: Complex Objects and Their Use in Design Transactions. In: [IEEE]

[Lor(2)] Lorie, R.: Issues in Databases for Design Applications. In: [Enc]

[Lor(3)] Lorie, R., Daudenarde, J.-J.P.: On Extending the Realm of Application of Relational Systems. Information Processing 86, 1986

[Phi] Phillips, R.J. et al.: AESOP: An Architectural Relational Database. Computer Aided Design, Vol. 11, No. 4, July 1979

[Sch] Schmidt, H.A., Swenson, J.R.: On the Semantics of the Relational Data Model. Proc. ACM SIGMOD Conf., 1975

[Smi] Smith, J.M., Smith, D.C.P.: Data Abstractions: Aggregation and Generalization. ACM TODS, Vol. 2, No. 2, June 1977

[Ulf] Ulfsby, S. et al.: TORNADO: A DBMS for CAD/CAM Systems. Computer Aided Design, Vol. 13, No. 4, July 1981

[Ull] Ullman, J.D.: Principles of Database Systems. 2nd ed., Computer Science Press, 1982

[Val] Valle, G.: Relational Data Handling Techniques in Computer Aided Design Procedures. In: Allan, J.J. (ed.): CAD Systems. North-Holland, 1977

[War] Warn, D.R.: VDAM – A Virtual Data Access Manager for Computer Aided Design. Proc., Workshop on Databases for Interactive Design, ACM, 1975

[Wel] Weller, D., Williams, R.: Graphic and Relational Data Base Support for Problem Solving. Computer Graphics, Vol. 10, 1976

[Wil(1)] Williams, R.: A Survey of Data Structures for Computer Graphic Systems. ACM Computing Surveys, Vol. 3, No. 1, March 1971

[Wil(2)] Williams, R.: On the Application of Relational Data Structures in Computer Graphics. Information Processing 74, North-Holland, 1974

[Wil(3)] Williams, R. et al.: Data Structures in Computer Graphics. Proc. Workshop on Database for Interactive Design, 1975

[Wil(4)] Williams, R., Giddings, G.: A Picture-Building System. IEEE Trans. Software Eng., Vol. SE-2, No. 1, March 1976

[Zin] Zintl, G.: A CODASYL CAD Data Base System. Proc. 18th Design Automation Conf., 1981

Chapter 3
First Order Theory and Its Interpretation

In this preparatory chapter, we present in a condensed manner the fundamental concepts and certain important results of first order predicate logic (FOPL) which we will use or reference in succeeding chapters. We have tried to state all definitions in a rigorous and self-contained manner, yet because of the limited space, we have omitted the proofs of theorems. Interested readers are referred to Shoenfield [Sho], Manna [Man] and Lloyd [Llo]. The word 'logic' is understood to mean the discourse that deals with a formal language, theories constructed in it and their interpretations. These are the respective subjects of the first three sections. Next, some of the major results on the undecidability and incompleteness of FOPL are summarized. The final section discusses a restriction of FOPL which is particularly relevant to computer science in that it forms the theoretical basis for theorem proving and logic programming.

3.1 First Order Language

(1) *Logical Symbols*

⟨variable⟩ ::= ⟨any of the finite or countably infinite number of symbols reserved for denoting variables⟩
⟨operator⟩ ::= ⟨negation (¬)⟩ | ⟨binary operator⟩
⟨binary operator⟩ ::= ⟨or (∨)⟩ | ⟨and (∧)⟩ | ⟨difference (−)⟩ | ⟨implication (→)⟩
⟨quantifier⟩ ::= ⟨existential quantifier (∃)⟩ | ⟨universal quantifier (∀)⟩
⟨equality symbol⟩ ::= =

(2) *Non-Logical Symbols*

⟨function symbol⟩ ::= ⟨n-ary function symbol ($n = 0, 1, 2, \ldots$)⟩
⟨n-ary function symbol⟩ ::= ⟨any of the finite or countably infinite number of symbols reserved for denoting n-ary functions⟩
⟨constant⟩ ::= ⟨0-ary function symbol⟩
⟨predicate symbol⟩ ::= ⟨n-ary predicate symbol ($n = 0, 1, 2, \ldots$)⟩

⟨n-ary predicate symbol⟩	::=	⟨any of the finite or countably infinite symbols reserved for denoting n-ary predicates⟩
⟨logical constant⟩	::=	⟨0-ary predicate symbol⟩

(3) *Punctuation Marks*

⟨punctuation mark⟩::=, | (|)

(4) *Terms*

⟨term⟩	::=	⟨variable⟩ | ⟨constant⟩ | ⟨n-ary function symbol⟩ (⟨term_1⟩, ..., ⟨term_n⟩) ($n \geq 1$)
⟨ground term⟩	::=	⟨term containing no variable⟩

(5) *Formulas*

⟨atomic formula⟩ or ⟨atom⟩	::=	⟨logical constant⟩ | ⟨term⟩ = ⟨term⟩ | ⟨n-ary predicate symbol⟩ (⟨term_1⟩, ...,term_n⟩) ($n \geq 1$)
⟨literal⟩	::=	⟨atomic formula⟩ | ¬⟨atomic formula⟩
⟨ground atom⟩	::=	⟨atom containing no variable⟩
⟨ground literal⟩	::=	⟨literal containing no variable⟩
⟨formula⟩	::=	⟨atom⟩ | (⟨formula⟩) | ¬⟨formula⟩ | ⟨formula⟩ ⟨binary operator⟩ ⟨formula⟩ | ⟨quantifier⟩ ⟨variable⟩ ⟨formula⟩

A variable preceded by ∃ or ∀ is said to be existentially or universally quantified respectively. The formula succeeding it is the 'scope' of quantification. Occurrences of a quantified variable within the scope of quantification are called 'bound'. Occurrences of variables which are not quantified are 'free'. For example:

$$\exists x(\;p(x)\;\wedge\;\forall y(\;q(y)\;\rightarrow\;r(x,\;y))\;\wedge\;s(x,\;y,\;z)) \tag{3.1}$$

```
∃x(  p(x)   ∧   ∀y(  q(y)   →   r(x,  y))  ∧  s(x,   y,   z))
|      |        |      |          |    |          |    |    |
|    bound      |    bound        |  bound        |    |   free
existentially   universally     bound             |   free
quantified      quantified                      bound
```

Formulas with no free variables are called 'closed'. Otherwise, they are 'open'. As we will see later, the meaning of a formula will remain the same if any quantified variable and its bound occurrences are replaced by any other variable that does not occur in the scope of quantification. It will hence be appropriate to abbreviate a formula to show only its dependence on free variables. For example, the above formula may be written as

$$\begin{aligned}\alpha \equiv \alpha(y,z) &\equiv \exists x(p(x) \wedge \forall y(q(y) \rightarrow r(x,y)) \wedge s(x,y,z)) \\ &\equiv \exists u(p(u) \wedge \forall v(q(v) \rightarrow r(u,v)) \wedge s(u,y,z)).\end{aligned} \tag{3.2}$$

where α and $\equiv$ should be understood not as symbols within the language but as belonging to metalinguistic notation.

That variables denote only terms but neither function nor predicate symbols is one of the characteristics of FOPL from which the phrase 'first order' derives. Indeed, function and predicate symbols are used to make statements about terms, hence any language that allows variables to denote these symbols and be quantified should be said to be second or higher order.

3.2 Formal Theory

Let it be understood throughout this section that:

- x, x_i and y_i $(i = 1, 2, \ldots)$ denote arbitrary variables,
- f and g denote arbitrary function symbols,
- t denotes an arbitrary term,
- p and q denote arbitrary predicate symbols, and
- α, β and γ denote arbitrary formulas.

(1) *Logical Axioms*

$$
\begin{aligned}
\langle\text{logical axiom}\rangle &::= \langle\text{propositional axiom}\rangle \mid \langle\text{substitution axiom}\rangle \\
&\quad \mid \langle\text{identity axiom}\rangle \mid \langle\text{equality axiom}\rangle \\
&\quad \mid \langle\text{equivalence axiom}\rangle \\
\langle\text{propositional axiom}\rangle &::= \neg\alpha \vee \alpha \\
\langle\text{substitution axiom}\rangle &::= \alpha(t) \rightarrow \exists x(\alpha(x)) \\
\langle\text{identity axiom}\rangle &::= x = x \\
\langle\text{equality axiom}\rangle &::= ((x_1 = y_1) \wedge \ldots \wedge (x_n = y_n)) \\
&\quad \rightarrow (f(x_1, \ldots, x_n) = f(y_1, \ldots, y_n)) \mid \\
&\quad ((x_1 = y_1) \wedge \ldots \wedge (x_n = y_n)) \\
&\quad \rightarrow (p(x_1, \ldots, x_n) \rightarrow p(y_1, \ldots, y_n)) \\
\langle\text{equivalence axiom}\rangle &::= (\alpha \wedge \beta) \rightarrow \neg(\neg\alpha \vee \neg\beta) \mid \neg(\neg\alpha \vee \neg\beta) \rightarrow (\alpha \wedge \beta) \\
&\quad \mid (\alpha \rightarrow \beta) \rightarrow (\neg\alpha \vee \beta) \mid (\neg\alpha \vee \beta) \rightarrow (\alpha \rightarrow \beta) \\
&\quad \mid (\alpha - \beta) \rightarrow (\alpha \wedge \neg\beta) \mid (\alpha \wedge \neg\beta) \rightarrow (\alpha - \beta) \\
&\quad \mid \forall x(\alpha(x)) \rightarrow \neg(\exists x(\neg\alpha(x))) \\
&\quad \mid \neg(\exists x(\neg\alpha(x))) \rightarrow \forall x(\alpha(x))
\end{aligned}
$$

(2) *Non-Logical Axioms*

$$
\langle\text{non-logical axiom}\rangle ::= \langle\text{closed formula}\rangle
$$

(3) *Inference Rules*

$$
\begin{aligned}
\langle\text{inference rule}\rangle &::= \langle\text{expansion}\rangle \mid \langle\text{contraction}\rangle \mid \langle\text{association}\rangle \mid \langle\text{cut}\rangle \\
&\quad \mid \langle\exists\text{-introduction}\rangle \\
\langle\text{expansion}\rangle &::= \alpha \vdash \alpha \vee \beta \\
\langle\text{contraction}\rangle &::= \alpha \vee \alpha \vdash \alpha \\
\langle\text{association}\rangle &::= \alpha \vee (\beta \vee \gamma) \vdash (\alpha \vee \beta) \vee \gamma \\
\langle\text{cut}\rangle &::= \alpha \vee \beta, \neg\alpha \vee \gamma \vdash \beta \vee \gamma
\end{aligned}
$$

$\langle\exists\text{-introduction}\rangle \quad ::= \alpha \rightarrow \beta \vdash \exists x(\alpha(x)) \rightarrow \beta$
where x has no free occurrence in β.

The formula at the left or right side of the inference symbol '$\vdash$' is called the premise or consequence of the inference respectively.

Definition 3.1. A (*first order*) *formal theory*, or simply a *theory*, is the combination of:

- the set of all logical axioms,
- the set of all inference rules, and
- a set of non-logical axioms

based on a given first order language.

We will let $L(\mathcal{T})$, $\Lambda(\mathcal{T})$ and $A(\mathcal{T})$ denote the language and the sets of logical and non-logical axioms respectively for a theory $\mathcal{T}$. $(\mathcal{T})$ will be omitted when there is no danger of confusion.

Definition 3.2. Let Σ be a given set of closed formulas. A formula σ is said to be a *logical consequence* of Σ if:

- $\sigma \in \Lambda \cup \Sigma$ or
- there exist logical consequences σ_1 and σ_2 of Σ such that

$$\sigma_1 \vdash \sigma \quad \text{or} \quad \sigma_1, \sigma_2 \vdash \sigma.$$

By extending the use of the symbol '$\vdash$', we express this fact as

$$\Sigma \vdash \sigma \quad \text{or} \quad \sigma_1, \ldots, \sigma_n \vdash \sigma$$

if Σ is a finite set $\{\sigma_1, \ldots, \sigma_n\}$. In particular,

$$\vdash \sigma$$

shows that σ is a logical consequence of Λ.

Definition 3.3. Given a theory $\mathcal{T}$, a *theorem* of $\mathcal{T}$ is a logical consequence of $A(\mathcal{T})$.

For each theorem α of $\mathcal{T}$, we can construct a tree whose root is α, whose leaves are axioms and whose branches represent instances of inference rules connecting intermediate consequences to their premises. Such a tree is called a *proof tree* for α. A simple example is illustrated below.

We readily see the occurrences of sub-trees of the same pattern at several places in Fig. 3.1. These and some other patterns which occur frequently are summarized in the following series of lemmas. Let us first introduce a combined operator '$\leftrightarrow$' (equivalence) such that

$$\alpha \leftrightarrow \beta \quad \text{stands for} \quad (\alpha \rightarrow \beta) \wedge (\beta \rightarrow \alpha).$$

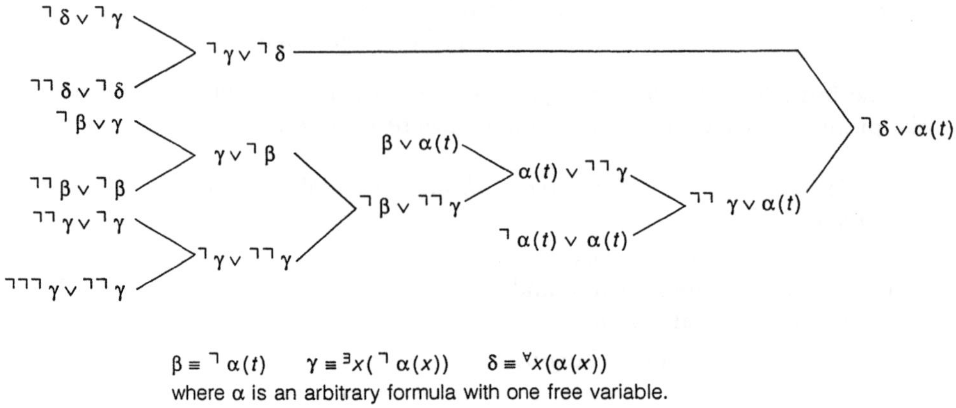

Fig. 3.1. A proof tree for a theorem

Lemma 3.1. Let $\omega(\alpha)$ be any formula containing α as a sub-formula. Then

$$\vdash (\alpha \leftrightarrow \beta) \rightarrow (\omega(\alpha) \leftrightarrow \omega(\beta)).$$

Lemma 3.2. (1) $\vdash (\alpha \vee \beta) \leftrightarrow (\beta \vee \alpha)$
(2) $\vdash \alpha \leftrightarrow \neg\neg\alpha$
(3) $\vdash (\alpha \rightarrow \beta) \leftrightarrow (\neg\beta \rightarrow \neg\alpha)$
(4) $\alpha, \beta \vdash \alpha \wedge \beta$
(5) $\vdash (\alpha \wedge \beta) \rightarrow \alpha$
(6) $\vdash (\alpha \rightarrow \beta) \rightarrow (\alpha \rightarrow \forall x(\beta(x))$ where x has no free occurrence in α
(7) $\vdash \exists x(\alpha(x) \vee \beta(x)) \leftrightarrow (\exists x(\alpha(x)) \vee \exists x(\beta(x)))$
(8) $\vdash \forall x(\alpha(x) \wedge \beta(x)) \leftrightarrow (\forall x(\alpha(x)) \wedge \forall x(\beta(x)))$
(9) $\vdash \langle q\rangle x(\alpha\langle op\rangle\beta(x)) \rightarrow (\alpha\langle op\rangle\langle q\rangle\beta(x))$ where x has no free occurrence in α, and ⟨q⟩ and ⟨op⟩ stand for either quantifier and any operator respectively.

Lemma 3.3. Let $\mathcal{T}$ be any theory and α be any formula of $\mathcal{T}$. Then,

$$\mathcal{T} \cup \{\alpha\} \vdash \beta \quad \text{if} \quad \mathcal{T} \vdash \alpha \rightarrow \beta$$

hence in particular,

$$\alpha \vdash \beta \quad \text{if} \quad \vdash \alpha \rightarrow \beta.$$

The converse implications also hold true if furthermore α is a closed formula.

Definition 3.4. A formula is said to be in *conjunctive form* or a *conjunction* if it combines its outermost constituent formulas by $\wedge$ (and). It is said to be in *disjunctive form* or a *disjunction* if it combines its outermost constituents by $\vee$ (or). It is said to be a boolean combination if both $\wedge$ and $\vee$ are used. *Conjunctive normal form* is that in which each conjunct is in disjunctive form. *Disjunctive normal form* is that in which each disjunct is in conjunctive form.

Lemma 3.4. Every formula can be converted equivalently into conjunctive and disjunctive normal forms in which each constituent is an original constituent or its negation.

Definition 3.5. A formula is said to be in *prenex form* if it is written as

$$\langle q\rangle x_1 \ldots \langle q\rangle x_n \quad \langle\text{formula containing no quantifier}\rangle$$

where ⟨q⟩ stands for either quantifier. The part $\langle q\rangle x_1 \ldots \langle q\rangle x_n$ and the formula following it are called the *prefix* and the *matrix* respectively.

Lemma 3.5. Every formula can be converted equivalently into prenex form in which each constituent literal of the matrix is an original constituent or its negation with possible renaming of bound occurrences of variables.

Corollary. Every formula can be converted equivalently into prenex form of which the matrix is in either conjunctive or disjunctive normal form.

Example. A formal theory designated by N for the natural numbers is defined as follows:

(1) *Function Symbols*
(1.1) ⟨constant⟩ ::= 0
(1.2) ⟨unary function symbol⟩ ::= *succ*
(1.3) ⟨binary function symbol⟩ ::= + | ×

(2) *Predicate Symbols*
(2.1) ⟨binary predicate symbol⟩ ::= <

(3) *Non-logical Axioms*
(3.1) $\forall x \forall y((succ(x) = succ(y)) \rightarrow (x = y))$
(3.2) $\forall x(x + 0 = x)$
(3.3) $\forall x \forall y(x + succ(y) = succ(x + y))$
(3.4) $\forall x(x \times 0 = 0)$
(3.5) $\forall x \forall y(x \times succ(y) = (x \times y) + x)$
(3.6) $\forall x \forall y(\neg(x = y) \leftrightarrow ((x < y) \vee (y < x)))$
(3.7) $\forall x \forall y(((x < y) \wedge (y < z)) \rightarrow (x < z))$
(3.8) $\forall x \forall y((x < y) \leftrightarrow ((succ(x) = y) \vee (succ(x) < y)))$
(3.9) $\forall x(\neg(x < 0))$

Note that the induction principle is not included. If we want to formulate it, we should systematically list all legitimate formulas $\{\alpha_1, \alpha_2, \ldots\}$ that contain exactly one free variable and postulate the (countably infinite number of) axioms

$$(\alpha_i(0) \wedge \forall x(\alpha_i(x) \rightarrow \alpha_i(succ(x)))) \rightarrow \forall x(\alpha_i(x)) \quad (i = 1, 2, \ldots).$$

3.3 Interpretation of First Order Theory

It is assumed throughout this section that an arbitrary formal theory $\mathcal{T}$ is given and fixed.

3.3.1 Fundamental Concepts and the Completeness of FOPL

Definition 3.6. An *interpretation* ϑ of $L(\mathcal{T})$ consists of a set $|\vartheta|$ called the universe of interpretation and the following assignments for non-logical symbols of $L(\mathcal{T})$:
(1) an element of $|\vartheta|$ for each constant,
(2) a function from $|\vartheta|^n$ to $|\vartheta|$ for each n-ary function symbol ($n \geq 1$),
(3) either TRUE or FALSE (a truth value) for each logical constant, and
(4) an n-ary predicate, that is, a mapping from $|\vartheta|^n$ to {TRUE, FALSE} for each n-ary predicate symbol ($n \geq 1$).

Given an interpretation, we can immediately expand the definition of assignments to include those for terms and atoms which may or may not contain variables. To further expand it to include those for arbitrary formulas, let us first define the (boolean) operators or truth functions to be applied to TRUE and FALSE:

$$\neg\text{TRUE} = \text{FALSE} \quad \neg\text{FALSE} = \text{TRUE}$$
$$\text{TRUE} \vee \text{TRUE} = \text{TRUE} \vee \text{FALSE} = \text{FALSE} \vee \text{TRUE} = \text{TRUE}$$
$$\text{FALSE} \vee \text{FALSE} = \text{FALSE}$$
$$tv_1 \wedge tv_2 = \neg(\neg tv_1 \vee \neg tv_2)$$
$$tv_1 \rightarrow tv_2 = \neg tv_1 \vee tv_2$$
$$tv_1 - tv_2 = tv_1 \wedge \neg tv_2 \quad \text{where } tv_1 \text{ and } tv_2 \text{ stand for a truth value.}$$

Definition 3.7. Given an interpretation ϑ of $L(\mathcal{T})$, let $\vartheta(\alpha)$ denote that which the interpretation assigns to α, where α may be a non-logical symbol, term or formula of $L(\mathcal{T})$. We need only to define $\vartheta(\alpha)$ as a predicate for a non-atomic formula α, which is done recursively as follows:
(1) $\vartheta(\neg\alpha) = \neg\vartheta(\alpha)$.
(2) $\vartheta(\alpha\langle\text{operator}\rangle\beta) = \vartheta(\alpha)\langle\text{operator}\rangle\vartheta(\beta)$.
(3) Suppose the free variables of α are $x_1, \ldots, x_n$. The *extension* of α in ϑ denoted by $\text{ext}_\vartheta(\alpha)$ is a subset of $|\vartheta|^n$ defined by

$$ext_\vartheta(\alpha) = \{(x_1, \ldots, x_n) : \vartheta(\alpha)(x_1, \ldots, x_n) = \text{TRUE}\}.$$

(4) Suppose the free variables of α are $x_1, \ldots, x_n, y$. For each element $(a_1, \ldots, a_n)$ of $|\vartheta|^n$, the value of $\vartheta(\forall y(\alpha(x_1, \ldots, x_n, y)))$ at $(x_1, \ldots, x_n) = (a_1, \ldots, a_n)$ is defined by

$$\vartheta(\forall y(\alpha(a_1, \ldots, a_n, y))) = \text{TRUE if } ext_\vartheta(\alpha(a_1, \ldots, a_n, y)) = |\vartheta|$$
$$\text{FALSE otherwise.}$$

(5) Under the same assumption as above,

$$\vartheta(\exists y(\alpha(a_1,\ldots,a_n,y))) = \begin{cases} \text{TRUE if } ext_\vartheta(\alpha(a_1,\ldots,a_n,y)) \neq empty \\ \text{FALSE otherwise.} \end{cases}$$

In particular, $\vartheta(\alpha)$ is a truth value for a closed formula α.

Definition 3.8. For a term or closed formula α, its *extension* $ext_\vartheta(\alpha)$ in ϑ is equal to $\vartheta(\alpha)$ which is an element of $|\vartheta|$ or a truth value.

Definition 3.9. A closed formula α is said to be ***valid*** in an interpretation ϑ if $ext_\vartheta(\alpha)$ is TRUE. An open formula $\alpha(x_1,\ldots,x_n)$ is said to be ***valid*** in ϑ if $\forall x_1 \ldots \forall x_n(\alpha(x_1,\ldots,x_n))$ is valid in ϑ or equivalently, $ext_\vartheta(\alpha)$ is equal to $|\vartheta|^n$.

Lemma 3.6. For any set of formulas Σ and any interpretation ϑ of $L(\mathcal{T})$, if every formula of Σ is valid in ϑ, so is every logical consequence of Σ.

Definition 3.10. An interpretation ϑ of $L(\mathcal{T})$ is said to be a *model* of $\mathcal{T}$ if all non-logical axioms (hence, all axioms) are valid in ϑ.

Definition 3.11. A formula is said to be ***valid*** in $\mathcal{T}$ if it is valid in every model of $\mathcal{T}$. It is said to be ***logically valid*** if it is valid in every theory in its language.

Lemma 3.7. Every logical axiom is logically valid.

Theorem 3.1. Every theorem of $\mathcal{T}$ is valid in $\mathcal{T}$.

It is fairly straightforward to prove the above lemmas and Theorem 3.1. Whether the converse of the latter holds or not is a profound question, however, and can only be answered through a much more sophisticated reasoning. While we will not elaborate on it any further, the introduction of a few more important concepts on which it is based is appropriate as they play fundamental roles in later discussions as well.

Definition 3.12. Let $\mathcal{T}'$ be another theory with the language $L(\mathcal{T}')$. $\mathcal{T}'$ is said to be an *extension* of $\mathcal{T}$ if

$$L(\mathcal{T}) \subset L(\mathcal{T}') \quad \text{and} \quad A(\mathcal{T}') \vdash A(\mathcal{T}).$$

$\mathcal{T}$ and $\mathcal{T}'$ are said to be *equivalent* if each is an extension of the other.

Definition 3.13. A formula of $L(\mathcal{T})$ is said to be ***satisfiable*** if it is valid in some interpretation of $L(\mathcal{T})$. Otherwise, it is said to be ***unsatisfiable***. A set of formulas is said to be satisfiable if every formula in it is satisfiable. $\mathcal{T}$ is said to be satisfiable if $A(\mathcal{T})$ is satisfiable, or equivalently, if there exists a model of $\mathcal{T}$.

Definition 3.14. $\mathcal{T}$ is said to be *inconsistent* if every formula of $L(\mathcal{T})$ is a theorem of $\mathcal{T}$. Otherwise it is said to be *consistent*.

Theorem 3.2. (Completeness of FOPL)
(1) A formula of $L(\mathcal{T})$ is a theorem of $\mathcal{T}$ if and only if it is valid in $\mathcal{T}$.
(2) $\mathcal{T}$ is consistent if and only if it is satisfiable.

Corollary. Let $\mathcal{T}'$ be another theory with the same language as $\mathcal{T}$. Then, $\mathcal{T}'$ is an extension of $\mathcal{T}$ if and only if every model of $\mathcal{T}'$ is also a model of $\mathcal{T}$. In particular, $\mathcal{T}$ and $\mathcal{T}'$ are equivalent if and only if they have the same models.

The significance of the completeness theorem is that it establishes the equivalence between the purely syntactic concept of provability (or consistency) and the semantic concept of validity (or satisfiability). Another implication is the impossibility of extending the class of logical axioms as characterized by Lemma 3.7 or the set of inference rules, hence the 'completeness' of FOPL. Indeed, any formula with the same characteristic will necessarily be a logical consequence of $\Lambda(\mathcal{T})$ under the current inference rules.

3.3.2 Morphism Between Interpretations

Definition 3.15. Let ϑ_1 and ϑ_2 be two interpretations of $L(\mathcal{T})$ and μ be a mapping from $|\vartheta_1|$ to $|\vartheta_2|$. μ is said to be a *morphism* from ϑ_1 to ϑ_2 if

$$\mu(f(e_1^{(1)},\ldots,e_n^{(1)})) = f(\mu(e_1^{(1)}),\ldots,\mu(e_n^{(1)})) \tag{3.3}$$

$$p(e_1^{(1)},\ldots,e_n^{(1)})) \rightarrow p(\mu(e_1^{(1)}),\ldots,\mu(e_n^{(1)})) \tag{3.4}$$

holds for any elements $e_i^{(1)}$ $(i = 1,\ldots,n)$ of $|\vartheta_1|$, function symbol f and predicate symbol p of $L(\mathcal{T})$.

Definition 3.16. An *imbedding* or *monomorphism* is a one-to-one morphism. When an imbedding from ϑ_1 to ϑ_2 exists, ϑ_2 is said to be an *extension* of ϑ_1. If furthermore ϑ_2 is such that an axiom α of $\mathcal{T}$ is valid in it which is not in ϑ_1, it is said to be an *extension of ϑ_1 with respect to α.*

An *epimorphism* is an onto morphism.

A *bimorphism* is a monomorphism and an epimorphism at the same time.

A *homomorphism* is a morphism such that

$$p(e_1^{(1)},\ldots,e_n^{(1)}) \leftrightarrow p(\mu(e_1^{(1)}),\ldots,\mu(e_n^{(1)})) \tag{3.5}$$

holds instead of condition (3.4) above.

An *isomorphism* is a bimorphism and a homomorphism at the same time.

Lemma 3.8. Let ϑ_1 and ϑ_2 be arbitrary interpretations of $L(\mathcal{T})$ and μ be an arbitrary morphism from μ_1 to μ_2. Then:
(1) The restriction of ϑ_2 to $\mu(|\vartheta_1|)$ denoted by $\vartheta_2|\mu(|\vartheta_1|)$ is itself an interpretation of which ϑ_2 is an extension.

(2) The derived interpretation $\mu^{-1}(\vartheta_2)$ can be defined on $|\vartheta_1|$ such that the same μ gives a homomorphism from $\mu^{-1}(\vartheta_2)$ onto $\vartheta_2|\mu(|\vartheta_1|)$.
(3) The identity mapping from $|\vartheta_1|$ to $|\vartheta_1|$ gives a bimorphism from ϑ_1 to $\mu^{-1}(\vartheta_2)$.

Thus, μ can be considered as the composition of two identity mappings and a homomorphism as depicted in Fig. 3.2.

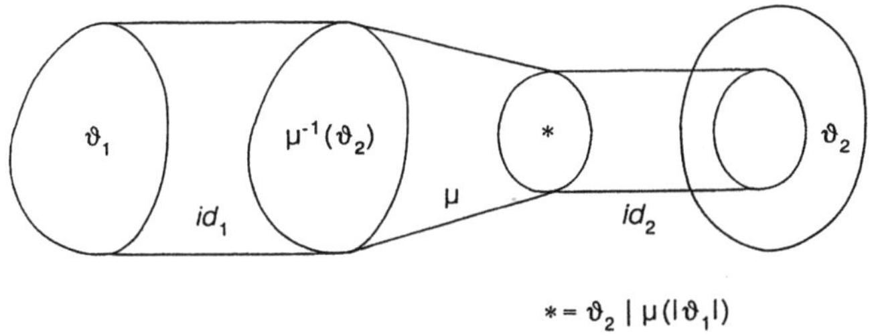

Fig. 3.2. Decomposition of a morphism

Lemma 3.9. Let ϑ_1, ϑ_2 and μ be as above.
(1) μ is a homomorphism if and only if ϑ_1 is identical to $\mu^{-1}(\vartheta_2)$.
(2) Suppose μ is a homomorphism and α is an arbitrary closed formula of $L(\mathcal{T})$:
(2.1) If α has no universal quantifiers and is valid in ϑ_1, then it is valid in ϑ_2.
(2.2) If α has no existential quantifiers and is valid in ϑ_2, then it is valid in ϑ_1.
(2.3) If μ is onto, then α is valid in ϑ_1 if and only if it is valid in ϑ_2.

Corollary. Suppose μ is an onto homomorphism. Then, ϑ_1 is a model of $\mathcal{T}$ if and only if ϑ_2 is a model of $\mathcal{T}$.

Definition 3.17. Two interpretations are said to be *elementarily equivalent* if the same formulas are valid in both of them.

Lemma 3.10. Two interpretations ϑ_1 and ϑ_2 are elementarily equivalent if there exists a homomorphism from ϑ_1 onto ϑ_2.

Definition 3.18. A *Herbrand interpretation* of a theory $\mathcal{T}$ denoted by $\vartheta_{\mathrm{H}}(\mathcal{T})$ (or simply ϑ_{H} when there is no danger of confusion) is an interpretation such that:
(1) $|\vartheta_{\mathrm{H}}|$ is the set called the *Herbrand universe* for $L(\mathcal{T})$ and denoted by $U_{\mathrm{H}}(\mathcal{T})$ which consists of all ground terms of $L(\mathcal{T})$.
(2) For every function symbol f, $\vartheta(f)$ is defined by

$$\vartheta(f)(t_1,\ldots,t_n) = f(t_1,\ldots,t_n)$$

for arbitrary elements t_i $(i = 1,\ldots,n)$ of $U_{\mathrm{H}}(\mathcal{T})$.

Note that no specific assignments are assumed for predicate symbols of $L(\mathcal{T})$. Hence, there will be generally more than one, possibly infinitely many Herbrand interpretations of the same theory $\mathcal{T}$, of which particularly important are those defined as in the following:

- $\vartheta_{\mathrm{H}}^{0}$ in which no ground atoms are valid.
- $\vartheta_{\mathrm{H}}^{\mathrm{L}}$ in which those and only those ground atoms that are theorems of $\mathcal{T}$ are valid.
- $\vartheta_{\mathrm{H}}^{\mathrm{G}}$ in which all the ground atoms are valid.

For a Herbrand interpretation ϑ, let us have the set of ground atoms that are valid in ϑ denoted by $V(\vartheta)$. Clearly, the collection of all interpretations of a given theory $\mathcal{T}$ constitutes a category in which $\vartheta_{\mathrm{H}}^{0}$ serves as the initial object, which we will denote by $Int(\mathcal{T})$. There are two sub-categories of $Int(\mathcal{T})$ worthy of attention, one denoted by $Int_{\mathrm{M}}(\mathcal{T})$ comprising models of $\mathcal{T}$ and the other denoted by $Int_{\mathrm{H}}(\mathcal{T})$ comprising those to which the morphisms from $\vartheta_{\mathrm{H}}^{0}$ are all onto. Although these sub-categories could be generally disjoint, we will see later that in fact they have a non-empty intersection for a certain important class of theories. Meanwhile, let us note that:

Lemma 3.11. If $\vartheta_{\mathrm{H}}^{\mathrm{L}}(\mathcal{T})$ belongs to $Int_{\mathrm{M}}(\mathcal{T})$, it will be the initial object of the sub-categories $Int_{\mathrm{M}}(\mathcal{T})$ and $Int_{\mathrm{H}}(\mathcal{T}) \cap Int_{\mathrm{M}}(\mathcal{T})$. More generally, if $Int_{\mathrm{H}}(\mathcal{T}) \cap Int_{\mathrm{M}}(\mathcal{T})$ is not empty, there is a model of $\mathcal{T}$ which is an extension of $\vartheta_{\mathrm{H}}^{\mathrm{L}}(\mathcal{T})$ with the same universe of interpretation.

The concepts of Herbrand interpretation and model are theoretically very important, as they apply generally to every formal theory. Yet they are syntactic in their essence rather than semantic, as they have nothing to do with any specific truth of the universe of interpretation. In fact, we will usually find ourselves subject to certain constraints when constructing a model for a given theory. Typically, we may have to take a given set as a part of the universe of interpretation and add elements and relations to it piecewise. If that is the case, it will be appropriate to expand the notion of morphism to allow those between partial interpretations. Definitions 3.15 and 3.16 remain the same. Lemmas 3.8 and 3.9 continue to hold. It should be noted the conditions (3.3) and (3.4) in Definition 3.15 entail that the destination of a morphism is structurally richer than the source. The notion of morphism is thus an abstraction of the act of extending an interpretation toward a more completed state.

3.4 Undecidability and Incompleteness

Definition 3.19. A formula α of a theory $\mathcal{T}$ is said to be *decidable* if α or $\neg\alpha$ is a theorem of $\mathcal{T}$. $\mathcal{T}$ is said to be *complete* when it is consistent and every closed formula of $\mathcal{T}$ is decidable.

Lemma 3.12. Suppose $\mathcal{T}$ is consistent. Then the following three conditions are equivalent:
(1) $\mathcal{T}$ is complete.
(2) Models of $\mathcal{T}$ are elementarily equivalent to each other.
(3) For each model ϑ, $\mathcal{T}$ is equivalent to the theory $\mathcal{T}(\vartheta)$ defined by

$$L(\mathcal{T}(\vartheta)) = L(\mathcal{T})$$
$$A(\mathcal{T}(\vartheta)) = \{\alpha : \alpha \text{ is a closed formula of } \mathcal{T} \text{ and valid in } \vartheta\}.$$

The completeness of a theory is intimately related to the existence of a mechanical method to decide if an arbitrarily given closed formula of that theory is a theorem or not. The precise definition of a 'decision method' is normally given in terms of the evaluation of recursive functions. Yet, in view of the well-known equivalence of the latter with the operation of Turing machines, let us be content with the following alternative which is rather intuitive.

Definition 3.20. Let S be an arbitrarily given countably infinite set. A (unary) predicate p on S is said to be *recursively enumerable* if and only if a Turing machine (that is, a program for a computer with unlimited memory capacity) TM_p exists such that:
(1) TM_p accepts elements of S as input and produces an answer YES or NO.
(2) For every element e of S:
(2.1) If $p(e)$ is TRUE, TM_p will halt and answer YES.
(2.2) If $p(e)$ is FALSE and TM_p halts, TM_p will answer NO.
p is said to be *recursive* if TM_p always halts.

Lemma 3.13. p is recursive if and only if both p and $\neg p$ are recursively enumerable.

Definition 3.21. Let $thm_{\mathcal{T}}$ denote the predicate that characterizes the set of theorems within the set of all closed formulas of a theory $\mathcal{T}$. $\mathcal{T}$ is said to be decidable if $thm_{\mathcal{T}}$ is recursive. It is said to be undecidable otherwise.

Theorem 3.3. Let $axm_{\mathcal{T}}$ denote the predicate that characterizes the set of axioms within the set of all closed formulas of $\mathcal{T}$. If $axm_{\mathcal{T}}$ is recursive, then $thm_{\mathcal{T}}$ is recursively enumerable.

The above theorem tells us that under the said premise, we can write a program which will halt and confirm that a formula is a theorem if it is indeed a theorem, but yet could run forever otherwise. Guaranteeing that the program always halts is tantamount to requiring that $\neg thm_{\mathcal{T}}$ too is recursively enumerable; an instance of Lemma 3.13. This will be the case when $\neg thm_{\mathcal{T}}$ characterizes the set of negated theorems (Lemma 3.15 and Fig. 3.3).

Lemma 3.14. Suppose $\mathcal{T}$ is an extension of N, the formal theory for natural numbers as defined in Section 3.2. If $\mathcal{T}$ is consistent, then it is undecidable.

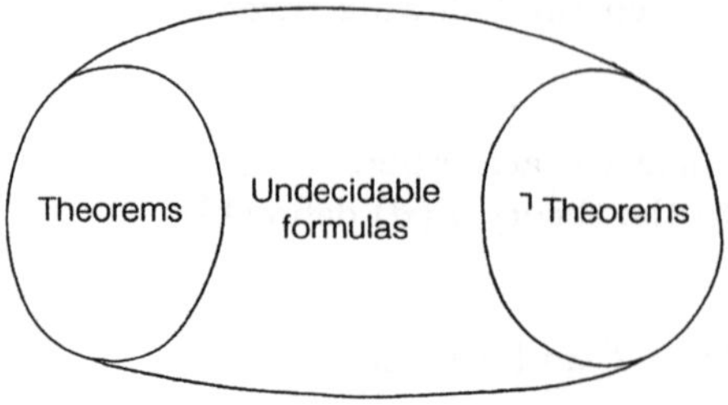

Fig. 3.3. Theorems, negated theorems and undecidable formulas

Lemma 3.15. If $axm_{\mathcal{T}}$ is recursive and $\mathcal{T}$ is complete, then $\mathcal{T}$ is decidable.

Theorem 3.4. (Incompleteness of first order theories) Suppose $\mathcal{T}$ is an extension of N. If $axm_{\mathcal{T}}$ is recursive, then $\mathcal{T}$ is not complete.

Definition 3.22. An interpretation ϑ is said to be *strongly undecidable* if every theory of which ϑ is a model is undecidable.

Let $\mathcal{N}$ denote the set of natural numbers, which of course is a model of N. It is trivial to see that $\mathcal{T}(\mathcal{N})$ is a complete extension of N, which implies by Lemma 3.14 or Theorem 3.4 that $thm_{\mathcal{T}}$ (which is equivalent to $axm_{\mathcal{T}(\mathcal{N})}$) is not recursive. One might imagine the possibility of formulating a theory for $\mathcal{N}$ which is not an extension of N and is decidable. The following theorem denies this.

Theorem 3.5. $\mathcal{N}$ is strongly undecidable.

Figure 3.4 depicts various categories of satisfiable formulas of a given theory and how they are related in the contexts of the completeness and incompleteness theorems.

3.5 Clausal Form and the Herbrand Model

Suppose a closed formula α is given in the prenex conjunctive normal form

$$\alpha \equiv \langle q \rangle x_1 \cdots \langle q \rangle x_n (\alpha^{(1)} \wedge \cdots \wedge \alpha^{(r)}) \tag{3.6}$$

$$\alpha^{(i)} \equiv \beta_1^{(i)} \vee \cdots \vee \beta_{s_i}^{(i)} \vee \neg\gamma_1^{(i)} \vee \cdots \vee \neg\gamma_{t_i}^{(i)} \quad (i = 1, \ldots, r)$$

where $\beta_j^{(i)}$ and $\gamma_k^{(i)}$ are atoms containing no variables other than $x_1, \ldots, x_n$.

Definition 3.23. α is said to be in *clausal form* if all quantifiers are universal. It is simply a *clause* if $r = 1$, that is, the matrix consists of a single disjunction

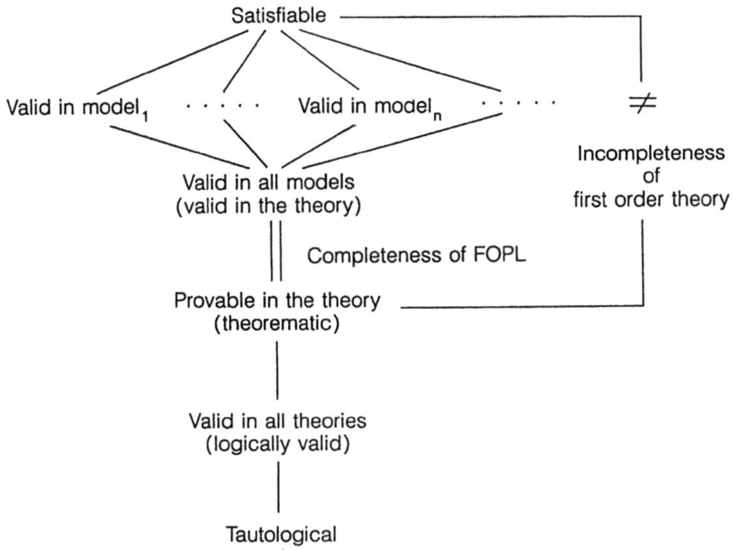

Fig. 3.4. Levels of validities of satisfiable formulas

of literals. Let us rewrite it as

$$\alpha \equiv \forall x_1 \cdots \forall x_n (\beta_1 \vee \cdots \vee \beta_s \vee \neg\gamma_1 \vee \cdots \vee \neg\gamma_t) \tag{3.7}$$

which is further classified as follows:

- *positive clause* if $s \geq 1$,
- *definite clause* if $s \leq 1$,
- *Horn clause* or *positive definite clause* if $s = 1$,
- *unit clause* if $s = 1$ and $t = 0$,
- *goal clause* or *negative clause* if $s = 0$.

Definition 3.24. Let α be as in equation (3.6) above. The *Skolemization* of α is the following process to eliminate existential quantifiers from α . Let $\tilde{\alpha}$ denote the formula in the middle as well as at the end of the process.
(1) Examine if there remains any existential quantifier in $\tilde{\alpha}$. If no, the process terminates.
(2) If yes, and if

$$\tilde{\alpha} \equiv \forall x_1 \cdots \forall x_h \exists y \cdots \langle \text{matrix of } \tilde{\alpha} \rangle$$

where $\exists y$ is the leftmost existential quantification and $\forall x_1, \ldots, \forall x_h$ are all the universal quantifications to the left of $\exists y$, introduce a new function symbol f_y, eliminate $\exists y$ and replace every occurrence of y within the matrix of $\tilde{\alpha}$ by $f_y(x_1, \ldots, x_h)$. Return to step (1).

The symbol f_y corresponding to each existentially quantified variable is called a *Skolem function symbol* for that variable. In case $h = 0$, it is called a *Skolem constant*. The formula $\tilde{\alpha}$ is called the *Skolem normal form* for α.

Lemma 3.16. Every closed formula α is satisfiable if and only if $\tilde{\alpha}$ is satisfiable. More precisely, every model of α can be regarded as a model of $\tilde{\alpha}$ and vice versa.

Definition 3.25. Given a first order language L, a *logic program* in L is a theory in L having as the axiom a formula in clausal form of which all conjuncts are Horn clauses.

Given a closed formula α, let us have the same α denote the theory that has α as the sole non-logical axiom.

Theorem 3.6. For every formula α in clausal form, the theory α has a Herbrand model if it is satisfiable.

Lemma 3.17. Every logic program λ is satisfiable, hence has a Herbrand model. More precisely, for an arbitrary set Σ of ground atoms of $L(\lambda)$, the set of all the ground atoms that are logical consequences of $\Sigma \cup \{\lambda\}$ provides a model of λ.

Corollary. $\vartheta_{\mathrm{H}}^{\mathrm{L}}(\lambda)$ is the least Herbrand model (LHM) in that it is imbedded in every Herbrand model of λ. $\vartheta_{\mathrm{H}}^{\mathrm{G}}(\lambda)$ is the greatest Herbrand model in that every Herbrand model is imbedded in it.

Lemma 3.18. For any set of Herbrand models $\{\vartheta_{\mathrm{H}}^{m}(\lambda) : m \in M\}$ for a given logic program λ, $\bigcap_{m \in M} V(\vartheta_{\mathrm{H}}^{m}(\lambda))$ defines another Herbrand model of λ.

Lemma 3.19. Let λ be a logic program and ρ be a closed formula. If $\lambda \wedge \neg\rho$ is unsatisfiable, then $\lambda \wedge \rho$ is satisfiable.

Definition 3.26. An *instance* of a formula α is that which is obtained by replacing some of the free variables with terms. A *ground instance* is that which is obtained by replacing all free variables with ground terms.

Theorem 3.7. A formula α in clausal form is unsatisfiable if and only if there exists a conjunction of ground instances of α's matrix which is unsatisfiable.

If α in clausal form is unsatisfiable, it will necessarily contain goal clauses by Lemma 3.17. Let us assume for simplicity that

$$\alpha \equiv \lambda \wedge \rho$$

where λ is a logic program and ρ is a single goal. By Lemma 3.19, $\lambda \wedge \neg\rho$ is satisfiable. Suppose ρ is written as

$$\rho \equiv \forall x_1 \cdots \forall x_n (\neg\gamma_1 \vee \cdots \vee \neg\gamma_t)$$

hence $\neg\rho$ as

$$\neg\rho \equiv \exists x_1 \ldots \exists x_n (\gamma_1 \wedge \cdots \wedge \gamma_t)$$
$$\gamma_i \equiv p_i(x_1, \ldots, x_n) \quad (i = 1, \ldots, t).$$

That $\lambda \wedge \neg\rho$ is satisfiable means that $\neg\rho$ is valid in a certain model (say ϑ) for λ. By actually refuting α, we may be able to find the set of tuples of ground terms (the answer set for α) $Ans(\alpha)$ such that

$$p_1(a_1, \ldots, a_n) \wedge \cdots \wedge p_t(a_1, \ldots, a_n) = \text{TRUE for any } (a_1, \ldots, a_n) \in Ans(\alpha).$$

If we take as ϑ the LHM $\vartheta^{\mathrm{L}}_{\mathrm{H}}(\lambda)$ which is the canonical interpretation of λ in logic programming, the answer to be found will be such tuples $(a_1, \ldots, a_n)$ that

$$\lambda \vdash p_1(a_1, \ldots, a_n) \wedge \cdots \wedge p_t(a_1, \ldots, a_n).$$

In a more general case, suppose the Skolem normal form $\tilde{\alpha}$ is written as

$$\tilde{\alpha} \equiv \mu \wedge \nu$$

where μ and ν are conjunctions of positive and negative clauses respectively. As far as our interest is solely in the satisfiability or unsatisfiability of α, it makes no difference by Lemma 3.16 whether we consider α or $\tilde{\alpha}$. Paraphrasing the unsatisfiability of α in terms of the negation of $\tilde{\alpha}$, what we want to see is whether

$$\mu \vdash \neg\nu$$

holds or not. Thus, Skolemization can be a useful way to solve the problem of whether a formula is a theorem or not in a given theory.

The incompleteness of the logic program λ is manifest in the separate existences of $\vartheta^{\mathrm{G}}_{\mathrm{H}}(\lambda)$ and $\vartheta^{\mathrm{L}}_{\mathrm{H}}(\lambda)$. The validity of any ground atom other than those in $V(\vartheta^{\mathrm{L}}_{\mathrm{H}})$ could only be decided from an extra-theoretical viewpoint. The canonical interpretation is based on the view that any elementary statement (ground atom) that cannot be logically inferred from the axioms is necessarily false. This view is often referred to as the *Closed World Assumption* (CWA).

Theorem 3.7 guarantees that we can actually refute a formula α in clausal form by systematically listing the ground instances of its matrix and testing the unsatisfiability of any of them that are indeed unsatisfiable. This procedure could be extremely inefficient, however, and logic programming normally relies on the unification algorithm, which does not deal with ground atoms but finds the most general instances that are unsatisfiable. While it goes beyond the scope of this book to fully elaborate on this algorithm, we will give a succinct explanation of it in Sect. 4.2.1.2.

References

[Gal] Gallier, J.H.: Logic for Computer Science. Harper & Row, 1986

[Llo] Lloyd, J.W.: Foundations of Logic Programming. 2nd ed., Springer-Verlag, 1987

[Man] Manna. Z,: Mathematical Theory of Computation. McGraw-Hill, 1974

[Sho] Shoenfield, J.R.: Mathematical Logic. Addison-Wesley, 1967

Chapter 4
Criticism of Logical Approaches

4.1 Pertinencies of Logic to Computer Sciences and Applications

The relationship between logic and computer sciences and applications is deep and broad. At the most basic end of the spectrum, we have the functional description of a circuit as a combination of AND, OR and NOT gates, and the boolean data type in most programming languages. The generation of sentences by repeated applications of production rules is equivalent to the enumeration of theorems of the form $S \rightarrow \langle$sequence of terminal symbols$\rangle$. A more specialized case is the formulation of the functional and multivalued dependencies of a relational database in terms of implications between the propositional variables, each corresponding to an attribute domain of a relation ([Fag], [Sag]). While these instances are all within the realm of propositional logic, the use of first order predicate logic (FOPL) has become prevalent in recent years. This trend is probably explained by the following reasons:

- the far richer contents of FOPL than those of propositional logic,
- the high degree of maturity and refinement it has attained as a theoretical discipline,
- the ease of the intuitive understanding of an interpretation for such a purely formal system of symbols, without impairing the rigor of the system, and
- that it has reached a sort of natural boundary as a framework for describing information systems in view of the precise and profound results concerning decidability and computability.

All these points should be clear from what was presented in the preceding chapter.

It might be considered only too natural to formulate an information system within some sort of logical framework since it belongs to the family of mathematical systems in a broad sense, all of which are built on the foundation of logic. The actualities are not that simple, however. Mathematically rigorous representations for information systems are not regularly given except in a few special cases such as the BNF grammars of programming languages and the algebraic specifications of abstract data types. It is notably rare in application fields, including CAD. This is due not only to negligence on the side of system builders but also to the lack of mathematical concepts and tools that are both theoretically sound and practically useful in dealing with many important problems. Of particular relevance for CAD are such problems as the uniform representations of finite spatial objects, specifications of man-machine interactions and descriptions of evolutional transformations of the database state. On the other hand, there also exists a distance between logic and mathematics. It cannot be expected, to begin with, to capture within the syntax of the universal language of logic all the specific meanings of each individual discipline of mathematics. A fundamental cause for this disparity lies in the fact that while mathematics sees the 'reality' of its objects behind their abstract forms, logic is primarily concerned with the formalities of inferences and hence inclines to nominalism. Jumping to the conclusion we will elaborate on later, the inevitable extension of logic programming to include constraint solvers clearly indicates the limitations of purely symbolic inference engines.

Actually, logical methods in computer sciences have been developed under the strong influence of artificial intelligence. Specifically, we have automatic theorem proving by refutation, and as one of its special cases, logic programming with Horn clauses. This reflects the fact that many (maybe the majority of) problems in AI can naturally be described as mathematical theorems with solutions obtained in the course of proving these theorems. More fundamentally, it appears to be generally agreed that human intelligence as opposed to senses, intuitions and other cognitive faculties inherent in any kind of living creature is characterized as the ability to use symbols and, with awareness, the mechanism of reasoning (see the item 'intelligence' in [Web] for instance). We may term such methods 'syntax oriented'.

Some methods, however, address themselves to the capturing of semantics rather than the manipulation of syntax. A well-established example is the verification of a program by interpreting it as a device for transforming the output assertion and confirming that the result of the transformation is logically implied by the input assertion ([Gri], [Man]). It should be noted that in defining the transformation for each statement type, we are (implicitly) exploiting our knowledge on the domain of all the possible states of the program variables and the transition from one state to another effectuated by executing that type of statement. Also that we are engaged in the dynamic aspect of the subject domain, which is not the usual case with the methods based on logical inferences.

An important source for logical methods other than AI is the relational data model. The correspondence between the dependencies and propositional logic was already mentioned. Speaking of the realm of FOPL, each relation can be seen as the extension of a predicate among the domains of attributes. The domain relational calculus mentioned in Sect. 2.4.3 (4) is nothing but the first order language with these predicates as the sole non-logical symbols. The important point is that a database is a concrete set defined as the union of those relations, hence any relational language operating on it, however formally it may be specified, is concerned not with the form but with the meaning of its operations. In fact, the relational algebra which carries out the evaluation of formulas as series of symbols is considered to be an interpretation of the relational calculus as a formal language.

The deductive database approach aims at an integration of the relational data model and logic programming by incorporating the capability for inference. That is, it considers some of the closed formulas which usually play the role of integrity constraints to be axioms or derivation rules and includes those tuples implied by them in the database in addition to the explicitly stored tuples. Any single change to the (finite) extensional part of the database could cause infinitely many effects. Answering queries or evaluating formulas cannot be accomplished any more by means of the relational algebra but require certain theorem proving techniques. Terms constituting tuples are uninterpreted constant symbols rather than denoting specific elements in the domains. The status of the database thus reverts to being syntactic instead of semantic.

There are methods that do not fix on specific interpretations from the beginning but conceive of a class (or 'category' in the mathematically more precise terminology) of all the 'natural' interpretations for each formal theory, then focus attention on particular members of this class with singular or distinct characteristics compared to the others. These are interesting in that they carry the methodological flavor of modern mathematics and give a clue as to how to integrate the syntax and semantics of formally described systems. The algebraic specification of abstract data types is one of the typical examples of such methods for which the mathematical foundations are provided by the logic of equality and many-sorted algebra.

Artificial intelligence has produced yet another family of devices which may be grossly termed knowledge representation schemes and have penetrated into other fields of applications including CAD. Associative language (of which LEAP discussed in Sect. 2.1.2 is an example), semantic networks and frame theory are among those better known and frequently referenced. What can roughly be said of them is first, that they are syntactically equivalent to certain subsets of FOPL with the arities of predicates being limited to one or two, second, that it is not clear what entities lie beyond syntax and are actually represented, and third, that there have been long and hot controversies concerning whether they (networks, in particular) are mere restricted and informal dialects of FOPL or not [Isr]. Those who contend that their networks stand in their own right tend to complain about logic as lacking in such capabilities as organizing objects (in hierarchical structures, for example), capturing semantics or deciding how to use the stored facts. While it is not necessary for us to examine the controversies in detail (which look quite confused and may not really be productive), it will be appropriate or even mandatory to take these voices as serious and constructive criticism of logical approaches and try to respond to the questions they raise. Only after that (and with a reasonable success), will our provisional selection of the predicate logic as the base language (see the discussion in Sect. 0.4) be justified.

Let us now summarize the above introductory remarks on the use of logic in computer sciences and applications, and make sure where we stand before proceeding further. First of all, there apparently exist several fundamental points of argument, and according to the positions taken on them, the class of methods that are commonly and popularly termed 'logical' is differentiated into a multitude of veins. Each of the examples mentioned above is committed to its own combination of positions, though not always in explicit terms, and is characterized by that. It will therefore be necessary and appropriate to do the following in order to understand rightly and clearly whether and how logical methods can be meaningfully applied to CAD. That is:

- to identify the methodological issues that are the causes for diversification,
- to explicate the positions on these issues assumed in each of the example cases, and
- to criticize these positions in the light of the fundamental requirements for describing the world of design objects and processes.

On the first point, at issue are:

(1) whether the system in question to be formulated in a logical language should be a formal theory or an interpretation,

(2) what logical formulas (particularly, those containing free or bound variables) should mean, whether integrity constraints, derivation rules or something else, provided that the system is an interpretation,

(3) whether it is required that in the system all the integrity constraints as well as derivation rules hold true,

(4) whether the system always comprises a single and fixed interpretation or whether multiple interpretations come into consideration, and if the latter, how the relations between different interpretations (which apparently transcend the logical status of each interpretation) should be formulated,

(5) what kind of set the universe of interpretation is, and

(6) in what kind of logic the system should be formulated, FOPL or some other, and if the latter, how it differs from FOPL.

The second and third are the subjects of the succeeding sections, each of which discusses each of the above issues in turn. We will select logic programming (with or without constraint), the relational data model, the deductive data model, abstract data types and semantic networks as the example cases, since in the author's view they exhibit distinct characteristics and/or raise serious questions with respect to these issues.

4.2 Theory or Interpretation?

4.2.1 Information Systems as Formal Theories

Let us take another and a slightly more serious look at the previously given naive justification for formulating information systems within some logical framework. The arguments against it are twofold. First, there seems to be opposition or at least reluctance among CAD researchers to accepting mathematics as the base language for modeling design objects and processes. It is contended that mathematics has nothing to do with reality and is more interested in generality than practicality, that doing things the pragmatic or empirical way can better solve the problems. Second, we know that the issues which separate logic and mathematics are more profound both theoretically and philosophically than largely subjective feelings, and have practical significance as well. In the following, we will try to substantiate the justification first in a general discussion, then by presenting specific methods taking these points of argument into consideration.

4.2.1.1 General Discussion

(1) *On Mathematical Languages*
No one will deny the vital roles mathematics plays in such fields as linear programming, finite element analysis, cutter path calculation for fabricating free surfaces, etc. These are however applications of computers to problems which have already been mathematically defined. Moreover, well established concepts, methods and tools have sufficed to define them and even to plan their solutions to some extent. The present issues we face are of quite a different nature. They concern not only the stage of solving but also the earlier stage of precisely defining the problems. Besides, these are kinds of problems which are not normally considered eligible for treatment by mathematics, at least by textbook mathematics. They typically include the functional specifications of parts and preliminary designs based on designers' empirical knowledge. Suppose, for example, the required function is the transformation of rotational to linear movement. The phrase 'to transform rotational to linear movement' should suffice for designers to conceive the idea of using one of the familiar mechanisms such as cam, link, vise, etc. A body of phrases similar to this exists in each engineering discipline, composed of well-defined technical terms and expressing unit functions. It serves as a common language for all designers and also as the language in which expert designers indoctrinate novices into their practices. All this is possible only because designers share an understanding of natural languages which in turn is made possible ultimately by appealing to human sensations and intuitive comprehension of shapes, motions and mutual interactions of physical objects. The situation is entirely different and essentially more difficult when trying to communicate to computers what the requirements are and how to meet them. Storage and retrieval of information for design alternatives off the shelf using the above-mentioned function phrases as the keys, while practically useful, is far from

the solution, as computers will simply play the role of a catalogue in such an environment and will not assist human designers in the creative and synthetic way which has been our central concern from the beginning. We cannot place realistic hopes on the advance of natural language processing either (that is the other extreme approach) because it presupposes the availability of methods and tools for effectively analyzing and making machine-intelligible much subtler and more extensive problems than requirements specifications and preliminary designs. In fact, several schools of modern linguistics tend to be strongly mathematically oriented, with all their concepts and terminologies defined formally and self-containedly (that is, without relying on anything outside a fixed system of symbols and rules). Such a tendency is quite understandable and perhaps inevitable considering the desire to make linguistics a truly rigorous science and natural languages eligible for computer processing as much as possible. This is yet more true of designers' languages and their methodical treatment. Let us then state our fundamental positions as follows:

- In order for computers to be able to support designers in their creative activities, it is not adequate to store individual pieces of knowledge concerning functions and their physical realizations in enumerative ways but is required to capture the principles underlying the individualities and postulate them in the form of general statements.
- It is strongly desirable to identify those portions and aspects of each designer's language that can be precisely and meaningfully communicated to computers and to organize them into a self-contained sublanguage. This sublanguage should necessarily be devoid of any impurities which are not intelligible to computers, though they might be indispensable for human understanding of meanings. Designers will not normally and consciously make such an isolation of a purely mechanical process from the rest when thinking and communicating in a natural language. It will significantly contribute, however, to making CAD software well-structured, manageable and intelligent.
- The said sublanguage should also be formal and abstract in order to be applicable to a wide range of applications. (See the discussions in Sect. 0.7.)
- All of these properties (generality, precision, self-containedness, formalness and abstractness) are very characteristic of mathematics. Conversely, mathematics in the present context should be understood as nothing but a linguistic framework in which to describe whatever can be clearly and precisely described.

Returning to the previous example, we know the movement of the rotating body is transmitted to another through their contact. Let us assume for simplicity that the relative positions of the two bodies are completely determined by any single point on the surface of contact (the simplest case being that the contact is indeed made at only one point). The movement of such a point will in turn be determined by specifying the surface (side or cross section) on which it lies and other constraints on the freedom of movement. Either this point or another connected to it moves linearly. If the latter, the line must be given as a further constraint. Since we talk about the movements of rigid bodies, these specifications must agree with physical laws and current technological limitations. This implies that we must stop elaborating purely mathematical arguments at some point, beyond which we must simply accept things and facts as they are. For example, we may specify the point of contact to be on the side of a cylinder and free to move along a line in the direction of the axis of rotation. Its locus relative to the cylinder is a spiral. The important point is that only in these 'mathematical' terms and phrases can humans and computers communicate with each other clearly and intelligently, although we may have to be content with merely naming 'screw' as a primitive element to realize the spiral movement (at least at the stage of preliminary design). Equally important is that a mathematical language provides a platform on which designers and engineers (software engineers in particular) with different backgrounds and expertise can cooperate with common understanding of some subject matter.

(2) *On Mathematics Versus Formal Theories*
We tend more or less to associate mathematics with the notion of mechanical reasoning and calculations. This tendency cannot be ignored simply as superficial and ungrounded. In fact, serious efforts have been made along a similar line of thought by leading logicians and mathematicians in each era of the history of mathematics. The results are profound, far-reaching, and have even brought practical consequences to computer sciences and applications. One of the classical and most ambitious cases was the conception of *Lingua Characteristica Universalis* or general mathematical language by G.M. Leibniz by means of which thinking and arguing could be replaced by calculations. Reconstructing mathematics on an absolutely sound and secure foundation was recognized as being of prime importance and vigorously pursued in the early decades of this century by various schools represented by such eminent figures as B. Russell, L.E.J. Brouwer and D. Hilbert. The motivation was to save mathematics from the paradoxes generated by excessively liberal interpretations of set-theoretic concepts and operations. Despite differences of philosophical position and certain technicalities, these schools essentially shared the formal-axiomatic method for formulating mathematical theories. With that method, whatever physical or intuitive meanings symbols may have originally carried are removed from them, encapsulated and frozen once and for all within the set of axioms where they are hidden and implicitized. What remains of a theory, its theorems in particular, should be derived purely logically from axioms through manipulations of meaningless symbols which observe a fixed set of rules. No creation of new symbols or meanings of symbols is allowed. It is then hoped that the process of deriving theorems can be carried out mechanically, that is, without requiring any truly human creativity or imagination. This is indeed the case if the logic is FOPL, as shown by Theorem 3.3 in Sect. 3.4.

There are crucial occasions in the development of mathematics when concepts and processes are applied in rather anomalous ways, and invention of new concepts or new interpretations of existing ones to straighten out the anomalies leads to major progress or even revolutions in established disciplines. It is interesting to note that mathematics restored to its normal state is again formal and transparent with all the tricks and mysteries removed, while the real fruitfulness, after all, lies in the law-breaking liberalism of the preceding stage. Let us cite a concrete example. Let $\mathcal{D}$ be the set of all functions that have derivatives of arbitrary orders and finite supports. (The *support* of a function is the set of points for which it takes non-zero values.) The desire that

$$\int_{-\infty}^{\infty} f(x)\psi'(x)dx = [f(x)\psi(x)]_{-\infty}^{\infty} - \int_{-\infty}^{\infty} f'(x)\psi(x)dx$$
$$= \int_{-\infty}^{\infty} f'(x)\psi(x)dx \quad \text{for all } \psi \in \mathcal{D} \tag{4.1}$$

holds for $f = Y$ (Heaviside function) where

$$Y(x) = \begin{cases} 0 & x \leq 0 \\ 1 & x > 0 \end{cases} \tag{4.2}$$

is fulfilled by defining its derivative δ (Dirac's delta function) as follows;

$$\int_{-\infty}^{\infty} \delta(x)\psi(x)dx = \psi(0) \quad \text{for all } \psi \in \mathcal{D} \tag{4.3}$$

or equivalently

$$\delta(x) = \begin{cases} 0 & x \neq 0 \\ \infty & x = 0 \end{cases} \tag{4.3$'$}$$

$$\int_{-\infty}^{\infty} \delta(x)dx = 1$$

which is of course not a legitimate function in the usual sense. However, if we consider each function f as representing a spatial distribution of some physical quantity such as mass, heat or electric charge, δ corresponds to the concentrated unit quantity which is quite normal. Guided by intuitive insights and physical meanings, physicists can perform calculations involving δ correctly. L. Schwartz's theory of distributions ([Schw(1)], [Schw(2)]) makes them not only correct but mathematically rigorous by reinterpreting δ and other pseudo-functions as being not functions but functionals, that is, mappings from $\mathcal{D}$ to the set of (complex) numbers. Y, δ and the derivative of arbitrary f will accordingly be defined so that

$$\left.\begin{aligned} Y(\psi) &= \textstyle\int_0^\infty \psi(x)dx \\ \delta(\psi) &= \psi(0) \\ f'(\psi) &= -f(\psi') \end{aligned}\right\} \quad \text{for all } \psi \in \mathcal{D} \tag{4.4}$$

Then, $Y' = \delta$ will immediately follow.

Even for computer programs that perform as well as mathematicians do in calculus, it will be next to impossible to extend the notion of functions to include δ as defined by (4.3)′ and to use physicists' insights in doing further calculations. However, (4.4) has nothing unclear about it and poses no problem when incorporated in such programs. Conceiving of the set of functionals on $\mathcal{D}$ and transplanting the intuitive meanings expressed in (4.3)′ and the like to definitions and axioms such as (4.4) may constitute mathematically the most substantial part of the present example. Once this has been done, the formal-axiomatic method will discard whatever remains of (4.3)′, clearing the way for 'computerization' of the theory.

Actually, no mathematical theory will be presented in its entirety within a purely formal axiomatic framework. That would certainly impose very undesirable and unnecessary constraints and restrictions on man's liberal exercise of his intellectual and imaginative faculties. But precisely this rigidness makes formal methods suitable for describing information systems. The privileged status of FOPL in this regard lies in Theorem 3.3 that guarantees the existence of an algorithm for deriving any formula from the set of axioms if it is in fact a theorem. On the other hand, the fact that we can construct theories in which any algorithm will run forever trying to decide that some formula is not a theorem (see comments following Theorem 3.3) seems to tell us that the class of first order formulas is rather abundant and that some extra-logical constraints are required to establish the falsehood of formulas. With this in mind, let us now proceed to present two typical methods from logical family based on FOPL, that is, logic programming and abstract data typing.

4.2.1.2 Logic Programming (LP)

Suppose we let terms designate a product, its constituent assemblies and parts, and their attributes, and let logical formulas express the relationships and constraints which hold among these terms. By considering each term or formula as a series of pure symbols, we will obtain a description of the product as a first order formal theory. Various behavioral characteristics and pieces of information necessary for manufacturing will be deduced as theorems. By providing a suitable query language interpreter and an inference engine, we will have in a sense realized an intelligent CAD.

The difficulty inherent in this approach is that generally we do not have an algorithm to decide whether a given formula is a theorem or not (undecidability), and there indeed exist such formulas that neither they nor their negations are theorems (incompleteness, Theorem 3.4). Speaking of the above example, this implies that however fine and precise we make the description, there always remains the possibility for us to encounter statements about the product which are significant but undecidable as true or as false according to the description. We will therefore have to question the adequacy and appropriateness of FOPL as the framework in which to describe physical objects. While we may be unable technically to prove or disprove certain statements from the given axioms, we nevertheless presume that the truth or falsehood of any statement has been decided in reality independently of our knowing it.

A natural way to alleviate this difficulty is to construct a set and define an interpretation on it instead of a formal theory. Sepcifically, LP takes the least Herbrand model (LHM) as the interpretation and restricts the set of permissible formulas to that of positive definite clauses in order to guarantee the existence of the said kind of model (Lemma 3.17). Thus, the description of a product will be given as a logic program, that is, a collection of logical formulas each of which is either a

$$\text{unit clause} \quad p \leftarrow \tag{4.5}$$

or more generally a

$$\text{program clause} \quad q \leftarrow (r_1 \wedge \cdots \wedge r_m) \tag{4.6}$$

where p, q, r are atomic formulas and all variables contained therein are universally quantified. Clause (4.5) represents an elementary fact which holds unconditionally. Clause (4.6) asserts a logical necessity or cause-effect relationship. Establishing the validity of an (existentially quantified) statement $s_1 \wedge \cdots \wedge s_n$ under this program is done by showing that its negation which is a

$$\text{goal clause} \quad \leftarrow (s_1 \wedge \cdots \wedge s_n) \tag{4.7}$$

in conjunction with the program leads to the

$$\text{contradiction} \quad (\text{null}) \leftarrow (\text{null}) \tag{4.8}$$

The refutation is done by unifying each s_i with the left-hand side of (4.5) or (4.6) and substituting it with the respective right-hand side where any variable can be instantiated by any other term containing variables and/or constants, and repeating this process until (4.8) is reached. The tuple of terms that are finally obtained makes the goal statement true when substituted for the corresponding variables. (Variables that are left uninstantiated may take any values.)

As a simple example, let us look again at the description of a lever ((1.3), (1.4)) restated in a slightly different form according to clauses (4.5) and (4.6):

$$\mathit{lever}(C, T, R) \leftarrow \tag{4.9}$$

$$(\mathit{movmnt}(x) = \tfrac{\overline{xz}}{\overline{yz}}\mathit{movmnt}(y)) \leftarrow \mathit{lever}(z, y, x) \tag{4.10}$$

The goal clause

$$\leftarrow (\mathit{movmnt}(R) = w) \tag{4.11}$$

will be unified with (4.10) resulting in

$$x = R$$

$$w = \tfrac{\overline{Rz}}{\overline{yz}}\mathit{movmnt}(y)$$

$$\leftarrow \mathit{lever}(z, y, R)$$

then with (4.9) producing

$$z = C$$
$$y = T$$

and finally establishing

$$movmnt(R) = \frac{\overline{RC}}{\overline{TC}} movmnt(T) \tag{4.12}$$

The above refutation procedure together with suitable methods to select an atom and a clause at each step from the goal clause and the program respectively provides a means to compute a correct answer. That the correctness of a statement in this procedural sense is equivalent to theoretical validity and also to validity in the LHM singles out the Horn language as a language that is both theoretically important and practically useful. Some fundamental results related to this are summarized in the following. Readers are referred to [Llo] for more details.

(1) The equivalence of theoretical validity and that in the LHM for any ground atom has already been mentioned (Corollary to Lemma 3.17). It will not necessarily hold, however, for formulas which are not ground or atomic. For example

$$\forall x \forall y \forall z (lever(x, y, z) \rightarrow ((x = C) \wedge (y = T) \wedge (z = R))) \tag{4.13}$$

is valid in the LHM of (4.9) and (4.10), but not a logical consequence thereof.

(2) If we strengthen a program P and turn it into its completion (comp(P)) by adding such axioms as (4.13) and equality axioms ([Llo], [Cla]), it will not affect the LHM. That is, the ground atoms (in fact, the unit clauses in general) that are logical consequences of comp(P) are also those of P. Moreover, the LHM of P is also that of comp(P).

(3) Suppose the goal is currently

$$\leftarrow (A_1 \wedge \cdots \wedge A_i \wedge \cdots \wedge A_n)$$

and the next goal is obtained by unifying one of the atoms, say A_i, with the head of one of the clauses in the program, say

$$A' \leftarrow (B_1 \wedge \cdots \wedge B_k)$$

and replacing it by the body. Selecting atoms and clauses at consecutive steps constitutes a derivation path. The tree of Fig. 4.1 depicts the totality of possible selections and derivation paths starting with the current goal.

Each derivation path may terminate at a

(S) *success node* (null) $\leftarrow$ (null)

or

(F) *failure node* $\leftarrow (C_1 \wedge \cdots \wedge C_m)$

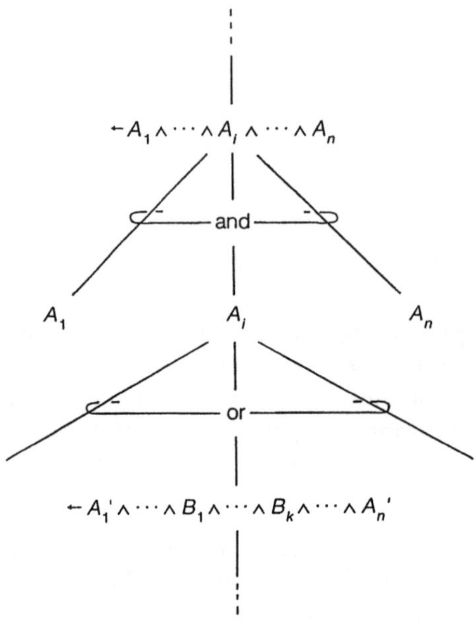

Fig. 4.1. The derivation tree

where none of C_i's is unifiable with any program clause's head, or may continue infinitely. The last case is further classified into

(FR) ***fair derivation path***

in which every atomic formula that appears at some step will eventually be selected, and

(UF) ***unfair derivation path.***

(4) Let P be a program and G ($\leftarrow A(x_1, \ldots, x_n)$) be a goal clause. Then,
(4.1) $\neg G$ ($\exists x_1 \ldots \exists x_n A(x_1, \ldots, x_n)$) is a logical consequence of P if and only if its derivation tree contains at least one success node (S). Specifically, for each (S), the instance of A with remaining variables being universally quantified is a logical consequence of P. Moreover,
(4.2) Any instance of A that is a logical consequence of P can be obtained by further instantiating the instance at some (S).
(4.3) G is a logical consequence of comp(P) if and only if its derivation tree contains at least one failure (F) node but neither an (S) node nor an infinite fair (FR) path. (UF) paths may or may not appear.

(5) Let W denote the set of all well-formed formulas the function and predicate symbols of which occur in P. These and only these have meanings in Herbrand interpretations. In any one of them, each formula is decisively true

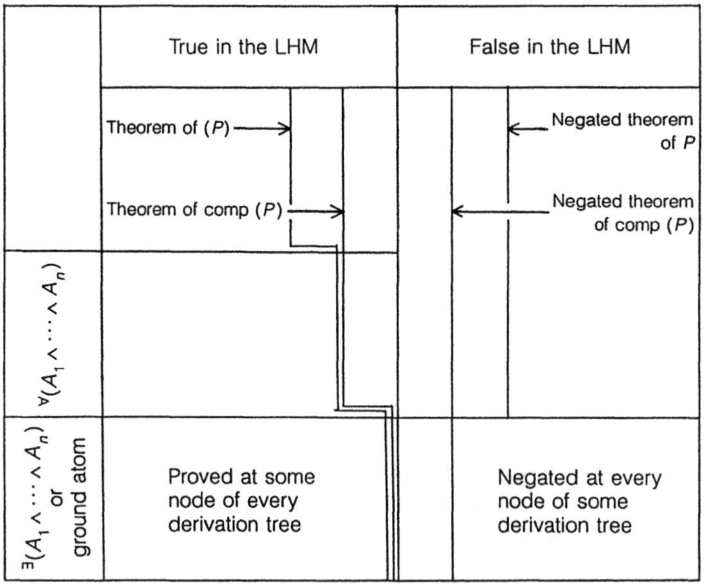

Fig. 4.2. Truth values and decidabilities of formulas in W

or false. We may or may not find a decision algorithm, however. Figure 4.2 depicts what we have stated above specifically for the LHM.

The so-called closed world assumption (CWA) expresses the view that the ground atoms that cannot be decided to be true are false [Rei(1)]. It is particularly relevant to logic programming, since existentially quantified conjunctions are the only legitimate questions to be asked, and those that are answered affirmatively are indeed positively decided by any fair derivation procedure. The same procedure will, if the formula in question is the negation of a theorem in comp(P) (a special case of falsity), detect this fact in a finite time. The problems are that, first, we do not know if the procedure will eventually stop (either in success or failure) or not, and second, it cannot be easily extended to more general classes of question.

4.2.1.3 Abstract Data Typing (ADT)

The principle of ADT is to view the target universe as a set of symbols on which a certain number of functions are defined and observe a given set of equality relations. Precisely, we define a first order theory comprising:

- unary predicate symbols ('sorts') $s_1, s_2, \ldots$,
- function symbols $f_{w,s}, f_{w',s'}, \ldots$ where $w, w', \ldots$ are (possibly empty) catenations of sorts and $s, s', \ldots$ are single sorts (the intended meaning being that $w, w', \ldots$ denote the sorts of domains and $s, s', \ldots$ denote those of ranges),
- variables $x_1, y_1, \ldots$ representing terms,

- binary predicate (equality symbol) $\approx$ and a set of equality relations (Γ) $\xi \approx \eta$, $\xi' \approx \eta'$, ... where ξ, η, ξ', η', ... are terms with both sides of each relation having the same domain and range,

and construct models (many-sorted algebras) out of the set of free terms in much the same way as we construct Herbrand models for a logic program. The theory and its models are considered to be the (algebraic) specification and implementations respectively of an abstract data type. Just as the set of all Herbrand models constitutes a complete lattice, so does the set of many-sorted algebras. Corresponding to the LHM, there exists the initial algebra in which all pairs of terms (α, β) such that the equalities between them $(\alpha \approx \beta)$ can be derived from Γ and only such pairs are identified. (The word 'initial' is used instead of 'least' because it is customary to consider not only free algebras but the totality of all models which together with homomorphisms between models constitutes a category.) Despite the apparent parallelisms in forms and theories, there seem to exist substantial differences in practices and motivations between LP and ADT. First, abstract specifications (theories) tend to redefine data types we already know and conceive of in terms of more concrete sets and structures while in the case of LP, proof-theoretic concerns prevail rather than constructing or investigating models. Second, every data type is usually a strictly structured and self-contained system, so that any small change to the specification could turn it into an entirely different entity. On the other hand, the body of knowledge represented in a logic program is allowed to increase or decrease piecemeal and more freely without losing its fundamental identity. In CAD as well as other applications, we can enrich our assets of software tools and components by identifying classes of objects which appear frequently, extracting the essential functions and rules governing them, and organizing them into abstract data type specifications. It is expected that this will contribute to describing models of design objects more systematically and economically. Among actual and potential works along this line are the specification of graphics programming language [Mal], definition of abstract data types for attribute domains of relational databases [Sto], formulation of presentation and manipulation of menus during man-machine dialogue, and construction of geometric relations and reasoning about them [Arb]. This last case was actually presented in a Prolog-like notation, but can be formulated as an abstract data type as shown in the following example.

Let us consider a very simple world comprising points, lines, line segments and polygons, and try to construct an abstract data type out of them. We focus our attention on how an entity is created from one or more other entities and how different ways come to produce the same result.

- *Sorts*
  In addition to those mentioned above, we require **integer** and **real**.
- *Functions*
  point : **real** $\times$ **real** $\rightarrow$ *point*

$xcoord$: $point \rightarrow$ **real**
$ycoord$: $point \rightarrow$ **real**
$lseg$: $point \times point \rightarrow line\text{-}segment$
$start$: $line\text{-}segment \rightarrow point$
end : $line\text{-}segment \rightarrow point$
$line$: $line\text{-}segment \rightarrow line$
extends a line-segment to an infinite line.
$inters$: $line \times line \rightarrow point$
defines the point of intersection.
$tringl$: $line\text{-}segment \times point \rightarrow polygon$
creates a triangle from a line-segment and a point.
$polygn$: $polygon \times point \rightarrow polygon$
replaces the last edge by two edges via a given point.
$order$: $polygon \rightarrow$ **integer**
returns the number of edges.
$edge$: $polygon \times$ **integer** $\rightarrow line\text{-}segment$
returns the edge at the designated order.
$clipls$: $line \times line\text{-}segment \rightarrow line\text{-}segment$
creates a new line-segment by clipping one with a line.
$clippg$: $line \times polygon \rightarrow polygon$
creates a new polygon by clipping one with a line.

– *Equality Relations*

$xcoord(point(x,y)) = x$
$ycoord(point(x,y)) = y$
$start(lseg(p,q)) = p$
$end(lseg(p,q)) = q$
$line(lseg(inters(l,m),\ inters(l,n))) = l$
$order(tringl(s,p)) = 3$
$order(polygn(g,p)) = order(g) + 1$
$edge(tringl(s,p),1) = s$
$edge(tringl(s,p),2) = lseg(end(s),p)$
$edge(tringl(s,p),3) = lseg(p,start(s))$
$edge(polygn(g,p),i) = edge(g,i) \quad (1 \le i \le order(g) - 1)$
$edge(polygn(g,p),\ order(g)) = lseg(start(edge(g,\ order(g))),p)$
$edge(polygn(g,p),\ order(g)+1) = lseg(p,\ end(edge(g,\ order(g))))$

$$xcoord(start(clipls(line(lseg(point(x_1,y_1),\ point(x_2,y_2))),\ lseg(point(x_3,y_3),\ point(x_4,y_4))))) = f(x_1,y_1,\ldots,x_4,y_4) \qquad (*)$$

It should be noted that the last equality relation contains an arithmetic function f. We have three other relations similar to this concerning *clipls* and still more concerning a polygon and its clipping by a line (Fig. 4.3a, b).

It is obvious that numerical evaluations of function f and the like are required to establish these relations. The inclusion of sort **real** implies that performing arithmetic on real numbers is allowed, which is not the case with the purely formal and syntactic method of ADT. However, if we are interested only in constructive or combinatorial aspects of geometric relations and are

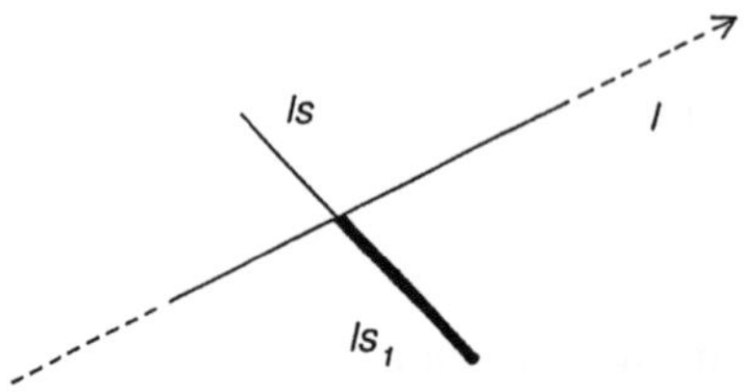

(a) *clip1s* : (*l*, *ls*) → ls_1

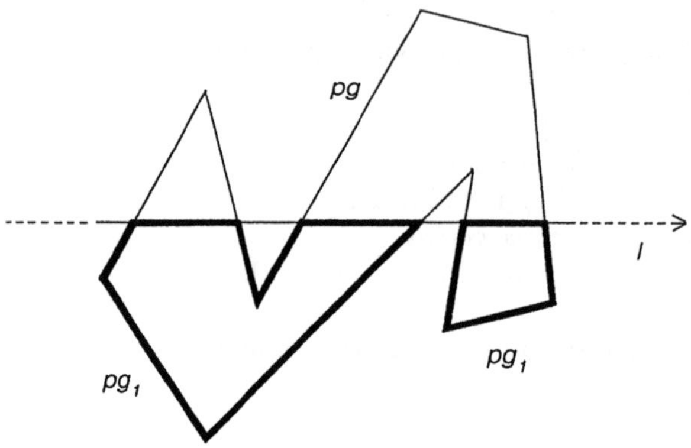

(b) *clippg* : (*l*, *pg*) → pg_1

Fig. 4.3. Clipping a line segment or a polygon with a line

prepared to discard precise coordinate values of points, we can dispense with arithmetic facilities and use instead the boolean function

$$rhand : line \times point \rightarrow \{\mathbf{true}, \mathbf{false}\}$$

which has the value **true** if and only if the point lies on the righthand half plane of the line. Then the equality (*) is simplified and integrated with the other three to form the following:

$$\begin{aligned} clipls(l, s) = \; & s && (P \wedge Q) \\ & lseg(start(s), inters(l, line(s))) && (P \wedge \neg Q) \\ & lseg(inters(l, line(s)), end(s)) && (\neg P \wedge Q) \\ & \text{null} && (\neg P \wedge \neg Q) \\ \text{where} \quad & P = rhand(l, start(s)) \quad \text{and} \quad Q = rhand(l, end(s)). && (*)' \end{aligned}$$

The case of *clippg* is slightly more complicated but should not pose any unmanageable difficulties. In a way similar to the above, the truth values of

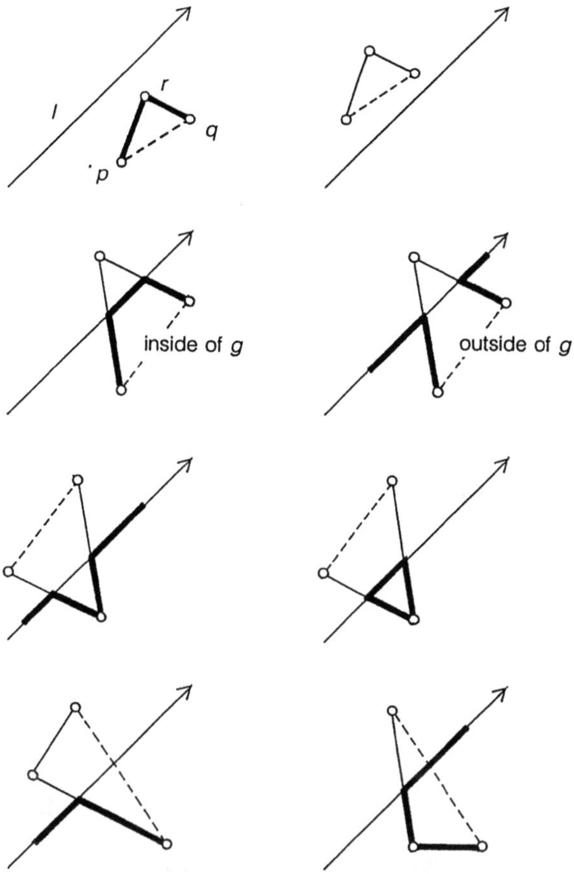

Fig. 4.4. Different cases of clipping a polygon

$P = rhand(l, p)$, $Q = rhand(l, q)$ and $R = rhand(l, r)$ play the key role in enumerating different cases of $clippg(l, polygn(g, r))$ where p and q are the start and end points respectively of the last edge of g (Fig. 4.4).

For example, the case of $P \wedge Q \wedge \neg R$ will produce

$$\begin{aligned}
&clippg(l, polygn(g, r)) \\
&= polygn(polygn(clippg(l, g), \\
&\qquad\qquad inters(l, line(lseg(start(edge(g, order(g))), r)))), \\
&\qquad inters(l, line(lseg(r, end(edge(g, order(g))))))).
\end{aligned}$$

4.2.2 Criticism

(1) *On Expressive Powers*
Although the LHM gives a natural, canonical and always constructible interpretation, the fundamental formal-theoretic character of LP cannot be denied. In the first place, the existence of Herbrand models presupposes the condition that every statement be in clausal form, which will in turn be fulfilled by Skolemizing all existentially quantified variables. Skolemization, however, does not necessarily transform a formula into an equivalent one but only maintains its satisfiability or unsatisfiability (Lemma 3.16). It will be justifiably applied only to those problems which are formulated as proving certain theorems within a given formal theory. Namely, that a formula Q is a logical consequence of another formula P (the conjunction of all axioms) is equivalent to the unsatisfiability of $\neg(P \rightarrow Q)$ or equivalently $P \wedge \neg Q$ which is to be converted to a conjunction of clauses to form a logic program and a goal. Second, program statements are restricted to definite clauses so that the canonical model for P may exist and every formula that belongs to a certain class and is valid in this model may be a logical consequence of P. Third, permissible goals are actually contained in this class and correct answer substitutions are obtained through logical inferences. In other words, within the framework of LP, we do not really deal with the structure of any interpretation, whether it be the LHM or not, as a separate matter from the syntax of a logic program.

Let us further elaborate on these points by presenting some simple examples. Suppose we have an axiom

$$\exists x \forall y (p(y) \rightarrow q(x,y)) \tag{4.14}$$

which is transformed to a clause by Skolemizing x and constitutes a logic program as follows together with a set of instances of p and q:

$$\left.\begin{array}{ll}\left.\begin{array}{l} p(a) \leftarrow \\ p(b) \leftarrow \\ q(l,a) \leftarrow \\ q(m,a) \leftarrow \\ q(m,b) \leftarrow \\ q(n,b) \leftarrow \end{array}\right\} & (4.15.1) \\ q(s,y) \leftarrow p(y) & (4.15.2) \end{array}\right\} \quad (4.15)$$

(Assertions and questions in the form of (4.14) frequently occur in various applications. For example, it may express that some factory (x) exists and produces all kinds of parts ($q(x,y)$) that meet a given criterion (p).) If we ask this program a question

$$\forall x \exists y (p(y) \wedge \neg q(x,y)) \tag{4.16}$$

or its Skolemized version

$$\forall x(p(f(x)) \wedge \neg q(x, f(x)) \tag{4.17}$$

which is the negation of the axiom (4.14), we will obtain the correct answer substitution

$$x = s, \quad y = f(s) \tag{4.18}$$

by the usual resolution (or by first asserting $p(f(s)) \leftarrow$ then asking $\leftarrow q(x, f(x))$ in the case of Prolog). However, the Skolem constant s does not carry any meaning, and the statement

$$\forall y(q(m, y) \leftarrow p(y)) \tag{4.19}$$

which we really expect to be given cannot be logically inferred from (4.15) though it is valid in the given set of instances (which is moreover just the LHM of (4.14)). Suppose further that we augment the theory by including the completion, equality and domain closure axioms [Gal(2)], which results in the following:

$$\begin{aligned}
&\forall x(p(x) \leftrightarrow ((x = a) \vee (x = b))) \\
&\forall x \forall y(q(x, y) \leftrightarrow (((x, y) = (l, a)) \vee ((x, y) = (m, a)) \vee \\
&\qquad\qquad ((x, y) = (m, b)) \vee ((x, y) = (n, b)))) \\
&\exists x \forall y(q(x, y) \leftarrow p(y)) \\
&\textit{equality axiom} \text{ (details omitted)} \\
&\forall x((x = a) \vee (x = b) \vee (x = l) \vee (x = m) \vee (x = n))
\end{aligned} \tag{4.20}$$

then (4.19) is valid in any of its models, hence a theorem of (4.20) (Theorem 3.2). (The latter conclusion can easily be inferred without appealing to the completeness theorem. But similar arguments are applicable to any logic program and will be more tractable than direct application of predicate calculus.) Completing theories with these supplementary axioms is strongly reminiscent of J.McCarthy's concept of 'circumscription' according to which:

- the objects that can be shown to have a certain property P by reasoning from certain facts A are all the objects that satisfy P (predicate circumscription), and
- the 'known' objects are all there are (domain circumscription) [McC(2)].

Both approaches try to strengthen the structures of theories to the extent that factual truths can be established also as logical truths, in particular that the CWA can be realized. Supplementation does not add anything to the structures of the canonical interpretation, however, in which the supplementary axioms trivially hold. Besides, the purpose will not always be fulfilled partly because the domain closure axiom simply cannot be written out for an infinite universe of interpretation. Without the domain closure axiom, theories may not be made complete. See Fig. 4.2 which indicates that some formulas may possibly be true (false) in the canonical LHM without being theorems (negation of theorems).

For example, if we add

$$\forall x(p(x) \leftarrow ((x \neq l) \wedge (x \neq m) \wedge (x \neq n)))$$

to (4.15.1), the first line of (4.20) is changed to

$$\forall x(p(x) \leftrightarrow ((x = a) \vee (x = b) \vee ((x \neq l) \wedge (x \neq m) \wedge (x \neq n))))$$

and the domain closure axiom becomes indispensable to establish the LHM as the unique model of (4.20) and to prove (4.19) as a theorem because without it, we may add $p(c)$ (c being a new constant) to the LHM, falsifying (4.19) in this model. If the additional axioms are:

$$\begin{aligned}
&\forall x(p(f(x)) \leftarrow p(x)) \\
&\forall x(p(x) \leftarrow \neg r(x)) \\
&\forall x(r(x) \leftarrow ((x = l) \vee (x = m) \vee (x = n))) \\
&\forall x(r(f(x)) \leftarrow r(x))
\end{aligned}$$

we will have to postulate an infinite domain closure axiom

$$\forall x((x{=}a) \vee (x{=}b) \vee (x{=}l) \vee (x{=}m) \vee (x{=}n) \vee (x{=}f(a)) \vee (x{=}f(b)) \vee \cdots)$$

Let us consider as another example the definition and evaluation of the transitive closure (r) of a binary predicate p. The usual definition of r runs:

$$r(x, y) \leftrightarrow (p(x, y) \vee \exists z(p(x, z) \wedge r(z, y))) \tag{4.21}$$

which is the completion of

$$\begin{aligned}
&r(x, y) \leftarrow p(x, y) \\
&r(x, y) \leftarrow (p(x, z) \wedge r(z, y))
\end{aligned} \tag{4.22}$$

The refutation of any goal of the form $\leftarrow r(a, y)$ where a is a constant is straightforward and will reach every object that is related to a via r in a finite number of steps by repeated backtrackings. If we let $p(z)$ and $q(y, z)$ represent $p(a, z)$ and $r(z, y)$ respectively, and rewrite the recursive part of (4.21) as

$$\exists z(p(z) \wedge q(y, z)) \tag{4.23}$$

and compare it with

$$\forall z(p(z) \rightarrow q(y, z)) \tag{4.24}$$

that characterizes the Skolem constants for (4.14), we will notice the existence of the same duality between them as we considered in Sect. 2.1.2 (4.2). That is, if we let $q[p]$ denote (4.23), then (4.24) will become $\neg(\neg q[p])$ and vice versa. We have seen that the evaluation of the extension of (4.23) is an easy task for standard Prolog while that of (4.24) is not possible with logical inferences. From the application standpoint, this lack of symmetry is indicative of the weakness of the base language (that is, FOPL).

Actually, the method of refutation as applied to this example is more powerful than the descriptive FOPL in that the transitive closure of p can be uniquely defined neither by (4.21) nor by any first order formula. That is, given any interpretation and the extension of p in it, there will generally exist more than one extension of r in it that satisfies (4.21). To single out the transitive closure among others will require some extra-FOPL apparatus, such as procedural methods, set-theoretic operations or higher order logics. The peculiarity of the refutation procedure is that while it generates each specific member of the transitive closure, it must exhaust the entire set before failing to establish $r(a, b)$ for any b not belonging to this set. It might therefore fall into an endless loop if the program contains function symbols that cause the Herbrand universes to be infinite sets or if the extension of p contains loops. This is an instance of the gap that exists between the set of false statements and that of negated theorems in Fig. 4.2.

In summary, FOPL, inference procedures and set-theoretic operations respectively have distinct levels of expressive power, in increasing order.

(2) *Not Theories but Interpretations*

As touched on above, one of the reasons for reliance on logical inference or theorem proving is the potential infinity of the canonical Herbrand interpretations. The question is whether both the reason and the conclusion will be justified from the viewpoint of modeling design objects. The author is inclined to a negative judgement on the following grounds.

First and simply, the objects to be represented in the computer model for an individual design task always constitute a finite set. Certainly, many of the domains such as integers, real numbers or more complex data types in terms of which the attributes and functions of objects are described are considered to be infinite. But we must distinguish clearly between the domains of objects and their attributes, or the target of modeling and the means. The logical or mathematical structures of the former are necessarily linked with those of the latter, but are not wholly contained therein. For example, while positions are fundamental attributes of center and end points, the relationship between them that defines the structure of a lever will not be imbedded in the Euclidean plane $\mathbf{R}^2$.

Second, the exclusive reliance on logic or logical inference is too restrictive both practically and theoretically. On the practical side, it should be possible to evaluate any FOPL formula (such as (4.24)) on a given interpretation and to define any of its subsets (such as a transitive closure and its complement). On the theoretical side, it should be noted that the set $\mathcal{N}$ of natural numbers, one of the most fundamental structures in mathematics and computer science, can never be completely formulated within the framework of FOPL (Theorem 3.5), but is precisely defined by the concise system of Peano's axioms. The point is of course that the law of induction contained in the latter is a typically set-theoretic concept.

Third, it is questionable whether the Herbrand universe generated by the mechanically repeated application of a small number of function symbols is suitable as the raw material for modeling the world of design in which the formation of each object and its relations with others is not that simple. Specifically, objects are differentiated according to their types and levels, for each of which only a limited set of functions are meaningful and applicable. In this regard, many-sorted algebras might be more appropriate.

In view of the fact that shapes, forms and topologies play fundamental roles in internal representations and man-machine communications in CAD, one may wonder if the method of ADT for graphic and geometric entities can be extended to define the models of design objects themselves. However, there exists a parallelism between ADT and LP, as was pointed out, hence the difficulties with LP just discussed, particularly the first and the second of them, apply also to ADT. Besides, we have seen that even the purely mathematical and abstract domain of geometric entities will not be completely described by an algebraic specification in the proper sense since it has to presuppose the availability of the sort **real**, which, unlike those of integers and rational numbers, cannot be constructed by manipulating a finite number of symbols. Likewise, LP will have to have built-in real arithmetic functions and incorporate constraint solvers that essentially lie outside the framework of symbol manipulations and logical inferences. Otherwise, its applicability in the CAD environment become much limited.

The above analysis clearly demonstrates the inherent weakness of formal-theoretic approaches and leads to the conclusion that we should conceive of models of design objects as interpretations in which arithmetic and set-theoretic operations as well as symbol manipulations and logical inferences may be employed.

4.3 Meanings of Logical Formulas

The next issue we want to discuss is what logical formulas do or should mean in information systems formulated within a logical framework. This issue has been raised primarily in the context of deductive data modeling (DDM) in which two of the main logical methods, that is, logic programming and relational data modeling (RDM) are to be integrated. The roles of logical formulas differ significantly between these two, however. In LP, all formulas are axioms, and moreover derivation rules, while in RDM, they are evaluated against explicitly stored facts (tuples and relations) functioning as queries, view definitions or integrity constraints. Let us therefore look at the way formulas are classified and understood in DDM, and examine whether and how it suits our purpose of modeling the world of design.

4.3.1 Deductive Data Modeling

Let a relational database D be the union of relations $R, S, \ldots, T$ such that

$$\begin{aligned}
&R = \{(a_1^{(i)}, \ldots, a_l^{(i)}) : 1 \le i \le I\} \subset A_1 \times \cdots \times A_l \\
&S = \{(b_1^{(j)}, \ldots, b_m^{(j)}) : 1 \le j \le J\} \subset B_1 \times \cdots \times B_m \\
&\ldots \\
&T = \{(c_1^{(k)}, \ldots, c_n^{(k)}) : 1 \le k \le K\} \subset C_1 \times \cdots \times C_n \\
&\{e_1, \ldots, e_N\} = \{(a_1^{(i)}, \ldots, a_l^{(i)}, b_1^{(j)}, \ldots, b_m^{(j)}, \\
&\qquad c_1^{(k)}, \ldots, c_n^{(k)}) : 1 \le i \le I, 1 \le j \le J, \ldots, 1 \le k \le K\}
\end{aligned} \tag{4.25}$$

Also, let the schema for D contain a set of general statements Σ (closed formulas in relational calculus). According to the conventional view, each relation is the extension of a predicate (also denoted by R, S, T, etc.) and each formula in Σ is an integrity constraint and must always be observed by D. The deductive view considers tuples as elementary facts, certain statements in Σ as derivation rules and others as integrity constraints. The first and second of these are then converted to completed forms. Furthermore, the equality axioms, the domain closure axiom and the unique name axiom are postulated ([Gal(2)], [Rei(3)]).

Suppose, for example, we have

$$R(t_1, \ldots, t_l) \leftarrow (R_1 \wedge \cdots \wedge R_r) \tag{4.26}$$

as a derivation rule, then the completed database will look as follows:

- *Completed elementary facts and derivation rules*

$$\begin{aligned}
&R(x_1, \ldots, x_l) \leftrightarrow ((\vee_i((x_1 = a_1^{(i)}) \wedge \cdots \wedge (x_l = a_l^{(i)})))\vee \\
&\exists u((x_1 = t_1) \wedge \ldots \wedge (x_l = t_l) \wedge R_1 \wedge \cdots \wedge R_r)) \\
&(u \text{ representing the free variables contained in } R_1, \ldots, R_r) \\
&S(y_1, \ldots, y_m) \leftrightarrow (\vee_j((y_1 = b_1^{(j)}) \wedge \cdots \wedge (y_m = b_m^{(j)}))) \\
&\ldots \\
&T(z_1, \ldots, z_n) \leftrightarrow (\vee_k((z_1 = c_1^{(k)}) \wedge \cdots \wedge (z_n = z_n^{(k)})))
\end{aligned} \tag{4.27.1}$$

- *Equality axioms*

$$\begin{aligned}
&(x = x) \leftarrow \\
&(y = x) \leftarrow (x = y) \\
&(x = z) \leftarrow ((x = y) \wedge (y = z)) \\
&P(y_1, \ldots, y_p) \leftarrow (P(x_1, \ldots, x_p) \wedge (x_1 = y_1) \wedge \cdots \wedge (x_p = y_p)) \\
&\text{for any predicate (relation or `=') } P
\end{aligned} \tag{4.27.2}$$

- *Domain closure axiom*

$$((x = e_1) \vee (x = e_2) \vee \ldots \vee (x = e_N)) \leftarrow \tag{4.27.3}$$

– *Unique name axiom*

$$((e_1 \neq e_2) \wedge (e_1 \neq e_3) \wedge \cdots \wedge (e_{N-1} \neq e_N)) \leftarrow \tag{4.27.4}$$

The purpose of postulating these axioms is to formulate the database as a complete first order theory and establish the collection of tuples logically derived as well as explicitly stored as its unique model. This requires, however, that the derivation rules be placed under certain restrictions such as the prohibition of indefinite clauses and recursive or cyclic definitions of predicates. Also, function symbols which cause the domain to be infinite are excluded. Otherwise, incomplete theories may result. (Refer to the discussions in the preceding sections.) In this view, a query which takes the form of a goal clause

$$\leftarrow \exists u_1 \ldots \exists u_k W(u_1, \ldots, u_k) \tag{4.28}$$

will be given an answer $(c_1, \ldots, c_k)$ if and only if $W(c_1, \ldots, c_k)$ is obtained as a theorem of the database as a formal theory. Likewise, integrity constraints are understood as those statements that remain theorems through whatever updates are made to the database. A question here is why we should cling to the formal-theoretic view when the theory is complete with its unique model and hence when both deal with virtually the same object.

The above concerned the special case of definite deductive data modeling (DDDM). Let us now turn to the more general case of indefinite deductive data modeling (IDDM). To begin with, it appears that the primary motivations behind the formal-theoretic view of database are:

- to precisely represent incomplete information such as unknown or null values, and
- disjunctive information, and
- to capture real world knowledge such as transitivity of ***part-of*** relations and inheritance through ***is-a*** hierarchies [Abi].

The selection of a logical framework with respect to these is natural and understandable, as is shown by the use of Skolem functions to denote unknown values, or set inclusions to implement inheritance. However, the use of logic does not necessarily lead to formulation in formal theories. We can always enjoy the power of expression or inference of logic, if we desire, in interpretations as well. Moreover, the interpretive view offers means other than inference for answering queries, as was seen in preceding sections. On the other hand, theories may no longer be complete now that disjunctive axioms are permitted. The deductive view will then face the same difficulties as LP does, or even more since the canonical LHMs may not exist, nor may the CWA be maintained. Besides, we cannot help but question how the imposition of the remaining restrictions on derivation rules and the provision of supplementary (domain closure and unique name) axioms can be justified, the primary purpose of which was to establish databases as complete theories. Another and more elusive question is whether and how it is possible to differentiate substantially between derivation rules (axioms) and integrity constraints (theorems), which is further elaborated next.

4.3.2 Derivation Rules, Integrity Constraints or What?

First, we note that certain axioms and theorems can possibly be interchanged while keeping the theory unchanged. Second, different suggestions are made by different authors for classifying derivation rules and integrity constraints. For example, Reiter [Rei(3)] seems to contend that disjunctive information such as

$$(p(a) \vee p(b)) \leftarrow \tag{4.29}$$

cannot be treated as an integrity constraint because to do so would require at least one of $p(a)$ and $p(b)$ to be explicitly stored. Also, formulas that contain Skolem constants for unknown values may occur as axioms. On the other hand, Gallaire/Minker/Nicolas [Gal(2)] suggest that derivation rules should be function-free in order for finite and explicit answers to be given to queries. This does not agree with Reiter's contention as (4.29) can be generalized to

$$\exists x(p(x) \wedge w(x)) \leftarrow \tag{4.30}$$

which will entail the introduction of a Skolem constant which in turn is only a special case of a Skolem function. Third, it is pointed out as the reason for putting purely negative clauses into integrity constraints that they do not produce any positive results [Gal(2)]. This will not apply, however, to IDDM since, for example, the clause

$$\leftarrow p(a) \tag{4.31}$$

in conjunction with (4.29) will produce

$$p(b) \leftarrow \tag{4.32}$$

More generally, when the extensional part of a database (that is, the collection of elementary facts) is updated, the propagation of updating may not be uniquely determined by derivation rules alone but only with the help of the further condition that integrity constraints be maintained. In this sense, integrity constraints may sometimes function also as derivation rules. The converse always holds, of course.

While the above discussion applies to any kind of application, there are several additional points that are more particular to CAD such as the following (see also Sect. 1.2 (3)).

- At the end of the design process, the database should describe the design objects completely, hence the extensional part should satisfy all the constraints that represent the design requirements.
- Certain elementary facts can still be stored implicitly if they are unequivocally derived from other facts and derivation rules. This should be under-

stood, however, as a convenience for the internal representation of facts rather than as a provision at the conceptual level.

- Since we have decided to see CAD databases as interpretations, integrity constraints and answers to queries need not be theorems but need only be valid formulas on the extensional part.
- In other words, we allow the theory (set of derivation rules and integrity constraints) to be incomplete and the (extensional) database to be one of its models. This reflects the possibility of alternate designs that meet the same given requirements.
- During most of design process, requirements represented in logical formulas will not be fully satisfied. Therefore, they do not really function as integrity constraints, the primary purpose of which is to protect the database from illegal updates. Instead, the database is updated so that one or more of these formulas that are currently false may become true.

In conclusion, logical formulas in a CAD database are neither derivation rules nor integrity constraints in their pure senses, but define the goal toward which the extensional part of the database is enlarged and/or enriched through the design process. They constitute the system of axioms of a formal theory, of which the extensional part is an interpretation that becomes a model when the design is completed. At any given instance during the design process, a certain set of axioms could function as a derivation rule if the (unique) minimum extension of the database to satisfy them exists. Such a case will be rather exceptional, however, and there will generally exist multiple minimal extensions for the same set of axioms.

Two major problems to be solved in constructing process models are how to select axioms to be satisfied at each step of design and how to find minimal extensions of a database for their satisfaction. These, together with the precise definition of minimal extension, will be the subjects of later discussions (Sects. 8.1 and 8.3).

4.4 Single or Multiple Interpretations?

The preceding section also answered the third and fourth questions raised in Sect. 4.1. In representing designs, multiple interpretations must be considered under a single formal theory. At least one, possibly more, but not all of these interpretations form models of the theory. Precisely speaking, this is a situation peculiar to the descriptions of design objects themselves which will not usually come up in defining attribute domains. This difference reflects that between objects and their attributes which we touched on in Sect. 4.2.2 (2) and shed further light on in the following.

(1) That we normally find multiple solutions for a given requirement implies the possibility for certain significant design decisions to be embodied only in specific solutions without being formally represented.

(2) Hence, it implies the necessity for the theory to be able to be dynamically expanded to incorporate such decisions, and for the extensional part and the theory to be equally accessible from applications.

(3) It also implies the necessity for a certain device through which alternate designs can be mutually related. Suppose, for example, designs A and B are identical except for their respective components a and b which are different but functionally equivalent. We may well want to write

$$alternative(A, B, a, b) \tag{4.33}$$

to express this fact. It should be noted however that A and B, each representing an entire universe of interpretation of the same theory, cannot both be treated under that theory. In other words, ***alternative*** does not really qualify as a predicate of that theory and should be understood as a statement on a metatheoretic level.

(4) Likewise, the transition of database state transcends the level of states themselves, hence cannot be defined within the theory. For example, suppose we rotate a face of a tetrahedral part around one of its boundary edges so that it makes a right angle with the adjacent edge (Fig. 0.8). We may represent this operation as a mapping (μ) from the old to the new tetrahedron that maps each element (body, faces, edges and vertices) to its corresponding one and preserves the structure that is defined by the winged-edge topology. We may formulate a theory which contains the winged-edge topology, the requirement statements (1)–(3) as in Sect. 0.6 and possibly others, and of which each of the tetrahedra and their elements constitutes an interpretation. There will be no function symbol in the theory, however, that denotes μ.

(5) On the other hand, each of the attribute domains in terms of which properties and functions of design objects are described concerns each specific aspect of design, hence is inevitably a particular and well-established structure. Changing attribute values means selecting different instances from the same (or possibly different) domains and entails changes of the structure of the object model. Yet, the structure of the attribute domain remains the same. Take, for example, the case of a triangular part (Fig. 2.4). What is changed is the correspondence between an abstract entity *edge* and a geometric entity (a line-segment or an arc), and the structure of the object *part* containing this edge. What remain unchanged are the identities of entities involved in these changes and indeed the entire domain of geometric entities. In particular, we will not consider the line-segment as being transformed into the arc. Clipping a polygon (Sect. 4.2.1.3) presents another example. The clipping of a polygon p with a line l to generate another polygon q does not really create a new entity q, nor transform p into q, but establishes the equality

$$q = clippg(l, p) \tag{4.34}$$

among existing and permanently fixed entities. We may of course take a more dynamic view of a geometric domain in which polygons may change their shapes, positions and structures. Then each polygon is conceived of as an abstract entity with the same identity before and after the change. Consequently, different polygons may coincide in their attributes. The domain thus viewed is closer to the database representing design objects than to the normal, static geometry. The distinction between static and dynamic views is subtle but decisive.

(6) Most attribute domains are strongly typed with limited sets of operations. They are defined as either primitive (**integer**, **real**, **boolean**, etc.) or more complex data types. Since primitive types are usually implemented as hardware or firmware functions, their meanings are established completely and independently of any particular implementations. Complex types are typically constructed as the canonical interpretations (initial algebras) of the formal specifications. In either case, we can and should give a formal theory to each attribute to define its meaning precisely and a unique interpretation to embody the meaning. Only the theories concern applications. Everything else should be discarded. Certainly, we conceive of the category comprising all the models and homomorphisms between them for a given specification in theoretical treatments of ADT. Such a conception belongs to metatheories however, and unlike the case of design alternatives, should not concern applications.

4.5 The Universe of Interpretation

We have decided that each model of a design object be an interpretation of a first order theory. But what is the universe of interpretation? This question concerns the semantics of the relational model discussed in Sect. 2.4 as a collection of ground atoms:

$$\left.\begin{array}{l} \{p(a_1^{(i)},\ldots,a_l^{(i)}) : a_\lambda^{(i)} \in A_\lambda\} \\ \{q(b_1^{(j)},\ldots,b_m^{(j)}) : b_\mu^{(j)} \in B_\mu\} \\ \ldots \\ \{r(c_1^{(k)},\ldots,c_n^{(k)}) : c_\nu^{(k)} \in C_\nu\} \end{array}\right\} \tag{4.35}$$

is syntactically indistinguishable from a relational database. Let us then briefly summarize previous discussions and state the conclusion that we can draw immediately from them.

(1) If tuples are to be taken as facts, the universe of interpretation will be the union of attribute domains.

$$(\cup_\lambda A_\lambda) \cup (\cup_\mu B_\mu) \cup \ldots \cup (\cup_\nu C_\nu) \tag{4.36}$$

However, tuple relational calculus (TRC) languages do not allow quantification of attribute variables while they do allow it of tuple variables.

(2) More fundamentally, the concept of key strongly indicates that tuples represent things (entities) rather than facts (relationships) since each of them maintains its identity throughout whatever changes are made to non-key attributes. By contrast, any two facts that show differences in their constituent elements cannot be considered identical, however slight the differences may be. It is this funtion of a key to identify things that enables the description of complex objects in multiple relations. Those tuples with a common key value are mere parts (dependent objects) of the single complex object which will be made explicit when they are joined.

(3) On the other hand, domain relational calculus (DRC) languages subsume the view in which the universe is as expressed in (4.36), hence allow quantification of attribute but not tuple variables. Manipulating tuples that denote objects is not so straigtforward as dealing with those that represent facts.

(4) In fact, there are relations in which tuples contain so-called foreign keys pointing to other tuples and hence are more appropriately interpreted as facts between referenced entities. The problem is that the semantic peculiarities of different types of tuple are not explicitly represented in the syntax of relations.

(5) Another problem with the relational model (or more generally record-based information models) as applied to CAD is that every entity is represented as an instance of a record type which has a fixed set of attributes whereas design objects should be allowed to change not only the values but the settings of attributes themselves without losing their identities. The effectiveness of the concept of object type definded by the homogeneity of attributes for the modeling of design is therefore questionable.

(6) Fixed record structures are not adequate to describe individual attributes either, which tend to be much more complex and elastic than in business applications.

(7) Moreover, multiple tuples of a relation are often required to describe the whole or a part of a single complex object. Specifying the condition for retrieving objects may involve an indefinite number of tuples of a relation, which goes beyond the powers of expression of either TRC or DRC languages.

Our conclusion from the above is as follows:

- The universe should comprise both kinds of elements, one to represent design objects and the other to denote values in attribute domains.
- Objects should not be represented as bundles of attribute values but be captured as more abstract or symbolic entities that exist and preserve their identities independently of whatever attributes and attribute values they may have.
- In particular, they should not be conceived of as instances of types, each of which is defined by a fixed and homogeneous setting of attributes.
- Rich and flexible structures should be allowed in defining attribute domains so that complex and macroscopic properties and functions of design

objects, the representations of which may otherwise require multiple tuples in one or more relations, can be described in a straightforward manner. Unnecessary details and implementation specificities should be hidden from applications. Practically, all the data types available in most of the widely used programming languages must be available while ideally, these should be specified as abstract data types.
- Each tuple should unequivocally represent an elementary fact in the said universe of interpretation.
- The data manipulation language should be able to treat complex objects as naturally and economically as simple objects. In particular, it should be possible to specify a condition for retrieval which involves the evaluation of functions of a variable number of objects related to the objects to be retrieved.

4.6 What Kind of Logic?

While opposition to the use of FOPL as the base language is rife, we will discuss in the following mostly the case made by proponents of semantic networks, primarily because the latter have had a significant influence on the development of semantic data models. We will also make passing remarks on the use of more sophisticated logic than FOPL, such as higher order, modal and intensional logic, etc.

4.6.1 Semantic Network (SN)

The common characteristic of those approaches that are grossly termed semantic network is the representation of interrelationships between objects (nodes) in terms of special binary predicates (links) each of which is supposed to carry a particular meaning. These usually include:

- *Classification* (*instance-of*)
  Plato is an *instance-of philosopher*.
- *Aggregation* (*part-of*)
  Tire is a *part-of car*.
- *Generalization* (*is-a*)
  Car is-a vehicle.

In some cases, *is-a* is further divided into *generalization* and *specialization* ([Abi], [Hull]), with an example of the latter being:

Philosopher is-a scholar.

Instance-of not only represents a class and its membership but subsumes the existence of a protoypical object for the class. The essence of the notion of a prototype lies in the stipulation that its properties be inherited by all instances unless overrides are explicitly specified. This also applies to *is-a* which may well be defined as the set inclusion relationship. In the case of aggregation, the dependence is in the opposite direction, that is, the properties of sub-types determine those of the super-type. While it is obvious that FOPL is more general and has greater expressive power syntactically, there are arguments that it is too flat or of low level, or lacks organizational principles such as explained above ([Bra(1)], [Isr], [Myl]).

Kowalski [Kow(2)] also advocates binary representations over general n-ary representations, though on different grounds. The former are said not to be special cases of the latter. For example, the table

supply	*ID*	*supplier*	*part*	*receiver*
	10	s	p_1	r
	11	s	p_2	r

should be represented as

$$is\text{-}a(10, supply) \wedge supplier(10, s) \wedge part(10, p_1) \wedge receiver(10, r)$$
$$is\text{-}a(11, supply) \wedge supplier(11, s) \wedge part(11, p_2) \wedge receiver(11, r)$$

rather than

$$supply(10, s, p_1, r)$$
$$supply(11, s, p_2, r)$$

Justifications for this run as follows. First, unknown information can easily be ignored and new information can easily be added. Second, properties which are independent of the order of columns can be represented more naturally. Third, certain derivation rules can be given in more concise forms. For example, the statement

$$\forall x(part(x, p) \rightarrow supplier(x, s))$$
("s is the sole supplier of part p")

cannot be expressed in a clause using the predicate *supply*.

4.6.2 Criticism

This first part of our criticism is directed to the gap between the meanings the representational formalism is intended (or claimed) to capture and those actually captured. In the next section, we will see to what extent this gap can be filled with FOPL and what remains unrepresented. Finally we will discuss the nature of the residue and the limitations of FOPL.

(1) *General Discussion*
Historically, SNs were invented and studied for symbolically representing the relationships between concepts embodied in grammatical elements (words, phrases, sentences) of natural language ([Fil], [Qui], [Rum], [Scha], [Sha(1)], [Sim]). This apparently had double-edged effects because on one hand, the meaning of any natural language as commonly understood by the people

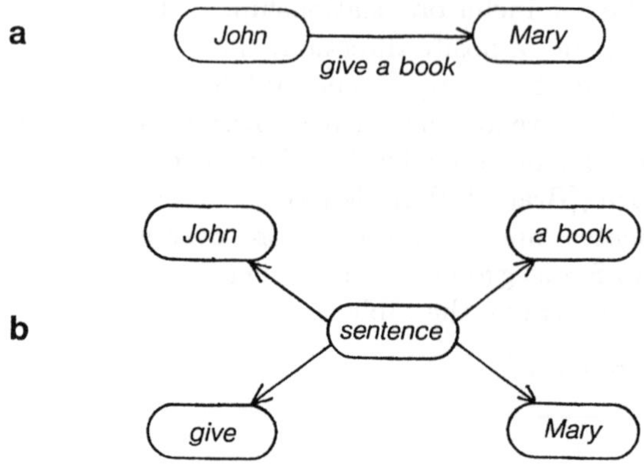

Fig. 4.5. Alternate network representations

who speak it as the mother language is so thoroughly established, yet, on the other hand, its precise formulation on a sound scientific basis seems almost hopelessly difficult. Consequently, the interpretation of nodes and links relies largely on the shared intuitions of authors and readers. Abstraction is not adequate either. A single network often contains objects and relations of different levels of abstraction without any explicit marks for differentiation. For example, a node which denotes an individual object may be connected to the node which represents its class by an ***instance-of***, to another node for a larger object by a ***part-of*** and to still others by links with application specific names such as ***verb***, ***agent***, ***right-of***, etc. Nodes may even represent phrases and sentences. Thus, we have at least two ways to construct networks for the statement

John gives Mary a book.

as shown in Fig. 4.5a,b.

That in Fig. 4.5b in turn can be considered as an instance of a type of sentence shown in Fig. 4.6

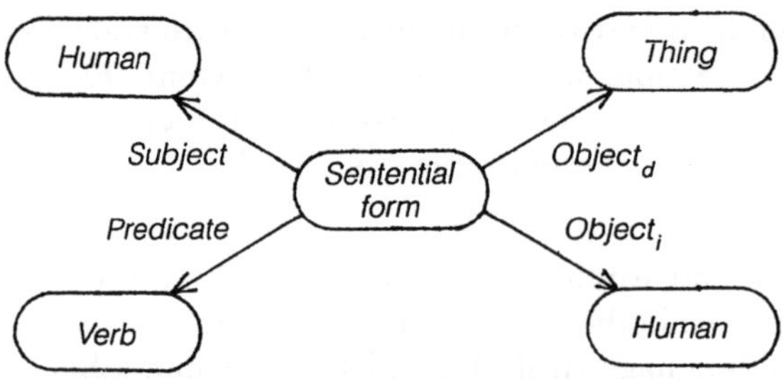

Fig. 4.6. Representing a type of sentence

Though the overall structure and the names of links are the same, these two representations carry fundamentally different meanings. That is, one asserts a particular fact while the other defines the kinds of constituents and the way they are organized into a sentence. In database terminology, these will be distinguished as being a schema and one of its instances. The distinct naming of head nodes (*sentence* and *sentential form*) which indicates the difference of meanings perfectly to human readers is not intelligible to computers at all. Yet, the treatment of varied cases in accordance with their correct meanings will not be given any clear definition and will usually be implemented inside the network interpreter. The network itself merely shows that such and such objects are related to each other in certain ways.

Natural languages are probably too subtle and complex to be dealt with in a naive formalism like nodes and links. We must either sharpen our conceptual tools or restrict the universe of discourse.

Ambiguities and difficulties with SNs including those described above and others have been widely recognized since a paper by W.A. Woods [Woo] pointed them out. Subsequent efforts to remedy the situation have tended to discern different levels and categories of nodes and links, ant to conceive of primitive entities at more abstract levels. Some authors take the extreme position which views SNs to be mere implementational mechanisms or restricted variants of predicate logic ([Schu(1)], [Schu(2)], [Sha(2)]). Others consider that logical languages are too neutral to be truly useful in representing knowledge and try to provide primitives at 'higher' levels ([Lev], [Bra(1)], [Bra(2)]). Particularly noteworthy among these primitives are the three basic relations mentioned in the preceding section and the notion of 'intensional object'. These have been abstracted out of more application dependent concepts, yet cannot be reduced to pure logical symbols and seem to constitute the core of the semantics of semantic networks.

Let us now turn to a few exemplary works and see how they succeeded in capturing the meanings of primitives within their formalisms.

(2) *The Case of KL-ONE*

Brachman ([Bra(1)], [Bra(2)] identified five separate levels of understanding networks and singled out as the most important what he called the 'epistemological' or 'knowledge-structuring' level. The names suggest that this is intended to provide a framework in which to define various sets of higher level primitives without committing to any specific one. In the example of the KL-ONE language, there are generic and individual types of elements which are called *Concepts*. Each *Concept* represents an intensional object and none denotes any external existence. These are also classified into *primitive* and *defined* groups. *Primitive Concepts* are those for which we cannot give sufficient conditions for unique identification but must simply accept as existing. Typical examples include natural kinds such as dog or male-animal, and undefined mathematical terms such as point or line-segment. *Defined Concepts* are those that are completely described in terms of other primitive or defined concepts.

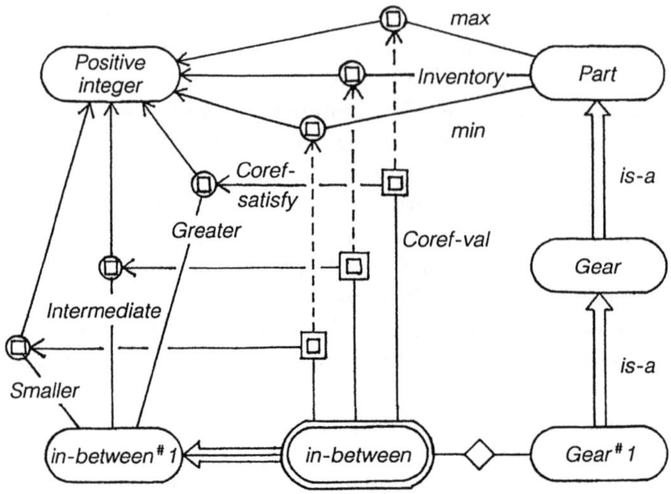

Fig. 4.7. Structural Description in KL-ONE

There are four kinds of relations holding between *Concepts*, that is, *Super-Sub*, *Individuates*, *Roles* and *Structural Descriptions* (*SD*s). The former two are the usual *is-a* and *instance-of* respectively. A *Role* is an arbitrary binary predicate between a *Concept* and its filler. An *SD* stipulates a constraint to be observed by two or more *Roles* of a *Concept* and is itself a *Concept*. For example, the condition that the inventory level of any part must be higher than the minimum and lower than the maximum will individuate a generic *Concept in-between* accompanied by *satisfies* links between corresponding *Roles* (Fig. 4.7).

While the proposed use of neutrality (toward higher levels), adequacy (for supporting higher levels) and (accountability for) semantics as the criteria for judging the utility and formality of a language is fully agreeable, we must question whether KL-ONE really meets these criteria, particularly whether it has succeeded in explicating its epistemology within its formalism. Our conclusion is that the degree of formalness and abstractness (ref. Sec. O.7) of the language is not satisfactory, with the meanings of terms not given rigorous definitions but relying more or less on readers' intuitions and interpretations as English words and phrases. This is elaborated below.

(2.1) What is a *Concept*? A *Concept* is said to represent an object, its attribute or its relation with another *Concept*. As we discussed previously, however (see Sects. 2.4.2 (2), (3) and 4.5.), things (objects) and facts (relations) demand fundamentally distinct treatments. Hence, as Woods [Woo] noted, structures with objects at their nodes must be interpreted differently from those with nodes representing predications. For example, the instance of the *in-between* relation

$$in\text{-}between(inventory(gear^{\#}1), max(gear^{\#}1), min(gear^{\#}1))$$

will wholly lose its meaning and must be deleted if $gear^{\#}1$ is deleted. The same applies also to the *supply* relation. The converse is not true, as a *gear* node may continue to exist quite legally even if its inventory moves outside the normal range or vendors stop supplying it.

(2.2) Generally, we understand a level of abstraction being supported by another below as implying that concepts at the former level must be able to be interpreted in terms of those at the latter. The interpretation must of course be unequivocal. This will not be the case with KL-ONE since what is uniformly represented as a *Concept* at the epistemological level will be normally formulated either as a term or a predicate at the logical level depending on whether it is an object or a relation.

(2.3) Distinctions seem mandatory among *Concepts* that represent relations as well. It is emphasized that names given to nodes are purely for human readers and hold no significance for KL-ONE. This position is certainly acceptable for such a case as *supply* where domain-specific knowledge of nodes and links will not be necessary for the network interpreter to perform epistemological functions. It will not be true of *in-between*, however, because the task of constraining relations among *Roles* of a *Concept* cannot be accomplished unless the label *in-between* stops being a mere character string and carries the complete semantic information for the interpreter. The information includes knowledge about the structure of attribute domains, which in the present case is the domain of natural numbers. There will generally be two approaches to 'formally' knowing paricular domains: define them within the given vocabulary or provide the interface to access those defined outside the language. Neither seems possible with KL-ONE or any network language at the strictly 'epistemological' level.

(2.4) What is a generic *Concept*? We seem to have to differentiate at least three cases in the above example. The first case is *part* for which individuation is done for each individual *Concept* independently of others of or any general conditions beyond conformity of its *Roles* to the specifications of the generic *RoleSet*. In one aspect, we may consider it as a unary predicate the extension of which is rather arbitrarily defined on the set of all individual *Concepts* (things). In another, it describes a binary relational schema in much the same way as Kowalski [Kow(2)] proposes. The second case is *quantity* (positive integer) which has a definite extension and structure. The characterization is given in terms of relations between individuals, however, rather than *Roles* of each individual or relations with other *Concepts*. In fact, what can we conceive as the *RoleSet* of a 'typical' *quantity*? It does not really represent any single object but a structured entirety, indeed an infinite number of objects. The third case is *in-between*. Again, there will not be any such thing as 'typical' *in-between*. Furthermore do we distinguish the occurrences of *in-between* in, say

$$in\text{-}between(5,10,3) \quad \text{and} \quad in\text{-}between(a+1,a+2,a)$$

Certainly not. The sense of *in-between* (or any predicate name) depends on the identity maintained through its occurrences, however different they may be.

The distinction will be more clearly understood if we imagine how these cases will be handled by the network interpreter. The first case will be implemented as an enumerative set. *Quantity* is a typical abstract data type on which *in-between* is defined as a ternary predicate. The point is that the distinction which is so clear to human readers and can undoubtedly be conveyed correctly to the interpreter must nevertheless be made explicit in the formalism of network language.

(2.5) How do primitive and defined *Concepts* differ? That a (generic) *Concept* A is defined in terms of *Concepts* $B, C, \ldots$ and local restrictions $\Gamma, \Delta, \ldots$ means that a given individual *Concept* α individuates A if only if it individuates $B, C, \ldots$ and satisfies $\Gamma, \Delta, \ldots$. Suppose A is also derived from $B', C', \ldots$ and $\Gamma', \Delta', \ldots$. Do we have to know that α individuates both sets of *Concepts* ($B, C, \ldots, B', C', \ldots$) and satisfies both sets of restrictions ($\Gamma, \Delta, \ldots, \Gamma', \Delta', \ldots$) before we conclude that it individuates A? While some pieces of knowledge suffice to decide, it is impossible to tell by examining the network which pieces qualify. The same problem arises whenever redundancy exists in defining a *Concept*. Or should we consider a network as representing a static organization of the world in which every individual *Concept* has been put to its right place with all of its attributes already known? Clues for individuations will then have to be given somewhere else, but where?

A variant of redundancy occurs in a recursive definition of *Concepts*. A very simple example is the case where two persons are related to each other by the *Role spouse*. An individual *person* is a *husband* if its *spouse* is a *wife* and vice versa. Are *husband* and *wife* both primitive and defined, or should either one be arbitrarily declared primitive and the other defined

So much for KL-ONE. Let us now examine another example.

(3) *The Case of IFO*

The IFO model by Abiteboul and Hull [Abi] has database schemas described in terms of object types, functions and two kinds of *is-a* relations: *specializations* (sub-types) and *generalizations* (unions). Object types are either one of the atomic kinds (*printable*, *abstract*, *free*) or constructed out of other types by means of a *product* or *power* operator. Printable types are predefined and include those usually provided in programming languages. Abstract types, like KL-ONE's primitives, allow partial descriptions in terms of functions but no complete definitions. Thus they cannot appear as products, power sets or generalizations (note that these constructors uniquely determine types from constituent types) though they can be specialized in one or more ways, or used as constituents to form more complex types. Different abstract types are much more mutually independent than different KL-ONE

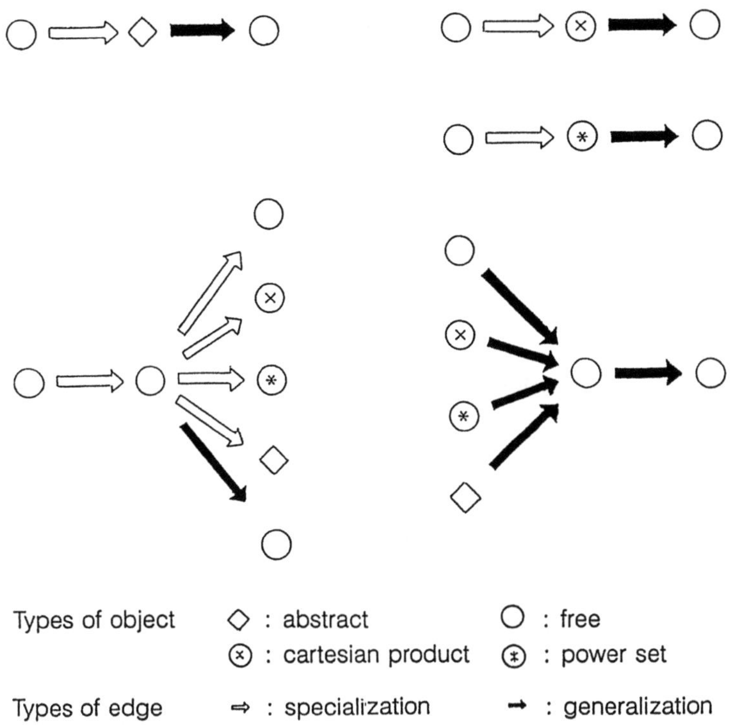

Fig. 4.8. Permissible chains of *is-a* relations

primitives in that neither kind of *is-a* relation is permitted among them. The primary role of free types is to represent those defined by specializing or generalizing one or more other types. Figure 4.8 depicts permissible cases of chaining two *is-a* relations.

Additional and more global constraints must be observed in specifying a database schema using elementary constituents and constructors explained above. These include:

- the prohibition of cycles consisting of either specialization or generalization links, and
- the requirement that two specialization paths originating at the same node be extended and converge again.

The extensional databases are obtained as instances of schemas (those in which *instance-of* relations do not appear explicitly).

The distinction of two kinds of *is-a* relation is the single most important feature, and at the same time the cause of some confusions and complications in the IFO model. Though the explanation that specialization defines 'roles' in a given type while generalization combines existing types into a union appeals well to human intuitions, it falls far short of giving a rigorous definition of either of them or of the distinction between them. The syntax (or rather syntactic constraints) of the hierarchical structures as we have seen above is clear but not very straightforward. Unless we have certain guiding principles established for building these structures for wide classes of applications, the questions of adequacy and neutrality as well as semantics will remain unanswered. Let us now be more specific and develop our questions on these points.

(3.1) To begin with, how do specialization and generalization differ from each other? If we write $spe(A, B)$ or $gen(A, B)$ to express the fact that B is a specialization of A or A is a generalization of B respectively, the only semantic conditions we must satisfy in forming a (partial) *is-a* tree are

$$\wedge_i\, spe(A, B_i) \rightarrow (\cup_i ext(B_i) \subset ext(A)) \qquad (4.37.1)$$
$$\wedge_i\, gen(A, B_i) \rightarrow (\cup_i ext(B_i) = ext(A)) \qquad (4.37.2)$$

where $ext(x)$ denotes the extension (set of instances) of type x. The so-called inheritance of properties (except default or exception handling) is a trivial consequence of these conditions, actually of (4.37.1) only, which is required by most network languages. How can a specialization that satisfies (4.37.2) of necessity be distinguished from a generalization? Suppose, for example, we have that *employee*(E) specialized into *single*(S), *married*(M) and *handicapped-employee*(HE). If the requirement comes up that the latter be a specialization of another type *handicapped*(H), we must make a few additions and modifications to the schema (Fig. 4.9a), namely:

- H and E nodes are changed to free types,
- these are specialized from the newly created *person*(P) node.

Suppose further that H has been defined as a generalization of some other types. The link connecting H and HE must then be a generalization, otherwise the global constraint on specialization paths will be violated (Fig. 4.9b). Looking at these alternatives, we note that two types of objects and two kinds of *is-a* relations respectively should be rather freely interchangeable. Exploiting this freedom will produce yet another alternative which is syntactically quite legal (Fig. 4.9c).

(3.2) The above example also indicated the ambiguity between abstract and free object types. Nevertheless, these are placed under different sets of syntactic constraints, which tend to cause local modifications to propagate globally. We question what the justification for this will be. Specifically, we question why

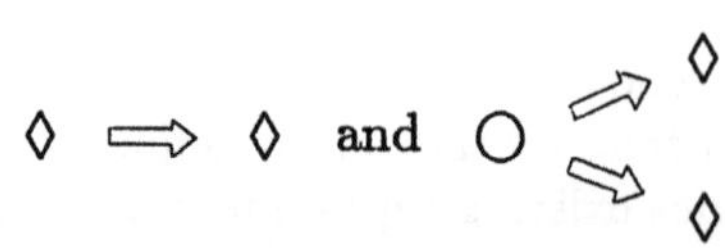

(4.38)

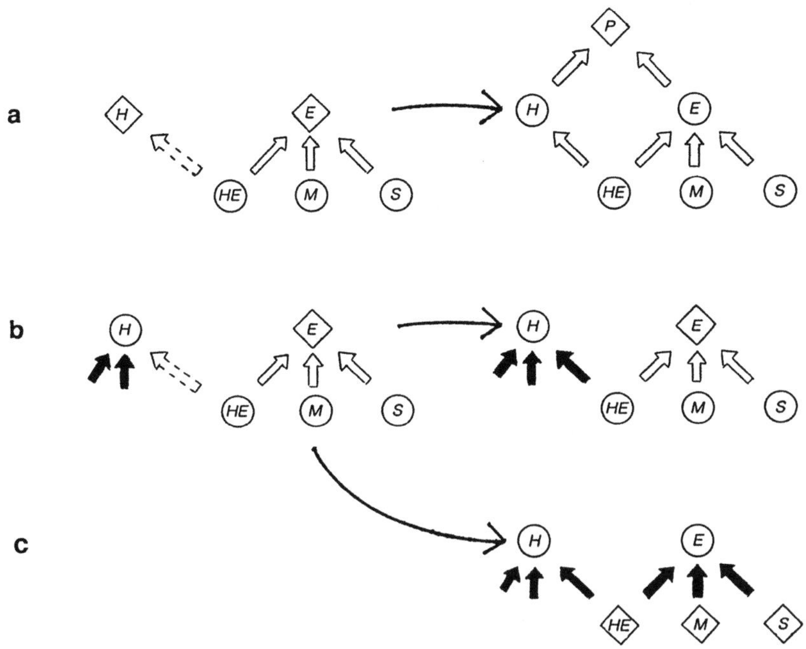

Fig. 4.9. Extending an IFO network

are not allowed while

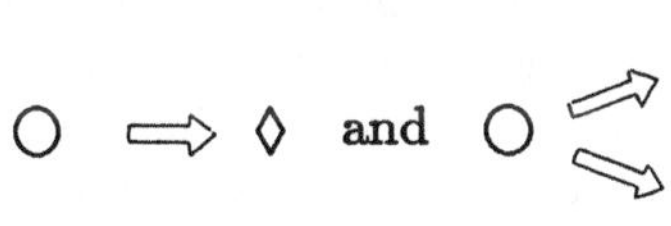

(4.39)

are allowed. That a specialization stipulates that the type at the head should be inherited by the tail is not simply accepted as the reason for rejecting (4.38) because:

- an abstract object is said to have no underlying structures though as many functions as desired may be defined on it, and
- sub-typings by two abstract objects in (4.38) will not necessarily cause conflicts provided functions defined on either of them can also be defined at the tail node with all constraints observed.

For example, we see nothing wrong or abnormal in specifying

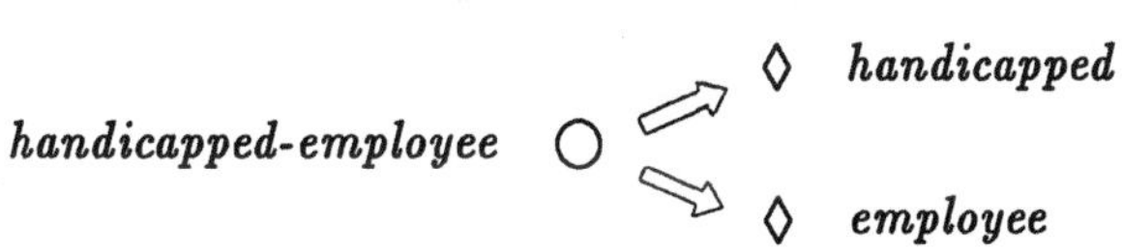

(4.38)′

but consider the requirement that two specialization links be extended to a common node to be rather artificial and unnatural.

(3.3) It might be that the true meaning of the hierarchical organization of IFO is to specify the order of determining types and structures of nodes in the completed network. Specifically, to start with, all abstract and printable objects are assumed to have been given fixed definitions regardless of any links connecting them to other objects in the network. These are 'top level' elements in the hierarchy. Second, free types are successively determined downward along the specialization paths. At each node where two or more links originate, types and structures inherited via different paths are combined into a union. Third, free and non-atomic types on the upward paths are determined by applying generalizations, product and power set operations in turn. The second step is then reentered and steps are repeated until all nodes are exhausted. It should be noted that this process involves hidden recursions, as functions may be defined whose ranges are determined only after their domains are, hence further conditions might have to be given in order to guarantee the existence of a fixed point.

Right or wrong, a procedural interpretation such as the above tends to obscure the essential nature of the structure, which is independent of any specific order of creation. For example, two networks in Fig. 4.10 for which typings proceed in different orders are intended to represent the same structure.

(3.4) The syntactic constraints on forming *is-a* networks have nothing to do with the semantic conditions. Their combination augmented by a procedural interpretation does not yet suffice for clearly discerning specialization and generalization. We are thus led to the conclusion that the formalism of the IFO model has not fully succeeded in capturing the intended semantics.

One last comment on IFO is that one of the causes for confusions and complications seems to lie in the complex and ambiguous roles a free type node is specified to play. This is obvious from Fig. 4.8 in which many a free node appears in various relative positions with respect to others. It may represent a definite sub-type of a super-type, or the union of several existing types, or serve as a mere passage for property inheritance from super- to subtypes. A clue may therefore be found to resolve the problem by distinguishing these different roles.

4.6.3 Network Represented as a Logical Structure

We do not attempt a systematic reconstruction of the concept of a semantic network but want only to suggest the possibility to represent within FOPL a few of the more characteristic features of network languages. These include:

- three (or four) basic hierarchical relations,
- prototype,
- inheritance of properties,

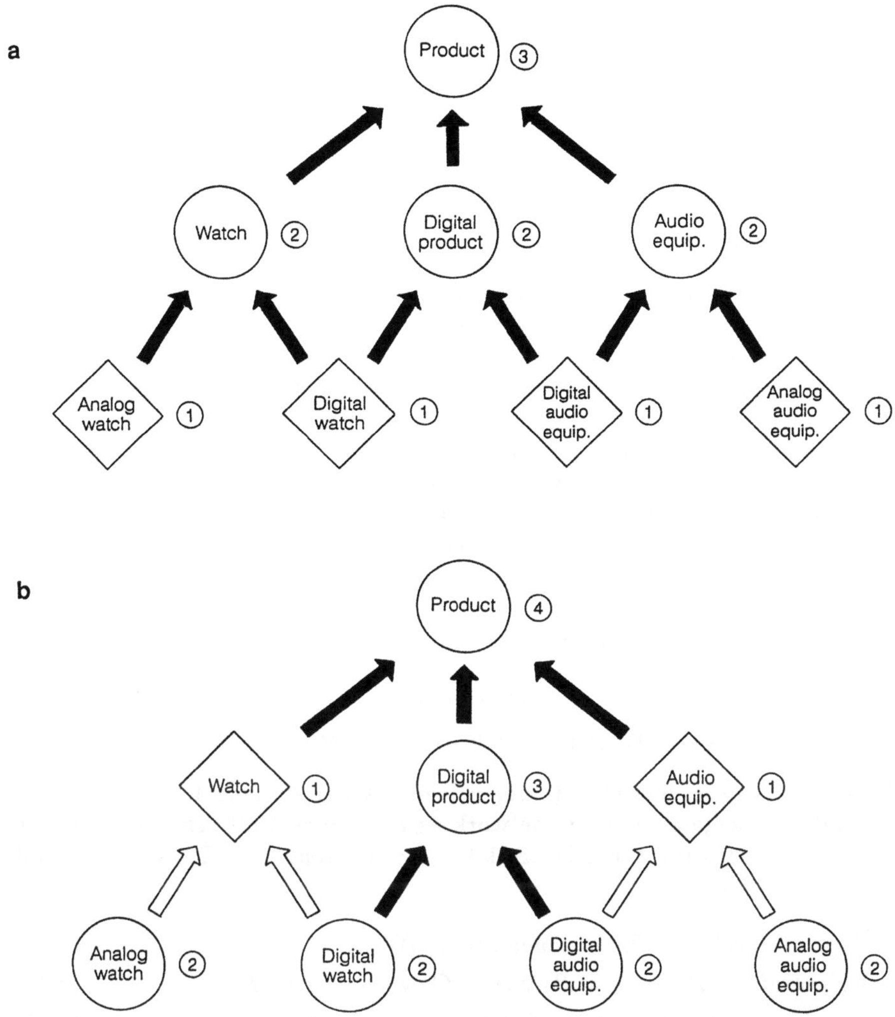

The number annexed to each box indicates the order of defining the type

Fig. 4.10. **Defining object types in networks**

- defaults or exceptions, and
- object orientation.

We will also see how the interpreted FOPL can introduce more precise distinctions among seemingly similar but essentially different notions.

(1) *Basic Hierarchical Relations*

(1.1) *Instance-of*

That A is an ***instance-of*** B can be expressed in at least three different ways as follows:

- Set membership $A \in B$ (4.40)
- Prototype instantiation $\mathit{instance\text{-}of}(A, B)$ (4.41)
- Unary predicate $B(A)$ (4.42)

The first of these raises the question of what is signified by '$\in$', the membership of a collection of individual objects or a formal as well as an interpreted predicate symbol. The difficulty with the former understanding is that B may not necessarily represent the 'class' of A but be an arbitrary set containing A. The latter demands that B is interpreted as a single object, though it may be identified with the set of A for convenience. The predicate '$\in$' in this sense is virtually the same as the binary predicate ***instance-of***. The notion of 'prototype' is confusing and complicated, however. The form of the assertion (4.41) in which it occurs as an argument indicates that B is just another object. If so, what distinguishes a prototype from ordinary objects and ***instance-of*** from arbitrary predicates must be explicitly stated. For example, the following seems both natural and inevitable:

$$\forall x \forall y(\mathit{instance\text{-}of}(x, y) \rightarrow \forall p(p(y) \rightarrow p(x))) \qquad (4.43)$$

$$\forall x \forall y \forall z((\mathit{instance\text{-}of}(x, y) \wedge \mathit{is\text{-}a}(y, z)) \rightarrow \mathit{instance\text{-}of}(x, z)) \qquad (4.44)$$

the first of which, however, is not a legal FOPL statement. Moreover, we normally envisage a semantic network as being essentially characterized by the given set of prototypes interrelated by ***is-a*** relations. This implies that we should postulate

$$\forall y(\mathit{prototype}(y) \leftrightarrow \exists x(\mathit{instance\text{-}of}(x, y))) \qquad (4.45)$$

$$\forall y(\mathit{prototype}(y) \leftrightarrow ((y = B) \vee (y = C) \vee \cdots \vee (y = D))) \qquad (4.46)$$

$$\forall y \forall z(\mathit{is\text{-}a}(y, z) \rightarrow (\mathit{prototype}(y) \wedge \mathit{prototype}(z))) \qquad (4.47)$$

as a part of the underlying theory for the network, a situation in which prototypes present themselves as constant symbols. Let us call them ***prototype symbols*** and their interpretive counterparts ***prototypical objects***. While prototype symbols are rightly contained in the theory, the question persists of what prototypical objects are and how they differ from instances. On one hand, we tend to conceive of a 'real' world as consisting solely of individual objects over which prototypes are placed as something 'ideal' or paradigmatic. On the other, we see little need to have prototypes interpreted as objects since the latter may add nothing significant to the descriptions within the theory.

These considerations lead us to the adoption of the third alternative which eliminates the notion of prototype as object and establishes it instead as a statement about objects. The most fundamental requirement of ***instance-of*** will be paraphrased as

$$\forall x(B(x) \to (\forall y(B(y) \to p(y)) \to p(x))) \tag{4.43$'$}$$

It should be noted that this is a legal FOPL formula and a tautology for any p while (4.43) is a non-logical axiom. The secret lies in attributing the properties to be inherited to the entire class rather than the seemingly independent prototypical objects.

(1.2) *Is-a*
"*B is-a A*" understood as a specialization will be given the expression

$$\forall x(B(x) \to C(x)) \tag{4.48}$$

which makes (4.44) and the inheritance via *is-a* also trivial.

A generalization of $B_1, \ldots, B_n$ into C will require an entirely different formulation. That is, we give the definition

$$C(x) \equiv B_1(x) \vee \cdots \vee B_n(x) \tag{4.49}$$

of C rather than make any assertion among previously defined predicates and/or formulas. The 'upward' inheritance is obvious as any object satisfying C will come from the extensions of some of the B_is. Constraints such as in Fig. 4.8 may be interpreted as warnings against potential contradictions caused by multiple definitions of the same object type. For example,

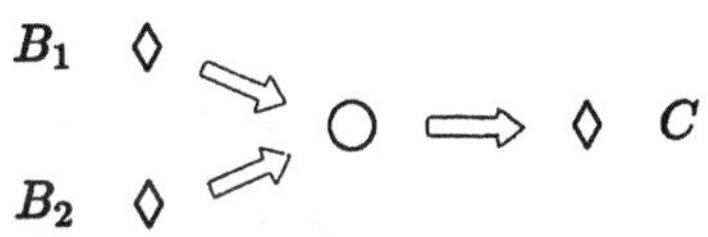

is allowed only when $\forall x((B_1(x) \vee B_2(x)) \to C(x))$ holds.

(1.3) *Part-of*
The connotation of *part-of* is twofold. First, there exists an object (the whole) which represents a collection of constituent objects (parts). Second, it carries the proper functions and meanings which are realized only through the interrelationships between parts. One way to incorporate these ideas in a suitable form will be to provide a set of predicates

$$\begin{array}{l} \textit{part-of}(\textit{whole}, \textit{part}) \\ p(\textit{whole}, \textit{part}_1, \ldots, \textit{part}_m) \\ \cdots \\ q(\textit{whole}, \textit{part}'_1, \ldots, \textit{part}'_n) \end{array} \tag{4.50}$$

where $p, \ldots, q$ together with appropriate constraints binding them define the internal structures of the whole. Additivity will be predicated of certain functions (e.g., weight, cost, number of sub-parts, etc.) defined on the set of all parts (transitive closure) belonging to the same root object. An alternate way is to have a ternary predicate

$$\textit{part-of}(\textit{whole}, \textit{part}, \textit{sub-part}) \tag{4.51}$$

which will enable the transitive closure to be evaluated in a much more straightforward manner.

(2) *Prototype, Inheritance*
We have probably oversimplified the notion of prototype. A more realistic case we often deal with in CAD environments is that of variational or parameterized objects which themselves define only 'patterns' and have to be instantiated to produce individual objects. These are typical complex objects with their constituents organized into ***part-of*** hierarchies. Moreover, each of them should be considered a small formal theory of which instances are interpretations. Inheritance should then be understood as being a model.

(3) *Object Orientation*
The heart of object orientation seems to be the encapsulation of all the information about each object within the node (or affiliation of nodes) representing it. There exists just too much freedom, however, to include pieces of information as pertinent: own properties, properties to be inherited, relations with other objects, functions and procedures to be executed, effects on behavior and state changes of the entire network, etc. With FOPL, each of these is given an appropriate form of representation according to its semantics rather than simply 'attached' to the node. Specifically:

- Properties are defined as (single variable) functions, the values (or their ranges) of which are given as part of the description of the object in question (or its class).
- Binary relations are represented by including one of the arguments in the description of the other.
- Relations of higher arities can be decomposed into a set of binary relations in the manner shown in Sect. 4.6.1. There are two cases which require separate treatments, however. First, we may naturally consider each tuple an independent object (such as *Course* in [Kow(2)]). Second, we may be actually describing a predicate or its instances in the disguise of an object as with *Supply* in Sect. 4.6.1.
- Defining functions and procedures that operate in the domains of attributes and including them in the description of objects should be quite possible, since we have decided that our logical model of design objects is an interpretation rather than a theory and that its universe of interpretation includes all the data types supported by the usual programming languages. Executing them should serve only to complete the definition of the static structure of the model, however, and not to cause any change in it. Logic neither can nor should concern itself with such a dynamic aspect.
- Inheritance may be regarded as a special case of default reasoning, which is discussed next.

It is interesting to note that so-called object oriented data models tend to rely on system-generated identifiers (surrogates) for identifying objects ([Cod], [Hei]). This reflects the conception of an object as something that exists and maintains its identity independently of any properties, relations with other objects or forms of representation. At the extreme end, there is the view of an object as completely detached from any particular piece of information about it, the exact opposite to that discussed above. Indeed it was this position that we took in discussing what the universe of interpretation should be (Sect. 4.5). The emphasis was of course on the abstract existence of object. The phrase 'object orientation' is multifaceted and multivocal.

(4) *Default Reasoning*
While default reasoning is a concept of much broader scope [Rei(2)], we limit the current discussion to the task of assigning values to attributes and predicates, which is a special case of defining functions and procedures. As the simplest case, the clause

$$A = \mathbf{default}$$

shall cause the procedure to be called that evaluates the default value of attribute A. For example, the inheritance which implicitly allows exceptions could be formulated as

$$\forall A \forall p \forall q(((A \textbf{ defined on } p) \wedge (A \textbf{ defined on } q) \wedge (p \rightarrow q) \wedge ((A|p) = \mathbf{default})) \rightarrow ((A|p) = (A|q|p))) \tag{4.52}$$

Although this is not a legitimate FOPL statement, it should be accepted as meaningful in any interpretation. The crucial point is that:

- the interpreter can access axioms as well as interpretations and interpret formal-theoretic as well as set-theoretic components,
- the statements are constructed solely of these components, and
- the constructions are formal and recursively interpretable.

(5) *Advantages of the Interpreted FOPL*
First and foremost, the achievement of a high degree of formality and rigor, the lack of which has plagued semantic networks so much, must be pointed out. More specific yet equally fundamental advantages include the following:

- Things (objects) and facts (relations) are categorically distinguished and represented as terms and formulas respectively.
- Concepts at different levels are given different levels of representation, namely schema and its instance or theory and its interpretation. Examples: sentential form and each sentence, prototype and individual object.
- The recursive definability of terms and formulas and the orthogonality between a relatively small number of elementary constituents enables the conceptually economical construction of highly structured objects. Example: complex object as prototype.
- Redundancies are allowed within a theory provided that appropriate axioms are asserted to ensure consistency.
- The decision to take interpretations rather than theories bestows the advantage of utilizing our knowledge about the universe of interpretation (including especially the usual data types) freely and legally for modeling

design objects. Examples: predicate *in-between* on integers or real numbers, default attribute values.

4.6.4 Limitations of FOPL

Like any other methods, FOPL has its own limitations. Indeed, an examination of the phrase 'interpreted first order theory' is enough to make us question once again the capability and applicability of such a framework for our purpose of modeling design objects. We must particularly consider the following questions, which are the most fundamental and difficult to answer:

- On 'theory': What methods and tools can and should we provide to facilitate formulating design requirements in a formal theory?
- On 'interpretation': How can and should we guide constructing and adapting interpretations for the theory?
- On 'first order': Why should we not go to modal, intensional or higher order logics to better handle such matters as beliefs, intensional objects or relations between relations?

The purpose of this section is not to answer these fully but to delineate the problems that should be given due considerations by our framework and/or its users. Let us first take another look at default reasoning, which may be regarded as an aid for defining theories, and proceed to discussing the above questions in turn in more general terms.

(1) *Default Reasoning*
Reiter [Rei(2)] has defined default reasoning as sanctioning the inference of a statement from another unless explicitly denied by a given set of negative conditions. Precisely, a *default* is an axiom of the form

$$(p \wedge \mathcal{M}(q_1 \wedge \cdots \wedge q_n)) \rightarrow r \tag{4.53}$$

where p, q_i, r are closed formulas. $\mathcal{M}$ indicates that the condition $\wedge_i q_i$ holds most likely. 'Holding most likely' means that its negation is not known to hold (is not proved). We may as well write

$$(p \wedge \neg(pr(\neg q_1) \vee \cdots \vee pr(\neg q_n))) \rightarrow r \tag{4.53$'$}$$

with $pr(s)$ expressing s is provable, which shows that default is a metatheoretic notion.

Given a (consistent) set of axioms A and a set of defaults Δ, we may extend A to include the consequents of a subset Δ' of Δ so that:

- the said extension A' of A will make a consistent theory,
- $\neg(pr(\neg q_1) \vee \cdots \vee pr(\neg q_n))$ under A',
- $pr(r)$ under $A \cup \Delta'$ for any default (4.53)$'$ in Δ', and
- Δ' is maximal in that any increase of it will cause an inconsistency.

There are generally multiple extensions for the same couple (A, Δ) which represent different sets of beliefs and are mutually incompatible.

Default reasoning as formulated above keeps intact the fundamental structure of FOPL. This strategy is quite agreeable in view of the inherent transcendency of the notion of default over any representation scheme, whether it be logic or network. The spirit is to formally establish a mechanism for identifying possible theories from a fuzzy set of pretheoretic alternatives. Once the decision is made on any specific alternative, fuzziness disappears and the selected theory contains nothing foreign to standard FOPL. Indications of this are twofold. First, the necessity to incorporate defaults should not at all impair the viability of FOPL as a representation framework. Second, we may probably exploit default reasoning at the stage of specification development or requirement definition which comes prior to designing. Preliminary ideas and general rules permitting exceptions together with more solid prerequisites will be given by product planners as a collection of defaults from which each alternate set of design requirements is produced as a formal theory and presented for approval.

(2) *Methodology*
The criticism of FOPL that it is too flat or lacks organizational principles is rejected, since the provision within a logical framework of much of the conceptual apparatus that semantic networks have long provided has been demonstrated to be just as good or even better. Defaults and exceptions which it is considered difficult to accommodate in logic have been shown also to be treatable in a way conforming to FOPL. Nevertheless, we admit that the collection of these fragments of means and devices is far from making a solid methodological base for systematic requirement specifications. We hasten to add, however, that the preparation of a framework alone, logical or not, will never produce any methodology for effectively applying it to a sufficiently general class of problems. An analogue is noticed in the contrast of programming language versus programming methodology. The needed task of methodology development belongs to a different level, and in our case of modeling design objects, should start with a thorough comparative study of alternate representations of certain fundamental structures, which include the following:

- shapes (topologies and geometries)
- form features
- physical constraints
- dimensions and tolerances
- spatial positionings and occupancies
- motions and transformations
- block structures of parts and assemblies

(3) *Interpretation*
The construction of an interpretation, like that of a theory, lies inherently beyond the descriptive capability of logical language. We will not follow the paradigm of object oriented programming in which the passing of messages invokes the execution of pertinent procedures which in turn causes the change

of states and subsequent passing of messages. The so-called procedural interpretation of a logic program [Kow(1)] together with such functions as ASSERT and RETRACT, makes logic appear to be capable of modeling dynamic aspects. This appearance is of course a mere illusion. The inference procedure does not change the state of the universe (of the Herbrand interpretation) at all. ASSERT and RETRACT, as database commands CREATE, DELETE and MODIFY, have nothing to do with logic. The task we must carry out is therefore to design a language in which to express the strategy and functions to effect a change of the state of the database in the desired direction. This language will necessarily differ from, yet accord with FOPL. We may compare it to a data manipulation language (DML) or Reiter's default reasoning as against a data (schema) description language (DDL) or FOPL respectively. More specifically, it will facilitate the following:

- prioritizing axioms so that one of the highest ranking among those currently false may be selected for sanctioning at any instance during the design process,
- specifying for each axiom the procedures to be taken to sanction it,
- evaluating the current state of the database,
- analyzing the adverse effects that may possibly result from sanctioning the selected axiom, and
- executing the specified procedures comprising CREATE, DELETE, MODIFY and other functions.

It should be noted that while providing the linguistic framework for expressing strategies is one thing, supporting actual strategy planning is quite another. The latter is analogous to the methodology problem for theory construction and will remain open.

(4) ***Modality, Intension, Higher Order***
We encounter only too often situations that we cannot properly formulate within FOPL. A typical example is Peano's axioms on natural numbers containing mathematical induction. The transitive closure of a binary relation is another. The inheritance of properties is yet another one. We can define all of these precisely if only we are allowed to quantify predicate symbols. In fact,

$$\Gamma(r,p) \wedge \forall s(\Gamma(s,p) \rightarrow (r \rightarrow s)) \tag{4.54}$$

where $\Gamma(r,p)$ is given by (4.21) will uniquely characterize the transitive closure r of p.

It will certainly be difficult to capture the mutually distinct meanings of the phrases 'Morning Star', 'Evening Star' and 'the planet second closest to the sun' under the view that the meaning of an expression is what it denotes which is associated with predicate logic. Hence the need for intensional representation has been emphasized ([Woo], [Bra(1)]). Some authors have gone to the extreme that every node in a network represents an intensional object [Mai]. A proposal by J. McCarthy [McC(1)] separated individual concepts from their reifications and introduced a special function symbol *denot* to relate them. Thus, the three phrases above represent three distinct objects of sort *concept* and have the same fourth object of sort *thing* as the value of *denot*.

While McCarthy upheld FOPL as the base structure to incorporate intensions and modalities, current trends are to rely on more sophisticated logic such as modal or intensional logic. A seemingly very exotic (at least from the classic FOPL standpoint) example is the definition of the adjective *every* by R. Montague [Tho]

$$every \equiv \lambda p \lambda q \forall x (p(x) \rightarrow q(x)) \tag{4.55}$$

abstracted out of the set of such sentences and their logical correspondents as

$$\left.\begin{array}{l} \mathit{Every\ dog\ runs} \equiv \forall x(\mathit{dog}(x) \rightarrow \mathit{run}(x)) \\ \mathit{Every\ star\ shines} \equiv \forall x(\mathit{star}(x) \rightarrow \mathit{shine}(x)) \\ \ldots \end{array}\right\} \tag{4.56}$$

In general, the semantics of a word will be given by formalizing the syntactical pattern in which it occurs.

Having admitted that there are conceptions that are not directly translatable into the first order language, we face a naive question which way we should go, whether to adopt one of these non-classical languages or cling to FOPL. The answer is that we will continue to rely on FOPL, though the reasons are not that naive. Let us state them briefly and tentatively as follows; a further elaboration will be given in the next chapter, particularly in Sect. 5.1.2.

- *On higher order or metatheory versus first order interpretation*
  - As we have already noted, the decision to take interpretations rather than theories enables us to incorporate in our framework many of the features that would usually require higher order logic.
  - Defaults and exceptions can also be incorporated without appealing to metatheoretic treatments to the extent that they are considered as conveniences for assigning attribute values.
- *On intensions or concepts versus extensional objects*
  - We question the value of conceiving of an intensional object or concept for each aspect or description of the same extensional object as a distinct entity. It may be more natural and appropriate to consider aspects and descriptions as attributes of a single object and to understand the 'meaning' of this object as the entirety of all the attributes and relations with others it may have. Thus, there will exist one entity V (a particular planet) which has such attributes as

$$\mathit{shine\text{-}in\text{-}the\text{-}morning}(V)$$
$$\mathit{shine\text{-}in\text{-}the\text{-}evening}(V)$$
$$\mathit{order\text{-}from\text{-}the\text{-}sun}(V) = 2$$

and whatever others may be predicated of it.

- The well-known 'telephone number problem' can be handled as follows. In trying to logically paraphrase the sentence "Pat knows Mike's telephone number", we should write

  $$know\text{-}telephone\text{-}number(Pat, Mike)$$

  instead of

  $$know(Pat, telephone\text{-}number(Mike))$$

  because first, what a person knows is not a thing but a fact and second, the occurrence of *telephone-number(Mike)* in the latter expression is equivocal. Indeed, we would never say "Pat knows *5025*" but instead "Pat knows *the fact that Mike's telephone number is 5025*". The distinction between the two underlined phrases is not that between object and concept, or extension and intension, but that between thing and fact, or entity and relation. The expression *telephone-number(Mike)* is deceptive in that it not only denotes the value of the function *telephone-number* for an object *Mike* but tends to connote (to human writers and readers) a functional correspondence between objects. This connotation of the correspondence, which is a fact, will completely disappear when we replace the expression in question by 5025 or whatever.

– *On the logic of natural language*
It is interesting to note that the logical correspondent of such a word as *every* that plays a very important role in our world of natural language takes a form which occurs frequently in information processing environments. Also interestingly, the dual of that form

$$\lambda p \lambda q \exists x (p(x) \wedge q(x)) \tag{4.57}$$

corresponds to another important word, that is, the indefinite article *a*. These make us speculate on the possibility of further exploration of correspondences between the structures of words in natural language and the pattern of logical formulas to which we may profitably pay special attention. This is not only speculative but points to another task at the metatheoretic level, however.

While we refrain from making any general argument for or against intensional logic, we feel that its justification, if it can ever be justified, will derive primarily from the fact that natural language is so self-contained and has an infinite degree of freedom, which results in its extreme complexity. The meaning of a word must be completely explained by the syntactic environment in which it occurs and which comprises strictly linguistic elements. Whatever psychological or empirical base the meaning may have may not be considered in formally expressing it. As the example of adjective *every* shows, words of different levels and categories occur in the flat surface structure. Hence the need arises to provide tools to discern them precisely and to organize them in accordance with their deep structure.

The situation is profoundly different with the universe of design objects, which is simply but fundamentally non-linguistic. A need for and expected benefit from going intensional which may overcome the technical risks and complications involved is not apparent. On the other hand, it is felt that there remains great scope for wider and deeper exploitations of FOPL, including the development of the methods and tools mentioned above.

References

[Abi] Abiteboul, S., Hull, R.: IFO: A Formal Semantic Database Model. ACM TODS, Vol. 12, No. 4, Dec. 1987

[Arb] Arbab, F., Wing, J.M.: Geometric Reasoning: A New Paradigm for Processing Geometric Information. In: Yoshikawa, H., Warman, E.A. (eds.): Design Theory for CAD. North-Holland, 1987

[Bra(1)] Brachman, R.J.: On the Epistemological Status of Semantic Networks. In: [Fin]

[Bra(2)] Brachman, R.J.: An Overview of the KL-ONE Knowledge Representation System. Cognitive Science 9, 1985

[Bro] Brodie, M.L. et al. (eds.): On Conceptual Modelling. Springer, 1984

[Cla] Clark, K.L.: Negation as Failure. In: [Gal(1)]

[Cod] Codd, E.F.: Extending the Database Relational Model to Capture More Meaning. ACM TODS, Vol. 4, No. 4, Dec. 1979

[Fag] Fagin, R.: Functional Dependencies in a Relational Database and Propositional Logic. IBM J. R & D, Nov. 1977

[Fil] Fillmore, C.: The Case for Case. In: Bach, E., Harms, R. (eds.): Universals in Linguistic Theory. Holt, 1968

[Fin] Findler, N.V. (ed.): Associative Networks: Representation and Use of Knowledge by Computer. Academic Press, 1979

[Gal(1)] Gallaire, H., Minker, J. (eds.): Logic and Databases. Plenum, 1978

[Gal(2)] Gallaire, H. et al. (eds.): Logic and Databases: A Deductive Approach. ACM Computing Surveys, Vol. 16, No. 2, June 1984

[Gri] Gries, D.: The Science of Programming. Springer, 1981

[Hei] Heiler, S. et al.: An Object-oriented Approach to Data Management: Why Design Databases Need It. Proc. 24th Design Automation Conf., 1987

[Hul] Hull, R., King, R.: Semantic Database Modeling: Survey, Applications and Research Issues. ACM Computing Surveys, Vol. 19, No. 3, Sep. 1987

[Isr] Israrel, D.J., Brachman, R.J.: Some Remarks on the Semantics of Representation Languages. In: [Bro]

[Kow(1)] Kowalski, R.A.: Predicate Logic as a Programming Language. Information Processing 74, North-Holland, 1974

[Kow(2)] Kowalski, R.A.: Logic for Data Description. In: [Gal(1)]

[Lev] Levesque, H.J., Brachman, R.J.: A. Fundamental Tradeoff in Knowledge Representation and Reasoning. Proc. CSCSI-84, 1984

[Llo] Lloyd, J.W.: Foundations of Logic Programming. 2nd ed., Springer, 1987

[Mai] Maida, A.S., Shapiro, S.C.: Intensional Concepts in Propositional Semantic Networks. Cognitive Science 6 (4), 1982

[Mal] Mallgren, W.R.: Formal Specification of Interactive Graphics Programming Languages. MIT Press, 1982

[Man] Manna, Z.: Mathematical Theory of Computation. McGraw-Hill, 1974

[McC(1)] McCarthy, J.: First Order Theories of Individual Concepts and Propositions. In: Hayes, J.E. et al. (eds.): Machine Intelligence 9. 1979

[McC(2)] McCarthy, J.: Circumscription – A Form of Non-Monotonic Reasoning. Artificial Intelligence 13, 1980
[Myl] Mylopoulos, J., Levesque, H.J.: An Overview of Knowledge Representation. In: [Bro]
[Qui] Quillian, M.R.: Semantic Memory. In: Minsky, M. (ed.): Semantic Information Processing. MIT Press, 1968
[Rei(1)] Reiter, R.: On Closed World Databases. In: [Gal(1)]
[Rei(2)] Reiter, R.: A logic for Default Reasoning. Artificial Intelligence 13, 1980
[Rei(3)] Reiter, R.: Towards a Logical Reconstruction of Relational Database Theory. In: [Bro]
[Rum] Rumelhart, D.E. et al.: A Process Model for Long-term Memory. In: Tulving, E., Donaldson, W. (eds.): Organization of Memory. Academic Press, 1972
[Sag] Sagiv, Y. et al.: An Equivalence Between Relational Database Dependencies and a Fragment of Propositional Logic. JACM, Vol. 28, No. 3, July 1981
[Scha] Schank, R.C.: Conceptual Information Processing. North-Holland, 1975
[Schu(1)] Schubert, L.K.: Extending the Expressive Power of Semantic Networks. Artificial Intelligence, Vol. 7, No. 2, Summer 1976
[Schu(2)] Schubert, L.K. et al.: The Structure and Organization of a Semantic Net for Comprehension and Inference. In: [Fin]
[Schw(1)] Schwartz, L.: Theorie des Distributions. Herman, 1950
[Schw(2)] Schwartz, L.: Methodes Mathematiques pour les Sciences Physiques. Hermann, 1961
[Sha(1)] Shapiro, S.C.: A Net Structure for Semantic Information Storage, Deduction and Retrieval. Proc. 2nd Int'l Conf. on AI, 1971
[Sha(2)] Shapiro, S.C.: The SNePs Semantic Network Processing System. In: [Fin]
[Sim] Simmons, R.F.: Semantic Networks: Their Computation and Use for Understanding English Sentences. In: Schank, R.C., Colby, K.M. (eds.): Computer Models of Thought and Language. Freeman, 1973
[Sto] Stonebraker, M.: Adding Semantic Knowledge to a Relational Database System. In: [Bro]
[Tho] Thomason, R. (ed.): Formal Philosophy: Selected Papers of Richard Montague. Yale University Press, 1974
[Web] Webster's Third New International Dictionary. G & C Merriam Company, 1969
[Woo] Woods, W.A.: What's in a Link: Foundations for Semantic Networks. In: Bobrow, D.G., Collins, A.M. (eds.): Representation and Understanding: Studies in Cognitive Science. Academic Press, 1975

Chapter 5
Philosophical Discussions

Often, as Brachman [Bra] has pointed out, semantic networks contain primitives of mixed levels. In fact, this is true not only of SNs but of any kind of model description language. Sometimes these are more naturally considered as providing sets of tools and means for implementing higher level constructs. Also, ease and efficiency for accessing information is emphasized. What remains of them after discarding such implementational aspects is that which may be termed 'cognitive', 'conceptual' or 'epistemological' and concerns us almost exclusively in the present context. Deciding on a specific language to describe models at this level presupposes committing to a specific position to view and understand the universe of discourse in question. The propriety of a language will therefore depend on that of the underlying epistemology or *Weltanschauung*. Carrying out a rigorous analysis and objective evaluation of various positions at such a depth will be next to impossible, however. The reason is that, besides the general difficulties in philosophical arguments, we face some that are particular to our case, namely:

- Few technical works in data modeling and knowledge representation refer explicitly to the *Weltanschauung* on which they are based.
- Many do not even associate themselves with any specific universe of discourse as their prime concern. However, there is no reason why we should believe that the same ontology and epistemology forms an equally good foundation for comprehending a great variety of domains which includes psychology, natural language, business information processing, design, manufacturing, and many others.
- Levels of primitives tend to be confused. For example, will links attached to a node in a network establish a 'semantic vicinity' [Sch] or provide a convenient access mechanism?

With this situation as background, what we intend to do is:

(1) to reinterpret in philosophical terms the conceptual nuclei abstracted out of the exemplar methods which we have already judged through the discussions in preceding chapters to have failed in properly modeling our universe of design objects and processes,

(2) to examine the result of the more fundamental causes of failure,

(3) to establish a better and more proper *Weltanschauung*, and

(4) to select paradigmatic methods or achievements which we may follow fruitfully in developing our own framework based on the said *Weltanschauung*,

of which the principal part is of course steps (3) and (4). The first two steps serve as preliminaries and will be handled rather freely and informally. They are nevertheless indispensable since only in contrast with them may the significance of the principal results be clearly understood. An apology is offered for the lack of rigor and objectivity in the following critical analysis: we hope it will be compensated by elucidating and explicating the philosophical foundation on which a linguistic framework may be constructed.

5.1 Philosophical Reinterpretations of Conceptual Frameworks

5.1.1 Entity-Centered and Subject-Predicate Weltanschauung

Let us consider, to begin with, what the conceptual substance of the residue of a 'semantic vicinity' will be if graphic notations and implementational considerations are discarded. Apparently, a universe which, in its pure form, consists of symbolic entities and binary relations among them is envisaged. By 'symbolic' is meant that each entity serves as a syntactic element by itself and is given its meaning only in terms of the entirety of relations with other entities in which it is embedded. Frege's *Principle of Compositionality* is thus violated [Mai]. The semantic density is not uniform throughout the universe, however. There exist entities of prime significance and of a higher level of independence around each of which the collection of directly related subordinate entities forms the semantic vicinity. Each enter of a vicinity may represent either a proposition (sentence) which holds among surrounding entities or an object (subject) of which surrounding entities are predicated. Vicinities may overlap and subordinate entities may have their own vicinities. Central entities may in turn be related to other entities at an equal or higher level of the semantic hierarchy.

Kowalski's binary decomposition of general n-ary relations (Sect. 4.6.1) exemplifies the above view by making each tuple and its surrogate into a semantic vicinity and its center respectively. Reflecting the ambiguity of the concept of relation, a tuple may be interpreted either as a proposition or an an object. If the former, the entity representing the relation will be contained in the vicinity as the predicate, while in the latter case, it will rank one level higher in the hierarchy and will serve as the class or prototype for the tuple in question (Fig. 5.1.). Other works in semantic networks and data models take similar views of worlds. In Schubert et al. [Sch], for example, semantic units are represented either as atomic propositions of the form [subject, predicate, additional arguments] or as functions of the form [subject, function, value] which may in turn be combined to represent more complex entities.

A few ontological biases behind these conceptions are manifest. First, we note a strong tendency to conceive of any object of thought as an 'entity',

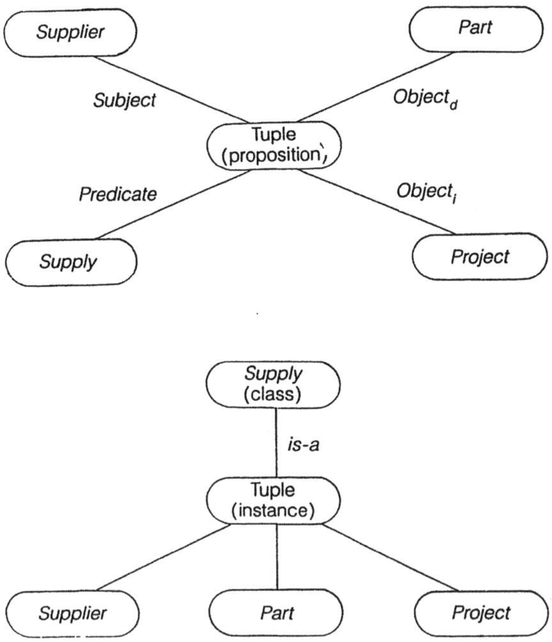

Fig. 5.1. Subject-predicate interpretations of n-ary relations

or something that exists independently of others, whether it is individual or generic, existential or propositional. Second, the meaning of each entity is made to depend on a (usually) small part of the universe which contains it and other entities closely related to it. Third, in this 'semantic vicinity', the central entity acts as the subject while others function as the predicate, properties, objects and other grammatical elements. The use of linguistic terminology is not a mere analogy. On one hand, natural language is often the universe of discourse of prime concern. On the other, it reflects the fundamental premise that an entity's holding the place of subject signifies its carrying a 'substance' which belongs only to it and that the nature of the universe is made known by knowing the substance of each constituent entity. Predicating properties, especially 'essential' properties, of entities constitutes the primary statements about the universe. The view based on these ideas will be termed the 'entity-centered' and 'subject-predicate' *Weltanschauung*.

Kowalski's contention that binary relations are not a special case of general n-ary relations even when $n = 2$ points to the asymmetry between the subject and the other argument intrinsic to this view. Let us also recall that asymmetries exist in many of the directed graph and associative structures even if implementation details are set aside. In ASP and CODASYL models, basic structures are binary relations in which a root entity 'owns' a number of member entities. The LEAP language which treats the three items in the triplet (A, O, V) completely symmetrically exhibits in its own way an adhesion to an entity-centered and subject-predicate *Weltanschauung*. Namely, an association's being ternary is an immediate result of its functioning as the predication of a property of the subject, and its defining an item allows a proposition to be treated as just another entity. An extreme of this tendency towards an entity-centered view is the object-oriented approach in which every bit of information concerning a particular object is encapsulated within that object, and interrelations among different objects are rather sparse and narrow, with message passing being the sole means of interaction. In contrast, the internal structure of an object which is invisible to the outside could have unlimited intricacy. The collection of objects thus makes up a 'federal' universe of small universes rather than a single and tightly organized universe.

Let us consider another aspect of the entity-centered organization. According to the ontology of the relational model, the universe consists of entities each of which is nothing but a set of attribute values and is uniquely identified by any one of its predefined subsets (sets of keys). The implications of this are that first, things are knowable by knowing their attributes, second, things with the same attribute values must be necessarily the same, and third, certain attributes constitute the essential and intrinsic part while others are accidental and dispensable. To know the key values is hence to know the things precisely as they are and nothing else. The creation of new relations by normalizing a given relation indicates that the existence of key attributes and dependency on them is considered to provide not only a necessary but a sufficient condition to recognize the existence of things as independent entities. These characteristics of the concept of key are strongly reminiscent of the Aristotelian concepts of 'substance' (Greek *ousia*) and 'essence' (Greek *to ti ēn einai*) as among the original causes of things. Indeed, we note in *Metaphysics* [Bar], for example, such statements as the following regarding these concepts, which may as well be quoted to characterize the functions of key:

The essence of each thing is what it is said to be in virtue of itself. [VII. 4]

'In virtue of itself' applies to ... whatever attributes belong to a thing alone and qua alone. [V. 18]

If there is to be any such thing as being essentially [something] ... there was some one thing which it meant, and this was the substance of something. And denoting the substance of a thing means that the essence of the thing is nothing else. [IV. 4]

Primary substance is that kind of substance which is peculiar to an individual, which does not belong to anything else. [VII. 13]

Metaphysics is full of vivid and stimulating examples to illustrate how we will think and discourse when faced with the problem of finding what makes things be as they essentially are. Let us then quote some further lines so that the abstract statements above may become rather more alive and understandable:

If a cup is mutilated, it must still be a cup. ... To be mutilated, things must be such as in virtue of their substance have a certain position. ... The parts removed must be neither those which determine the substance nor any chance parts, irrespective of their position; e.g., a cup is not mutilated if it is bored through; but only if the handle or a projecting part is removed. [V. 27]

People do not realize that it is impossible to define in the case of eternal things, especially those which are unique, like the sun or the moon. For they err not only by adding attributes after whose removal the sun would still exist, e.g., 'going round the earth' or 'night-hidden' (for from their view it follows that if it stands still or is visible, it will no longer be the sun; but it is strange if this is so; for 'the sun' means a certain substance); but also by the mention of attributes which can belong to another subject; e.g., if another thing with the stated attributes comes into existence, clearly it will be a sun; the formula therefore is general. But the sun was supposed to be an individual, like Cleon *or* Socrates. [VII. 15]

Of course, the concepts of substance and essence are deeply and inseparably connected to the subject-predicate *Weltanschauung*, the origin of which is also attributed to Aristotle. It is said that a substance is that which is not predicated of a subject, but of which all else is predicated [VII. 3]. It is also held to exist apart and be able to be indicated as a 'this'. The question is raised what the substance of an individual will be which is the compound of matter and form, or substratum and universals. When all the universals are taken away which are predicated of the ultimate substratum and hence can not be a substance, evidently nothing but matter remains. It is impossible, however, for matter to be a substance because it lacks both separability and individuality. Quite naturally, individuals themselves such as this man or that horse are cited examples of substance in the most strict and primary sense of the word (*Category*, V). Accidental attributes should be dismissed, of course. Aristotle remains noncommittal as to the ultimate answer to the question and even sounds contradictory when he takes up and scrutinizes one alternative after another, changing his viewpoint and drawing different conclusions. Yet on our side, we need to have the position, right or wrong, stated clearly to the extent that it may be translated into computer intelligible languages, at least theoretically, and allow comparisons with other positions. So, at the risk of misinterpretation, let us dare to define as follows what we understand to be the Aristotelian conception of substance and essence:

- The *substance* of a thing is the formula which results from accumulating successive differentiae of the topmost species (genus) and gives a necessary and sufficient condition for uniquely distinguishing the thing in question. The *essence* of a thing lies in its satisfying the said formula.

Note that the exclusion of accidental and superfluous attributes follows automatically. Now, our observation is that the above formulation of the substance of things may run precisely parallel to the provision of an identification mechanism for entities in a relational database with generalization/specialization (*is-a*) hierarchies. Specifically, the most abstract relation may contain the second level species as tuples with key values denoting specific differentiae. Each of these tuples may in turn be expanded in one of the next level relations and so on until individuals occur in the most specific relations. At each step, non-key attributes can be added as desired, and may be treated as non-essential ones in that they do not affect the identity of any entity, whether generic or individual [Smi].

As we have seen (Sect. 4.6.3), prototypes or generic entities could have been formulated as unary predicates, or more generally, small formal theories. As unary predicates correspond to adjectives (and theories to adjective clauses), conceiving of prototypes as 'entities' implies substantivizing adjectives or treating them as nouns. Furthermore, each of these nouns is considered to name the one thing of which the quality denoted by the corresponding adjective is purely and typically predicated. For example, 'Plato is aristocratic' will be expressed as

$$instance\text{-}of(Plato, aristocrat) \tag{5.1}$$

instead of

$$aristocratic(Plato) \tag{5.2}$$

where *aristocrat* names the entity that has all the properties necessarily belonging to every aristocrat but no others. In other words, these nouns name universals conceived as individuals which exist apart from concrete instances and yet in their own right. Hence they are nothing but what Aristotle understood as Platonic *Ideas* (*Metaphysics* X III. 4). There is indeed an abundance of evidence in the *Dialogues* to show the close affinity between *Ideas* and prototypes as conceived as above. To give a few quotations [Ham]:

The world has been formed in the likeness of that which is apprehended by reason and mind and ... must be a copy of something. [Timaeus 29]

These forms are as it were patterns fixed in the nature of things. The other things are made in their image and are likenesses. [Parmenides 132d]

We are in the habit of posting a single idea or form in the case of the various multiplicities to which we give the same name. ... For example, there are many couches and tables ... but, only two ideas or forms, one of a couch and one of a table. ... The craftsman who produces either of them fixes his eyes on the idea or form, and so makes in the one case the couches and in the other the tables. [Republic 596a–b]

Far more frequent references are made and a greater importance is given to the *Ideas* of adjectives, verbs or their substantivizations (i.e., abstract nouns) such as *good*, *beauty*, *existence*, which are considered to be even more typical unary predicates than common nouns. The existence of absolute beauty or goodness for example is assumed, and each individual thing partakes of the absolutes in order to be beautiful or good. Thus we have two ways for expressing 'Socrates is beautiful';

$$\textit{partake-of}(\textit{Socrates}, \textit{beauty}) \tag{5.3}$$

$$\textit{beautiful}(\textit{Socrates}) \tag{5.4}$$

in parallel to (4.41) and (4.42), also (5.1) and (5.2).

Just as prototypes are related to each other through generalizations and specializations forming a hierarchically organized network, *Ideas* are said to be woven together through blending or combining enabling discourses (such as 'Human is gregarious mammal') on them (*Sophist* 251c–260, *Republic* 511b–c). Two *Ideas* I_1 and I_2 (e.g., *gregarious* and *mammal*) can blend if and only if there exists an individual that partakes of both of them. That is,

$$\begin{aligned} \textit{blend}(I_1, I_2) &\leftrightarrow \exists x(\textit{partake-of}(x, I_1) \wedge \textit{partake-of}(x, I_2)) \\ &\leftrightarrow \exists x(I_1(x) \wedge I_2(x)) \end{aligned} \tag{5.5}$$

if we use the same symbol to denote a predicate and its corresponding *Idea*. *Blend* is symmetric with respect to its arguments, unlike a semantic network's binary relations, but can be paraphrased in terms of the latter as follows;

$$\begin{aligned} \textit{blend}(I_1, I_2) &\leftrightarrow \exists x(\textit{instance-of}(x, I_1) \wedge \textit{instance-of}(x, I_2)) \\ &\leftrightarrow \exists I(\textit{is-a}(I, I_1) \wedge \textit{is-a}(I, I_2)) \end{aligned} \tag{5.6}$$

Plato's theory of *Ideas* goes further, since it posits the existence of *Ideas* for relative terms such as *same*, *different* or *greater-than*. For example, it is said that:

Whatever is taller than something else is so simply by tallness and what is shorter is so simply by shortness. [Phaedo 101]

*(*Simmias*) is intermediate between (*Socrates *and* Phaedo*), and allows his shortness to be surpassed by the tallness of the one while he asserts his superior tallness over the shortness of the other.* [Phaedo 102d]

The apparently conflicting *Ideas* of smallness and tallness are in fact blendable, as indeed they are blended physically in the same individual *Simmias*. Presumably, two Ideas I_3 and I_4 for binary relations in general blend if there exists an individual that participates in both of the relations. That is:

$$\textit{blend}(I_3, I_4) \leftrightarrow \exists x \exists y \exists z(I_3(x, y) \wedge I_4(x, z)) \tag{5.7}$$

or a definition of the form (5.5) or (5.6) applies if unary predicates I_1 and I_2 are defined from I_3 and I_4 as follows:

$$I_1 = \lambda x(\exists y I_3(x, y)) \quad I_2 = \lambda x(\exists z I_4(x, z)) \tag{5.8}$$

Such redefinitions deprive binary relations of symmetry in significance between the participating terms to the effect that they become properties which are predicated of only one of the terms. As has been pointed out by many authors (e.g., Cornford [Cor], pp. 282–284), relations are not given their own distinct status but are considered as a special class of (unary) predicates in Plato's epistemology. The same is true of Aristotle's except that universals do not carry any substances, whether they be predicates or relations.

In our turn, we have noted the tendency of the relational model not to recognize the proper signification of general n-ary relations as such, but to treat them as a mere representation formalism and rely on other apparatus (e.g., data typing, *is-a* hierarchy, binary relations) in trying to capture the meaning of the world. Even with the Entity-Relationship model [Che] which purports to distinguish conceptually between relationships and entities, 'relationship' actually defines another subclass of more general 'relations' allowing predications of attributes just like 'entity' and spawning intermediate concepts like 'weak entity' and 'weak relationship' (Sect. 2.4.3 (2)).

A cursory look at an assortment of exemplars from data modeling should be enough to make us note the fundamental roles played by the concepts of 'entity', 'attribute' and 'relation' as they commonly appear in diversified approaches. The above analysis may as well be taken as a quest for the philosophical origins and implications of these concepts. We were particularly led to the classical notions of 'substance' and 'essence'. It will not be irrelevant then for us to have recourse to the dictionary explications of these key words in order to see how they relate to each other and reinforce our arguments. Thus, what follow are excerpts made from one of the most comprehensive and frequently referenced sources [Web]:

- *entity*
  - independent, separate or self-contained existence
  - the existence of something as contrasted with its attributes or properties
  - the essence, fundamental nature or real being of something.
- *attribute*
  - a characteristic, either essential and intrinsic or accidental and concomitant
  - an object closely associated with and thought of as belonging to a specific person, thing or office
  - (logic) any quality or characteristic that may be predicated of some subject
  - (philosophy) a necessary or essential quality or characteristic of substance.
- *relation*
  - an aspect or quality that can be predicated of two or more things taken together
  - the mode in which one thing or entity stands to another, itself or others.
- *substance*
  - [translation of Greek *ousia*] something that underlies all outward manifestations: the abiding part of existence or an existing thing as distinguished from what is accidental to it.

- *essence*
  - basic underlying or constituting entity, substance or form
  - the permanent as contrasted with the accidental and variable phases or foundation of being
  - condition or fact of being or existing: existence considered as a property of a thing
  - entity, especially an abstract entity
  - the/a most significant element, attribute, quality, property or aspect of a being.

As expected, interactions and interdependencies are obvious, with the meaning of each word being decided by the relative position it occupies in the network of cross-references. What is crucial is where to start tracing the thread of signification of those words. The entity-centered view starts with 'entity', assuming that the primary and basic meaning of the word (or rather the common meaning of the objects collectively denoted by that word) has been understood. The meanings of 'attribute', 'substance', and 'essence', which partially overlap, contribute to augment, modify and refine it. The interpretation of 'relation' is based on it. As will be seen later, an entirely different view with an opposite direction of interpretation is possible – and deemed more appropriate for our purpose.

5.1.2 Intensions and Extensions

Next, we want to examine the issue of intensions versus extensions, continuing and elaborating on the discussions in the last part of Sect. 4.6.4. First, we note that the use of the term 'intension' which may be simply taken as an equivalent to 'meaning' or certain of its components (e.g., cognitive or designative component [Car]) tends to emphasize the contrast with 'extension'. While an individual (physical) object, a class of objects and a statement's being true or false are extensions, the individual concept, the property that characterizes the class of objects and the proposition expressed by the statement are said to be the corresponding intensions. Thus, the intension of 'the sun' is that which makes a particular celestial body be nothing but the sun, and the intension of 'bright' is the property 'brightness' which is commonly predicated of all bright things. Two expressions

$$\textit{telephone-number}(\textit{Mike}) \quad \textit{department-code}(\textit{Mary}) \tag{5.9}$$

which could have the same extension will necessarily have different intensions. Now, these examples are again reminiscent of Aristotelian substance or essence. Indeed, the expressions in the last example are meant to serve as the defining formulas for their respective entities, and hence are considered to constitute the substances of these entities.

A crucial question is what sort of ontological status should be given to intensions, and whether they should ever be considered as entities that 'exist' separately from extensions. Varied approaches are taken according to varied degrees of separation. They are roughly classified into three different kinds as discussed in turn in the following.

First, some authors do not countenance the duplication of entities of comparable standing, while they admit or even emphasize the existence of distinct objects of thought as something objective and extra-linguistic ([Car], [Sch]). The same designator is to represent both an intension and an extension. The intension is given primacy because:

- it is completely determined in a logical manner, that is, by the designator and semantic rules alone, and
- it uniquely determines an extension in each 'possible world' but not vice versa.

Which of the possible worlds is the actual one will in turn be decided only through observing relevant facts. A natural and mathematically rigorous definition can be given to the extension for each concept type. On the other hand, it is not so simple to formalize intensions which are intuitively understood and usually referred to in popular terms such as mentioned above. Consequently, contextual definitions come up as more appropriate alternatives, which tends to make the notion of intension become still more virtual and less comparable to that of extension. If a further step is taken, either one of the two will be reduced to the other, or both to a third one. For example, an individual expression could be identified with the assignment of an individual object (extension) to each possible world.

Second, there are approaches in which distinct intensions are treated as entities distinct from one another as well as from extensions ([Woo], [McC]). Predicates and functions are provided to establish relations between different intensions and between intensions and extensions. These include among others:

- the equivalence relation *equiv* which relates one intension to another with the same extension,
- the function *ext* which maps every intension to the corresponding extension, and
- one of the inverse functions of the above *int*, that is, that which assigns to each extension its 'standard' intension.

Several ontological peculiarities associated with such conceptions can be observed. First, all the types of logical constructs – terms, function symbols, predicate symbols, and formulas alike – that represent intensions are eligible as the arguments of *equiv* and *ext*, hence must be treated as terms. A natural way to implement this is to conceive of these constructs as purely syntactical elements somewhat similar to those used in constructing a Herbrand universe. While distinct expressions generally carry distinct intensions, we may introduce an equality relation '=' to denote pairs of intensions which are the same. The usual equivalence law and equality axioms must hold. For example, the axiom schema

$$(X = Y) \to (A(\cdots X \cdots) = A(\cdots Y \cdots)), \tag{5.10}$$

where X, Y and A stand for any intensions of legal sorts, expresses the principle of interchangeability in an intensional context. As the function *ext* preserves equalities, the conclusion part of each axiom can be replaced by its extension. Thus, the above axiom schema will be rewritten as

$$\forall X \forall Y((X = Y) \to (\alpha(\cdots X \cdots) = \alpha(\cdots Y \cdots))) \tag{5.10$'$}$$

where α is the predicate or function that denotes the extension of A. Note that $\alpha(\cdots)$ and hence the whole matrix of (5.10)$'$ is a legitimate first order formula, hence permits quantifications on its free variables, whereas $A(\cdots)$ should be viewed not as having any internal structure but as being a primitive entity. If furthermore α is extensional with respect to X or Y, we will obtain

$$\forall X \forall Y(\mathit{equiv}(X, Y) \to (\beta(\cdots \mathit{ext}(X) \cdots) = \beta(\cdots \mathit{ext}(Y) \cdots)))$$
$$\beta(\cdots \mathit{ext}(\cdot) \cdots) = \alpha(\cdots) \tag{5.10$''$}$$

completing the extensionalization. If instead α defines an opaque (non-extensional) context, the antecedent may not be extensionalized. For example, it is said that $KNOW(P, X)$ (P knows X) is extensional with respect to P but not X, that is, that

$$\mathit{equiv}(P, Q) \to (KNOW(P, X) = KNOW(Q, X)) \tag{5.11}$$

holds for any X, but

$$\mathit{equiv}(X, Y) \to (KNOW(P, X) = KNOW(P, Y)) \tag{5.12}$$

will not [McC]. However, if we take $TELEPHONE\text{-}NUMBER(X)$ as X for example, the extensionality will be revived since we apparently have

$$\begin{aligned}\mathit{equiv}(X, Y) \to (&KNOW(P, TELEPHONE\text{-}NUMBER(X))\\ &= KNOW(P, TELEPHONE\text{-}NUMBER(Y))\end{aligned} \tag{5.13}$$

The cause for this is twofold. For one thing, as we previously noted (Sect. 4.6.4 (4)), anything we know and can speak of must somehow carry along with it the pertinent information on the facts involving it and related entities. For the other, the representatives of facts are nevertheless treated as if they themselves are simple entities. Opacity will result when this double role is demanded of the same designator. Indeed, $TELEPHONE\text{-}NUMBER(X)$ does not really represent the number itself but the fact that it is the telephone number of a certain person X. By contrast, intensional entities such as persons' names that are undecomposable and cut off from relations with other entities will not be distinguished from their extensions. If we augment the domain of extensions by including all the possible facts, every intensional entity will have in this extended domain a faithful representation which involves its extension, but is not generally equal to it in the usual sense. Figure 5.2 depicts this.

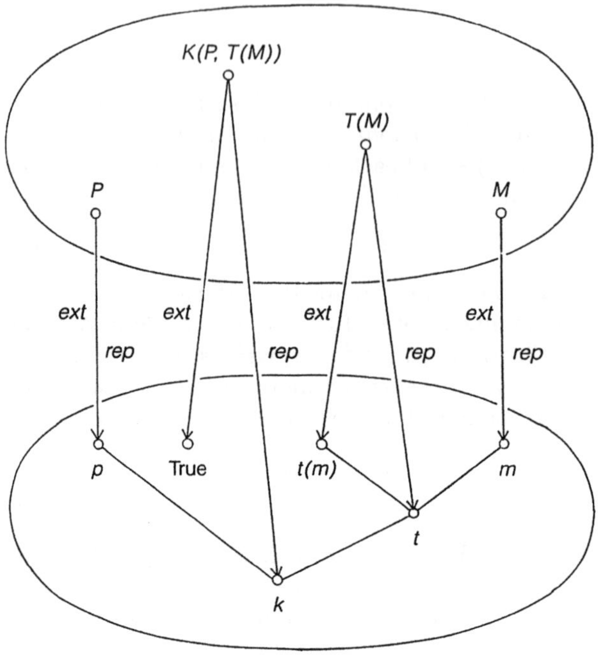

Fig. 5.2. Correspondences between intensions and extensions

To sum up, the world in this approach is divided into two levels, one representing observable things and facts and the other comprising substantivizations of facts. Viewed formally and abstractly, intensions are just another sort of extensions with structures formed through particular functions and predicates.

The third approach considers only intensions ([Bra], [Mai]). Every designator that appears in the system represents a concept but does not directly denote any real world entity. Since intensions are carried by facts, the system will be defined as a network of interrelating facts with individual concepts placed at terminal positions. Facts are represented either as the predications of properties or in the form of propositions. If the former, *is-a* and *instance-of* provide the apparatuses for interrelations. If the latter, individuals form the constituents of propositions, which are imbedded in other propositions in turn. If the system is completely severed from the reality and becomes self-contained, however, it may be considered to stand itself as a real world with designators serving as tangible objects. This is not the usual case. On the contrary, assertions are made for sets of individual concepts to share the same extensions (via *equiv* relations mentioned above) though the latter will not be referred to anywhere explicitly (that is, the function *ext* is not provided).

For example,

$$equiv(Morning\ Star, Evening\ Star) \tag{5.14}$$

should hold if the system is to reflect reality, without requiring the existence of an entity that denotes a particular planet. The intension of an individual concept is the sum total of those of the facts involving that concept. Being equivalent to another is among such facts. It is moreover a very special and crucial one in that it gives the ultimate key to completing the body of knowledge concerning the concept. Indeed, if a person knows (5.14), any of his knowledge concerning the Morning Star will automatically concern the Evening Star and vice versa. Otherwise, he will simply entertain two separate bodies of knowledge. On the other hand, knowing (5.14) conveys nothing meaningful by itself.

The primary cause for the distinction of intensions and extensions, and for the peculiarities of the *equiv* relation, lies in the epistemological position which treats opaque operators such as 'know', 'believe', 'have opinion', etc., as if they are normal first order predicates. Intensional entities may occur as their arguments. Knowing an entity is expected to enable one to answer any questions that may be asked regarding that entity except those involving *equiv* relations. This will not be the case, however, since questions could refer to accidental facts. Thus opacities exist among intensional entities as well. A revised conception of 'know-what' will then be that what is known is the definition of an entity; other facts can only be known through logical inferences and/or with the aid of domain-specific knowledge.

As the definition of an entity is nothing but the statement of the most fundamental facts that pertain to the entity, the revised interpretation of know-what suggests the possibility of organizing the world of intensional entities by means of 'know-that' (know-whether). Simple examples are:

$$know(A, have\text{-}phases(Morning\ Star)) \tag{5.15}$$

$$know(A, Morning\ Star = Evening\ Star) \tag{5.16}$$

$$know(A, \forall x \forall y(p(x) \land ((x = y) \rightarrow p(y)))) \quad \text{for any } p \tag{5.17}$$

where A represents a cognitive agent. If we stipulate appropriate inference rules such as

$$know(A, \forall x(p(x))) \rightarrow know(A, p(t)) \quad \text{for any } p \text{ and } t \tag{5.18}$$

$$(know(A, v) \land know(A, w)) \leftrightarrow know(A, v \land w) \quad \text{for any } v \text{ and } w \tag{5.19}$$

$$(know(A, v) \land know(A, v \rightarrow w)) \rightarrow know(A, w) \tag{5.20}$$

we can derive

$$know(A, have\text{-}phases(Evening\ Star)) \tag{5.21}$$

from the preceding propositions. Knowing someone's telephone number may be formulated as follows without referencing the number itself:

$$\exists x(know(A, telephone\text{-}number(Mike) = x)) \tag{5.22}$$

A's knowing Mary's telephone-number does not follow even if the two numbers are the same because

$$(know(A,v) \wedge (v \rightarrow w)) \rightarrow know(A,w) \tag{5.23}$$

will not generally hold. (Compare this with (5.20).)

While we do not intend to expound this theory any further, let us mention a few more axioms and inference rules just to make us feel a little more secure about its soundness. Here, A and B represent cognitive agents, $p(x)$ a formula containing a free variable x, and v and w arbitrary formulas.

$$\forall x(know(A,p(x))) \leftrightarrow know(A,\forall x(p(x))) \tag{5.24}$$

$$(know(A,v) \vee know(A,w)) \rightarrow know(A, v \vee w) \tag{5.25}$$

$$know(A,\neg v) \rightarrow \neg know(A,v) \tag{5.26}$$

$$\exists x(know(A,p(x))) \rightarrow know(A,\exists x(p(x))) \tag{5.27}$$

However, the converses of the latter rules do not hold, which reflects the incompleteness of human knowledge. The following axiom schema can be asserted for all the logical axioms (including equality axioms) and inference rules, here represented by w:

$$know(A,w) \quad \text{for any agent } A \tag{5.28}$$

The rationale for this is that all cognitive agents are intellectual beings, hence can be assumed to know logical truths and necessities. 'Know-that' contexts may be layered; that is, knowing a fact may itself be an object of knowing. Then,

$$know(A,know(B,w)) \rightarrow know(A,w) \tag{5.29}$$

If the converse of (5.27) is admitted, a correct chain of inferences will be

$$\begin{aligned} know(A,\exists x(know(B,p(x)))) &\rightarrow \exists x(know(A,know(B,p(x)))) \\ &\rightarrow \exists x(know(A,p(x))) \end{aligned} \tag{5.30}$$

contradicting the common-sense admission of a person A's claim to knowledge of another person B's knowledge of the answer to a question when A himself does not know the answer. (We know that Ben Jonson knew who it was who wrote 'Hamlet' though we ourselves may not be sure.) This partly supports our rejection of the converse of (5.27).

The 'know-that' operator is of course not a first order predicate but functions as the constructor of a local world for each cognitive agent on the common underlying set of entities. All entities are extensional, with their intensions as captured by an agent embodied as the totality of facts known in his world. Portions of a world may be imbedded in others with possible transformations such as negation ('We know Mozart did not know who brought him the order for a requiem'), disjunction ('Pat knows that either Mike knows Mary's telephone number or he knows her address'), etc. Opacities may exist between two worlds unless one is wholly contained, as is, within the other.

We might posit the existence of the supreme agent (the 'system') who knows all the facts that other agents know, provided that all of the worlds as a whole are compatible. His is the global world, which can be identified with a normal first order interpretation if 'know-that' is replaced by 'is-true'.

The last statement indicates that the concept of intension is inevitably associated with the human act of knowing, which in turn can only be crystalized in the form of language. Indeed, seemingly simple expressions such as 'Morning Star' or 'Mike's telephone number' actually reflect a multitude of human experiences, observations and recognitions. The issue of intensions versus extensions will come to be raised only when the meaning of linguistic expressions is the primary subject of discourse. This seems also to be the case with the various approaches discussed above.

5.2 Criticism of Underlying *Weltanschauungen*

We now proceed to the second stage of discussion, namely, the criticism of *Weltanschauungen* that underlie typical and well-known conceptual frameworks. To start with, let us summarize as follows the result of the preceding analysis:

(1) *Entity-Centered Conception of World*
A world is conceived of as a totality of entities. Its meaning is the composite of those of constituent entities which have been determined in advance. Counted among entities are not only individual objects but substantivized classes, properties and propositions. The world of Platonic *Ideas* gives a classic example. The possibility of 'know-what' is based on the same conception.

(2) *Subject-Predicate Dualism as an Organization Principle*
A strong tendency exists to relate entities in binary and asymmetrical forms. Propositions concerning multiple entities are transformed so that either one of the entities is given the status of subject (as with the views of Plato and Aristotle, also with some data models) or an artificial entity is created to represent the proposition itself (as with Kowalski [Kow] and some semantic networks). While this might have an origin in a common-sense prejudice of a linguistic nature, it certainly causes an ontological effect that entities are classified into two sorts, one the real being (*Ideas* with Plato, individuals with Aristotle) and the other something secondary (imperfect particulars or mere abstractions).

(3) *Individuation Based on Substance or Essence*
For each individual object, what makes it be nothing other than what it is must be found within the facts that concern it. With the subject-predicate *Weltanschauung*, this core of real being must be identified among the attributes that are predicated of that entity. A consequence of this is that individuals are conceived of as bundles of attributes belonging to them, some essential and others accidental.

(4) *Types (Genuses or Species) as an Organization Principle*
Moreover, individuals are often considered to exist only as instances of certain kinds, each of which is provided with or even characterized by a fixed set of attributes and/or attribute values. A kind tends to gain in substantiality as it is given further specific attribute values.

(5) *Primacy of Intensions as against Extensions*
Intensions are posited as entities belonging inherently to linguistic expressions and are conceptually independent of whatever extensional entities might happen to be denoted by the expressions. No canonical ontology is available for establishing intensional entities, however. They might coexist with extensions on a comparable standing in the same world. Or they might equally well be incarnated at a metalevel above extensions. Or extensions may simply be ignored, at least as explicit existences, with the net result that intensions present themselves as just another kind of extensional entity.

(6) *Human Mental Behavior as the Primary Subject of Discourse*
Any attempt to model knowledge or belief structures will entail consideration of opacities. Mysteries arise to haunt opaque contexts when operators such as 'know' or 'believe' are treated as if they are normal first order predicates which may take only terms (substantives) as arguments. Natural languages are versatile enough to allow different categories of concepts, such as things and facts, to be easily confused. A world that contains cognitive agents and their acts of knowing could be more naturally modeled as a structured assembly of multiple worlds, each of which comprises facts known by the corresponding agent.

The distinction between things and facts is so fundamental, however, that the first of our principal objections is directed to the tendency to blur it. Things are what exist and are identified. Facts are true statements (propositions) regarding things. That a thing is recognized to exist requires some facts, but none in particular, involving it to be recognized. In turn, a statement ceases to be a fact when any of the things involved in it ceases to exist. (Precisely, the statement becomes meaningless as its negation, which involves the same set of things, cannot qualify as a fact either, or rather the meaning of the statement becomes too indeterminate for a truth-value assignment.) While all this can be said of the everyday use of both terms, mathematics and logic carry the contrast between them to the farthest point. There, things are deprived of all attributes except the pure 'thinghood' of mutually distinct existence, and an unlimited variety of propositions may be asserted on the same set of things. This may possibly make a paradigm for us to follow in constructing an information model, which is just a mathematical system when viewed formally and abstractly.

We have attributed the responsibility for the confusion to an excessive versatility of natural language. The statement 'Pat knows Mike's telephone number' is perfectly normal and understandable practically, yet is a sheer nonsense theoretically. We would never say 'There exists a certain number, of which Pat knows that it is equal to Mike's telephone number.' though this is the logically correct way of saying it. 'He knows the Morning Star' is ambiguous even as an ordinary English sentence, while 'He sees the Morning Star' and 'He dials Mike's telephone number' are completely clear. We may question what puts two phrases 'Mike's telephone number' and 'Morning Star' on somewhat similar, yet not fully equivalent standings. It can be said of each of them that it is an individual expression and carries its own meaning (individual concept) apart from the extensional entity it happens to denote, hence this will not explain the difference. The crucial difference lies between the varieties of the facts (not things) these expressions imply. While 'Mike's telephone number' almost pinpoints the single specific fact, 'Morning Star' is involved in a multitude of facts such as its shining in the eastern sky before dawn, periodically changing its position relative to the sun, being the same planet as the Evening Star, etc. We simply do not know which of these is meant when we are told 'He knows the Morning Star'. We accept and exploit the versatility of natural language expressions in everyday practices, which itself presents an interesting linguistic problem. This should not be extended to the theoretical language, however, with which we analyze the problem. What is to be blamed is the inadequacy, inappropriateness and lack of precision of semantic analysis.

We have already noted (Sects. 4.6.4 (4) and 5.1.2) and would like to reemphasize that the meaning of a thing (entity) is the sum total of those facts (relations, propositions) in which it takes part. To maintain this contention, however, we have to consider common-sense objections such as those based on the *Compositionality Principle* (Frege). We seem indeed to derive our understanding of a sentence from our understanding of its clauses, that in turn from our understanding of smaller clauses, phrases and so on, and eventually of the words contained therein. In looking up the latter in a dictionary, however, we will find that the meaning of a word is explicated only by means of sentences which contain yet other words. The process of crossreferencing could continue ad infinitum. The ultimate recourse which enables us to terminate it is an appeal to our experiences and observations in the actual (non-linguistic) world, or the protocol statements expressing them which we accept as being immediately understandable without any further explanations. It should be noted that we will thus end up facing facts again in this pursuit of meanings.

Having admitted that the meaning of a thing is carried by the facts involving it, a question is whether or not different degrees of significance should be recognized among these facts in forming the meaning. An affirmative position tends to distinguish the most significant or necessary facts from contingent ones, and to regard the former as giving the definition for the thing. This is typical of the entity-centered view, which furthermore considers the definitional statements as belonging to the thing and constituting a core of real being or a substance. Thus, a substance is a self-sustaining entity in that all the conditions (facts) that necessitate its existence are contained in it. Other facts are either logically derivable or contingent and negligible. Hierarchies are introduced also among entities (things) as we have noted.

This is a grotesque and unacceptable view. It is indeed questionable whether we can ever differentiate more significant and less significant facts concerning any thing. It is even more questionable whether we can have any fact, substantial or not, 'belong to' a thing. It seems more plausible that a fact belongs equally to all the things that it involves. Take, for example, the case of 'Morning Star'. The periodic appearance in the eastern morning sky which might be considered superficially to belong solely to this one entity is derivable mathematically, not to say logically, from the relative positions and movements of it, the sun and the earth, a fact which stands clearly equidistant to all three. The latter in turn could be derived from the former with the help of some other observable facts. It is certainly possible that some facts logically precede others under a given condition or are more directly pertinent to a particular problem than others. Precedences and pertinences as well as significances are all relative to viewpoints, value judgements and/or situations, however, and no a priori or permanent status should be assigned to any fact or entity.

Aristotle would not have admitted relative position within the solar system as being a part of the substance of a planet known as the Morning Star. His (and also Plato's) failure to understand multi-term relations as such and his bias toward the entity-centered and subject-predicate *Weltanschauung* are reflected on each other. 'Substance' and 'essence' are hopelessly difficult and confused notions, and unworthy of saving at the cost of general relations. It took many centuries and stages, however, in the history of philosophy for them to be eventually abandoned [Rus]. Let us not linger here on the details of this process, but let us at least spend a few lines on the case of Leibniz, which is not only historically important but particularly relevant to our current concern. First, the distinction between 'essential' and 'contingent' facts is dismissed, or more precisely, replaced by that between logically valid (non-negatable) and satisfiable (negatable) propositions. Second, the concept of substance is retained or even reinforced, in that a substance now comprises the body of all the facts, necessary or not, concerning an entity. Third, since a fact which involves a specific entity could involve any other entities as well, any substance 'mirrors' the entire universe in it. Suppose, for example, propositions

$$\alpha(\cdots a \cdots b \cdots) \quad \text{and} \quad \beta(\cdots b \cdots c \cdots)$$

hold. The latter, which does not directly involve a, is reflected on the substance of a via b and the two entities' common involvement in proposition α. Moreover, these facts (and similarly others) may be made to 'belong to' a by setting

$$P(x) \equiv \lambda x(\alpha(\cdots x \cdots b \cdots) \wedge \beta(\cdots b \cdots c \cdots))$$

and asserting $P(a)$. Fourth, all the small universes must be 'compossible' with each other, which is equivalent to requiring that the global universe as a whole be consistent. In fact, we have

$$P(a) \wedge Q(b) \wedge R(c) \cdots \leftrightarrow \alpha(\cdots a \cdots b \cdots) \wedge \beta(\cdots b \cdots c \cdots)$$

where $Q, R, \ldots$ are defined like P. Now, looking at the above, one can't help but feel a desire to apply *Ockham's razor* to get rid of the encapsulation of facts in entities which entails the introduction of superfluous predicates. Once this is done, we will come close to conceiving of entities as purely symbolic and absolutely meaningless by themselves, completing the dismissal of the concept of substance.

The ontology of L. Wittgenstein [Wit] seems to be in line with what we have just stated, for we are told (under a rather liberal interpretation) that:

- The world is the totality of facts, not of things,
- It is essential to things that they are constituents of facts,
- Only a fact can represent a state of affairs, a set of things cannot,
- Facts are indicated by propositions, things by names, and
- Only propositions have meanings, things can only be denoted.

We may as well take the above as a manifestation of an alternate *Weltanschauung*, that is, the fact-centered *Weltanschauung*. A much more concrete statement of the same position was given by D. Hilbert which is reported to run as follows:

'It must be possible for us to construct a geometry in terms of tables, chairs and beer mugs instead of points, lines and planes.'
(Quoted from [Rei] with slight changes.)

It must be possible of course to use whatever words or symbols we like to denote geometric entities. 'Points', 'lines' and 'planes' will be actually used but will mean absolutely nothing by themselves. Their sole purpose is to identify certain entities as distinct from each other. All the meanings of the world of geometry go into the propositions that relate entities. We will return to Hilbert's doctrine a little later: meanwhile, let us make one last quotation from the classics in order just to show how far we have come both historically and philosophically:

'A point is that which has no part.'
'A line is breadthless length.'
[Euclid, *Elements*, Book 1, Definition 5]

Let us now face the issue of intensions versus extensions. As we have seen, the nature of the issue is inherently linguistic, and yet the intension of an expression can be known only through an ultimate recourse to our experiences in this real, physical world. Understanding the meaning of the phrase 'Morning Star' may be possible by understanding those of 'Morning' and 'Star', but the latter understanding can be attained only through living in the regular recurrences of days and nights, and watching stars from time to time. To this, we add the following observations:

- Obviously, watching a particular celestial body (that is, an extension) in particular environments forms the base of our understanding of what is meant by 'Morning Star'. This background belongs to the same coherent body of physical experiences as those associated with the component words, and there is no reason why we should reject it while accepting other experiences for incorporation into the intension of the phrase.
- If the intension of a compound expression contains nothing but the semantic combination of the intensions of its component expressions, how is it possible for the expression to denote anything extensional [Mai]? In fact, what enables one to say 'That is the Morning Star', pointing to a particular heavenly object, consists in the phrase itself. Contrary to this, understanding what 'Mike's telephone number' means as a linguistic expression is of little use in knowing what his number actually is. It contains only the knowledge of the concept of 'telephone number' and the information that 'Mike' names a person (that is, a potential owner of a telephone). The functional relationship between a certain number and 'Mike' which constitutes the true significance of the phrase has totally disappeared. Either the intension of 'Morning Star' is superfluous or that of 'Mike's telephone number' is deficient.
- Defining intensions as logical equivalence classes [Car] will give them an indisputable precision, yet deprive them of all their empirical actuality.
- The fog around the notions of opaque operator and context, which are intimately related to that of intension, comes primarily from two sources: first, the misconception of 'know-what', second, the failure to recognize different persons' ranges of knowledge as different worlds (interpretations). In particular, the question of interchangeability of equivalent expressions is raised when trying to apply

  $$\forall x \forall y((x = y) \rightarrow (know(A, x) \leftrightarrow know(A, y)))$$

  to the case where '$x = y$' is not known to A. (Its being true is tantamount to its being known to the supreme agent or 'system'. Thus the above inference confuses two agents' ranges of knowledge.)
- Except in the extreme case where the authors ([Bra], [Mai]) explicitly deny it, each individual expression is to denote an extensional entity. In this correspondence, which is generally many-to-one, each extensional entity is viewed as lying at the pivotal point of a set of expressions and fastening them together. Why, then, should we not consider the intensions of these expressions as constituents of the complete meaning associated with the extensional entity they denote? Treating them as something separate from and independent of extensions will fail to capture the meaningfulness of such a fact as the Morning and Evening Stars being the same, which can only be attributed to the common physical entity.

While each linguistic expression certainly carries an intension of its own, the meaning of the world as a whole is not equal to but far richer and more complex than the simple sum of these intensions. Many important meanings cannot be captured without referring to denoted entities and facts in the actual world. Actualities and extensionalities must not be too lightly dismissed.

We have occasionally referred to the 'meaning' of a 'thing', which may not be a normal use of the words. The reason for this is that the universe of our discourse indeed comprises physical things and that we want to speak of the totality of meanings pertaining to a thing as a single and consolidated meaning. It should be noted that, for any physical existence, any finite collection of constituent meanings will never exhaust the entirety, and that meanings generally pertain to other entities as well. On this ground, the method of classifying individuals according to certain fixed sets of attributes is not accepted.

5.3 An Alternate *Weltanschauung*

5.3.1 Characteristics of Our Universe of Discourse

Although the preceding criticism could benefit a more general class of problems, let us focus our attention on its relevance to the problem of modeling the world of design. As previously observed (Sects. 0.3, 0.9 and Chap. 1), the single most fundamental characteristic of this modeling project is that what is intended is to model real, physical entities and phenomena. Through refining and elaborating on this observation, taking the preceding criticism into consideration, we will arrive at specific positions on key issues which form the core of an alternate *Weltanschauung*.

(1) *Real Versus Linguistic*
'Reality' refers to that which is recognized as 'existing' prior to, independently of and beyond any linguistic expressions. For example, a drawing is interpreted not simply as a collection of lines, curves and symbols but as representing a real design object. This is possible not because the drawing is semantically complete (actually it is far from being complete) but because the object has been comprehended beforehand by human designers. The contents of the comprehension, that is, the way the object exists and behaves, or the existence itself, has already been established and will remain unchanged regardless of the kind of representation apparatus we may choose, drawing or not. Physical existence and sensory accessibility are the primary causes of perceived reality. Linguistic expressions, on the other hand, presuppose abstraction and conceptualization of human perceptions and recognitions. Not all abstract concepts are unreal, however. Lines, curves and mathematical entities in general which are conceptual and serve as linguistic components are themselves 'real' in the above sense (see the discussions in Sects. 4.1. and 4.2.1.1). A very simple example which illustrates this is a sphere and a plane

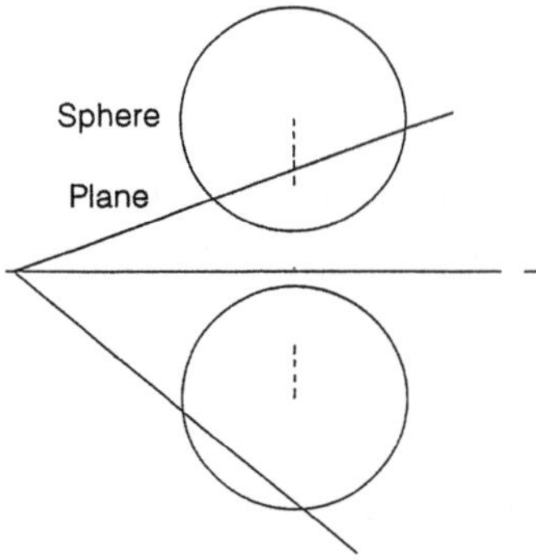

Fig. 5.3a. A sphere and a plane

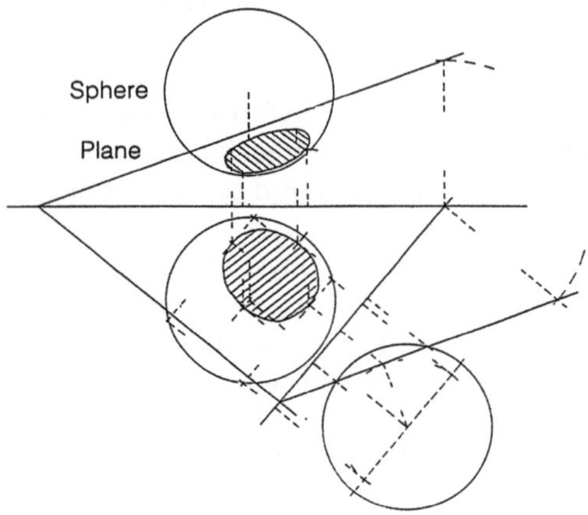

Fig. 5.3b. The intersection

represented as two circles and two half-lines (Fig. 5.3a). What enables us to draw the intersection (Fig. 5.3b) is not the collection of these 2-dimensional entities and annotations (SPHERE and PLANE) but the unrepresented understanding that what is to be truly represented is a sphere and a plane.

The situation is quite different with other universes of discourse, particularly business data processing where there is no such thing as 'reality' and the language is all that matters. Indeed, when we develop a payroll program, a customer information file, or a reservation system, the result is the creation of a new and self-contained world (of language) but not the modeling of one that already exists outside. Certainly, pieces of information are produced, manipulated and consumed in the real environment, by human beings, in physical media and by physical means. However, the realities and physical characteristics of these have nothing to do with the information structure, which is fully and precisely defined in a linguistic system.

A philosophically very profound question is how we can be confident of the correctness of a linguistic representation of a real world structure, or whether any correct representation will ever be possible. Without getting stuck in a metaphysical argument in trying to answer this almost unanswerable question, let us be content to note the following:

- Only a comparative evaluation of representation languages is possible with respect to naturalness, rigor, simplicity, well-structuredness and fertility.

Two major implications of the reality of what is to be modeled which should influence the specification of language are:

- the open-endedness and incompleteness of the model, and
- the objectivity and transparency of the represented facts,

which will be discussed next.

(2) *Open-Endedness and Incompleteness of the Model*
Real objects allow themselves to be placed at will in unpredicted environments, exposed to extraordinary experiments and/or observations, or studied under new theories, with the result that new phenomena are discovered and propositions asserted, which in turn require additional devices for representation. Therefore, any finite and fixed representation scheme will be inevitably incomplete and must be open-endedly expandable. In particular, an object cannot be represented as an *instance* of a certain *type* which is defined by a set of properties and manipulators. In the realm of mathematical discourse, the strong undecidability of the structure $\mathcal{N}$ of natural numbers (Theorem 3.5) most clearly and strikingly exemplifies the limitation of attempts to represent realities finitely and yet precisely, however simple and obvious they may appear to human cognitive and intellectual faculties.

(3) *Objectivity and Transparency of the Represented Facts*
Meant by this is that we are primarily interested in the structures and meanings of the world (of design) as commonly known and understood by all the parties concerned, that is, designers, engineers and information processing specialists, and moreover, conveyable to an artificially intelligent agent. Belief structures and other human psychological aspects which may necessitate the introduction of opaque operators and contexts will not be considered. Intensions conceived of as entities separate from and independent of extensionality are dismissed.

(4) *Multi-Facetedness or the Coexistence of Multiple Theories*
Trying to model a world of physical entities as a whole entails multi-faceted representations. In our case, design objects are to be described in terms of shapes, functions, physical properties, manufacturing processes, inventory regulations, environmental characteristics, marketing and support plans, etc. This is a situation where:

- multiple formal theories coexist which are more or less interrelated with each other, and
- the same physical structure serves as a common interpretation of these theories.

Hence, the (physical) existence itself should be the ultimate recourse for consistencies when describing what already exists. When creating a new design, we always face the problem of redundancies and inconsistencies. A more fundamental problem is whether and how we can provide a common (conceptual and linguistic) platform on which different theories are correlated. We will return to this problem shortly; meanwhile, an auxiliary remark is that each theory represents a piece of knowledge belonging to a single abstract domain (such as solid geometry, strength of materials, stochastic process, etc.) whereas a model integrating these pieces is considered as something interdisciplinary and yet representing a concrete world.

(5) *Distinction Between Essentials and Non-Essentials Negated*
It is true that some properties, functions or aspects are given relative precedence over others according to particular views, purposes or value judgements. This will be implemented typically by selecting a set of facts or valid propositions and postulating them as axioms. Others might be logically inferred, or constitute disjunctive propositions, or validate one or more axioms. However, there may exist a number of mutually equivalent, hence interchangeable, sets of propositions any of which qualifies as the system of axioms. Given two propositions, therefore, no definite statement can generally be made as to the distinction of significance between them in forming the meaning of the world. The notions of 'essential' and 'substantial' as something categorical and a priori are dismissed.

(6) *Fact-Centeredness*
While a real object may not be completely described within any finite scheme, it remains invariant throughout diversified descriptions and even when any properties or relations with other objects change. Thus, an identification is isolated from descriptions, an identity from meanings. An object standing by itself is an absolutely meaningless existence and will be represented as a pure symbol. Its meaning is nothing but the totality of all the propositions in which it takes part. As propositions generally pertain to more than one object, meanings of objects overlap each other and are embedded in that of the entire universe in a coherent manner. Coherencies are required also between different theories. The rationale for this again lies in the actual or expected existence of the same (physical) entity behind varied theories and descriptions. A fundamental and necessary condition for meeting this requirement is a representation of objects which is neutral between any particular domains of theory or description, hence which is purely symbolic.

(7) *Physical Versus Conceptual Versus Linguistic*
A physical object is by nature unlimitedly multi-faceted, which perhaps provides a psychological basis for us to view it as remaining identical despite any changes in this or that aspect. The case of mutilating a cup as discussed by Aristotle (Sect. 5.1.1) indicates this; what is lost by boring through is the function of containing liquid, but not the identity of the cup. We may not be able to call it a cup any more but are entitled to recognize it as the same object as before. On the other hand, each conceptual object is by definition conceived of as something that belongs to a particular abstract domain where all aspects other than those of current interest are discarded. Such an object is characterized by the relative position, properties and functions it is given within that domain. We will never say that the integer '1' has become '2' while can rightly speak of the position of a point moving from '1' to '2'. When Mike's telephone number is changed, say from '5025' to '0993', it does not affect the identity of the physical object 'Mike' at all, but may more or less change its meaning. These facts have nothing to do with the meaning of '5025', '0993' or any other number, however. The structure and meaning of the domain of integers or any conceptual domain is decided and fixed once

for all. So is the characterization of each object. The notion of an identity that is sustained through the changes of the meaning of the identified object is vital to physical objects but does not apply to conceptual objects.

Combining the arguments of this paragraph and (1) above, we have divided the world into three realms:

the realm of infinite, real, physical structures,
the realm of infinite, real, conceptual structures, and
the realm of finite, linguistic structures.

Those in the latter are to serve as the models for real structures. Besides, we have seen that some conceptual structures, mathematical structures in particular, may as well provide languages for describing other real structures. By imposing finitudes, however, we have essentially restricted the linguistic structures to those that are realizable on computers. Thus, infinite structures such as $\mathcal{N}$ are excluded from the linguistic realm whereas a finite system of axioms such as N (given in Sect. 3.2) is included in it.

In the case of design, the reality is what a designer envisages during and at the end of his work. Though not immediately tangible, it is fundamentally physical, as it is a mental image of a physical object that exists or comes to exist. The definition of properties and functions necessarily involves conceptual entities (such as forms, dimensions, motions, etc.). A linguistic representation of designer's image comprises a formal theory and an interpretation of it. The former specifies the requirements to be satisfied by the final state of the design and the latter represents the current state.

(8) *Indeterminacy of the Evolution Process*
So far, we have been considering static structures only. The world of design, however, is inherently and distinctly dynamic in that the contents of a design will grow very rapidly in a relatively short period of time, both in quantity and structural complexity. Static structures are deterministic, that is, at any given instance of time, the truth value of any proposition is uniquely determined, hence the closed world assumption is upheld. Indeterminacies will arise when steps are taken to advance the state of design. The cause for this is twofold. First, the requirements and designer's intentions may not be completely specified in a finite formal theory. In other words, interpretations could contain facts which are consistent with but not necessitated by the theory. Second, given an interpretation, there will generally be more than one way to expand or enrich it to interpret the theory more fully.

Indeterminacy does not necessarily mean a total lack of laws or principles to guide the evolution process, however. With the help of a sufficiently powerful and formal linguistic framework, it should be possible to address at least certain aspects of some fundamental problems in a rigorous and systematic manner, such as the following:

- to capture and precisely represent features, functions and designers' intents,

- to compare theories (specifications) and recognize similarities and commonalities between them so that an interpretation (design) for one may be exploited as a first approximation for another,
- to elucidate in particular what lies behind human judgements seemingly based on intuitions, empirical knowledge and other unspoken presuppositions,
- to prioritize the satisfaction of competing or cooperating requirements, and
- to determine the set of alternate plans to satisfy the requirement of the highest priority at each step of the evolution.

Given a requirement specification, we may envisage the class of all the full or partial interpretations of it organized into a directed graph structure along the evolution paths (see Fig. 0.4). Dynamic indeterminacies then correspond to the branching of paths. What we would actually like to deal with is a subclass in which laws and principles are applied to reduce the multiplicity of branching and priority considerations invoked to help select paths.

5.3.2 The Paradigm: A Methodological Basis

We want to be more confident of the legitimacy of our approach before proceeding to construct a linguistic framework. For that, we want to make appeal to the fundamental community of underlying *Weltanschauung* between our approach and those which we may regard as paradigms in more established fields of scientific endeavors. A 'paradigm' here has two senses. One refers to a concrete and exemplary achievement, and the other is a collection of elements shared by a research community, including laws, theories, beliefs, preconceptions, etc., originating in exemplars ([Kuh(1)], [Kuh(2)]). Our primary interest is in the methodological abstraction, and not in technical details or in social or historical relevances.

The two most prominent features of our discourse are the objective to model a part of the real physical world and the adherence to logico-mathematical formalism in carrying out the task. It is only too natural, then, that we turn our attention to the (meta) theories concerning the structure of natural-scientific theories, particularly those in the logical positivist tradition. Under the influence of new developments in modern physics, mathematics and logics, intensive efforts were made to emancipate philosophy from metaphysics, to interpret the philosophical issues raised by sciences in clear and rigorous terms, and to found the significance of propositions on logical calculus and empirical verifiability. The efforts culminated in the 'Received View' of scientific theories according to which each scientific discipline is to be formulated as a first order theory of which the real world is considered to be a model. The general construction of such a theory and its interpretation is as follows [Sup(1)]:

(1) The set of non-logical symbols is divided into two disjoint subsets, Σ_o comprising observational and Σ_t comprising theoretical symbols.

(2) The universe of interpretation consists of concrete physical objects, their properties (functions) and relations (predicates), all of which are 'directly observable'.

(3) Each observational term or proposition (formula) which does not contain any theoretical symbols has an explicit and full interpretation in this universe.

(4) Theoretical propositions which contain at least one theoretical symbol are only implicitly and partially interpreted in that only the observational propositions derived from them (in conjunction with the axioms) are immediately interpreted. The interpretations of theoretical terms are made through the propositions in which they occur, hence are also implicit and partial.

(5) The theory may possibly be augmented by modal operators, primarily in order to incorporate counterfactual conditionals, cause-effect relationships and dispositional properties, for example.

There have been severe and decades-long attacks on the view thus formulated of scientific theories. While we cannot and need not examine pros and cons in detail, we should pay due attention to the issues that are more important and closely relevant to our own case. Included among these are the following:

- The question whether the axiomatic method in formulating scientific theories is appropriate, fruitful, adequate and illuminating both practically and philosophically.
- Failure to capture the evolutionary aspect, the 'logic of discovery' in particular.
- Difficulty in maintaining the distinctions between observational and theoretical terms, and analytic and synthetic propositions.
- Inadequacy of partial interpretations of theoretical terms and propositions for capturing those aspects of their meanings which are not exhausted by observable manifestations.
- Oversimplification of the relevance to and reliance on other theories and experimental procedures which are not part of the theory in question, yet which play a vital role in connecting it to observable phenomena.

Because of these and other reasons, the Received View and positivistic positions in general in philosophy of science are said to be almost dying. Nevertheless, it seems that the Received View maintains its applicability to a certain extent, in the physical sciences in particular, and to that extent, we have yet to see the advent of alternate methodologies which are more (or at least equally) plausible. Thus, it should be able to serve as a paradigm (in the second sense) for our discourse, since the world of design objects in the discrete manufacturing environment with its structures and governing laws is naturally considered as the kind of world about which physical sciences theorize. To remedy the ailing condition indicated above, however, significant modifications and reinterpretations of some of the fundamental concepts in conjunction with a reasonable restraint on their applications are called for.

First, we note that the restraint on application should be accepted for any metatheoretic principle in view of the wide diversity of what may be called scientific theories, and that any theory that may be appropriately formulated under the Received View must inevitably be of a mathematical nature.

Second, it is almost by the definition of 'mathematical' that any such theory deals rather with a mathematical structure which is an abstraction than with physical and concrete objects themselves. Terms, functions and propositions in the theory will be interpreted accordingly as mathematical entities. It is of the utmost importance that different structures may possibly be devised, with each providing an explicit and full interpretation for the same theory. Different theories may in turn be formed for a given physical world or mathematical structure. The entire complex of correspondences and interrelations between the physical reality, mathematical abstractions and formal theories may be depicted as in Fig. 5.4.

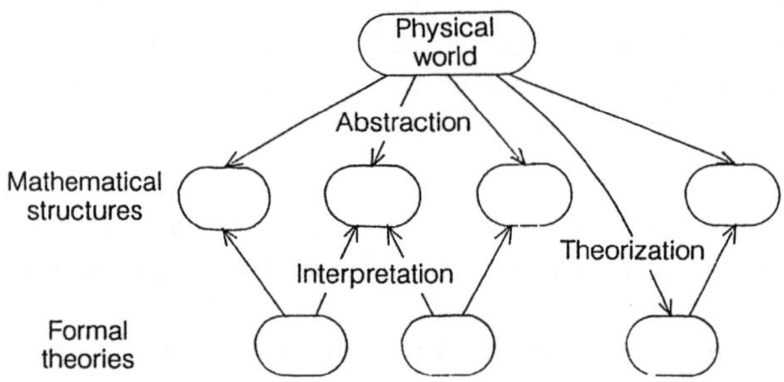

Fig. 5.4. Physical world, mathematical structures and formal theories

An immediate consequence of the above is the dissolution of the observational-theoretical distinction for terms and propositions as far as their interpretability is concerned. Of course, the ultimate recourse must still be to the observations in the physical, sensible world for establishing the empirical truth of mathematically true propositions. We cannot conceive of observability as something primitive and categorical, however, but will have to be content with its being relative to a prescribed body of experimental procedures, instruments, qualifications of observing persons, other theories and structures.

Another consequence is that the necessity for including modal operators is eliminated, or at least substantially decreased, as all the possible states of affairs can somehow be realized on a mathematical structure so long as they are precisely defined and consistent with the theory, hence counterfactual conditionals can be reinterpreted as material implications. For example, the statement

'Anything fragile will be broken if struck hard.'

which remains counterfactual for whatever is not actually struck may as well be reformulated as

$$\forall x((fragile(x) \wedge hard(impact(x))) \rightarrow excessive(deformation(x))) \qquad (5.31)$$

where functions *impact*, *deformation* and predicates *hard*, *excessive* are given rigorous definitions in either quantitative (continuous) or qualitative (discrete) terms. The state 'struck hard' is realized (or simulated, precisely speaking) for any object x for which these functions and predicates are defined, and the resulting value of *deformation* can be presented for observers' examination whether the object is 'broken' or not.

Next, the analytic-synthetic distinction for propositions gives way to a division into three well-defined categories, namely logically valid propositions, theorems of a particular theory and those that are valid in a particular structure for the theory. The distinction between the latter two reflects an extra-factual value judgement that underlies the selection from generally multiple possibilities of axiomatizing in a given structure.

Finally, while we admit the significance of the evolutionary dimension for scientific theories and the Received View's ignorance of it, we consider it to require an essentially separate treatment than that of the static structure, and we do not see any reason why the axiomatic method for the latter will particularly jeopardize the forming of dynamic theories. It is indeed quite possible to study the evolutionary processes of mathematical theories and structures from the axiomatic viewpoint.

Having adapted the Received View to better reflect the actual situation of scientific endeavors, let us pause here to reassure ourselves that it shares its *Weltanschauung* with our approach. For this, it will be sufficient to compare what has just been stated with the contents of the preceding section and note the correspondences with respect to the following:

- the physical-conceptual-linguistic (or physical-mathematical-logical) division of the universe of discourse,
- the necessity for multi-faceted conceptual modeling of physical reality,
- the recognition of incompleteness of any finite, linguistic representation (theory) of physical or conceptual reality, hence the necessity for the co-existence of multiple theories and structures,
- the fact-centered principle for capturing the meaning of the world, which entails purely symbolic representations of objects,
- the dismissal of any a priori and categorical distinction among entities, such as essential and non-essential or observational and theoretical,
- the emancipation from the spell of subject-predicate formalism and the use of the more general framework of predicate logic in constructing theories, and
- the primary interest in representing the static and objective (transparent) aspect rather than the dynamic and subjective (opaque) aspect.

Unfortunately, we do not find any exemplary work in the field of physical sciences that is constructed in strict accord with the principles laid down in the Received View. On the other hand, as we have seen, what should primarily concern physico-mathematical theories are not physical, sensible entities but their abstractions. The latter in turn are considered by mathematicians as being 'real' in their own sense. It is probably the case that many fundamental concepts in mathematics (e.g., sets, integers, spaces, spatial objects) have psychological as well as historical origins in our experiences in real, physical environments. The rationale for postulating specific non-logical axioms may be sought ultimately in these experiences. When considered at the level of metatheoretic structures, the seemingly great difference between abstract mathematics and empirical sciences will diminish to the degree that an analogue of Fig. 5.4 can be drawn for the field of mathematical disciplines. It will therefore be reasonable if we concentrate our attention on this field which is abundant in exemplars. We thus come to the selection as our paradigm the formal-axiomatic method of constructing mathematical structures due to D. Hilbert, as exemplified particularly in *Grundlagen der Geometrie* [Hil], which is one of the most typical of rigorous sciences, the most perfected and distinguished of logic-based achievements, and has had a decisive influence on the succeeding history of modern mathematics. In the following, we intend only to outline the construction of Hilbert's work very briefly and point out a few of the most significant features which are also of particular relevance to our task of modeling design objects.

Grundlagen der Geometrie begins by envisaging three different sorts of undefined terms (concepts). The terms belonging to the first sort are called *points* and denoted by $A, B, C, \ldots$. Those of the second sort are called *lines* and denoted by $a, b, c, \ldots$. The third is the sort of *planes* $\alpha, \beta, \gamma, \ldots$. *Points*, *lines* and *planes* are conceived of in the context of interrelationships called *on*, *between*, *congruent*, *parallel*, etc. The precise and mathematically complete description of these interrelationships is given by the 'axioms of geometry' which comprises five distinct groups.

Group I postulates *incidence* relations between terms of different sorts:

I_1 For any points A and B, there exists a line that both A and B lie on.
I_2 For any two distinct points A and B, there exists precisely one line (denoted by $A \cup B$) that both A and B lie on.

It should be noted that every term or relation occurring in these axioms is deprived of any mathematical meaning, not to speak of intuitive meaning (such as a *line* being a set of *points* and *lying on* interpreted as set membership), that is usually given to it. Terms are just symbols which can only be distinguished from each other, and relations simply stand for bundles of terms. We have six more incidence axioms.

I_3 For any line, there exist at least two distinct points that lie on it. There exist at least three distinct points that do not lie on any single line.
I_4 For any points A, B and C, there exists a plane that all of A, B and C lie on. For any plane, there exists at least one point that lies on it.

I_5 For any three distinct points A, B and C that do not lie on any single line, there exists precisely one plane (denoted by $A \cup B \cup C$) that all of A, B and C lie on.

I_6 If there exist two distinct points lying on a line a that lie on a plane α, any point lying on a also lies on α, and we simply say a lies on α.

I_7 If there exists a point A that lies on both planes α and β, there exists at least one point B that is distinct from A and lies on both α and β.

I_8 There exist at least four distinct points that do not lie on any single plane.

Several fundamental theorems can be derived from these axioms such as:

Theorem 5.1. Relations between two lines or planes.
(1) For any two distinct lines, there exists at most one point that lies on both of them.
(2) For any two distinct planes, either there exists no point that lies on both of them or there exists a line such that every point lying on it lies also on both of them and vice versa.
(3) For any line and plane, either the former lies on the latter or there exists precisely one point that lies on both of them.

Theorem 5.2. For either of the following two cases, there exists precisely one plane that every indicated point and line lies on:
(1) Any point and line such that the former does not lie on the latter,
(2) Any two lines such that precisely one point lies on both lines.

The derivations are carried out in a completely formal and symbolic manner without relying on any mathematical substance in the usual sense. For example, Theorem 5.1 (2) is derived by successively applying I_7, I_1, I_6, and I_5 on its premise. After the incidence axioms, the following groups of axioms:

II Axioms of *order*,
III Axioms of *congruence*,
IV Axioms of *parallels*, and
V Axioms of *continuity*

are postulated from which all the theorems of (3-dimensional) elementary geometry are obtained. While groups II–IV like group I are purely formal, group V has a completely different character in that it presupposes and relies on the domain of natural numbers and the standard notion of sets. It includes the following two axioms:

V_1 (Archimedes' axiom) For any line segments AB and CD, there exist a number n and points $A_1, \ldots, A_n$ on line $A \cup B$ such that each of the line segments $AA_1, \ldots, A_{n-1}A_n$ is congruent to CD and B lies between A_{n-1} and A_n (Fig. 5.5).

V_2 (Axiom of completeness) The set of points, lines and planes is maximal in that any expansion of it cannot satisfy axioms I–V_1.

Of course, V_1 and V_2 are not legitimate first order axioms, and because of their existence the whole system of axioms is changed from a formal theory

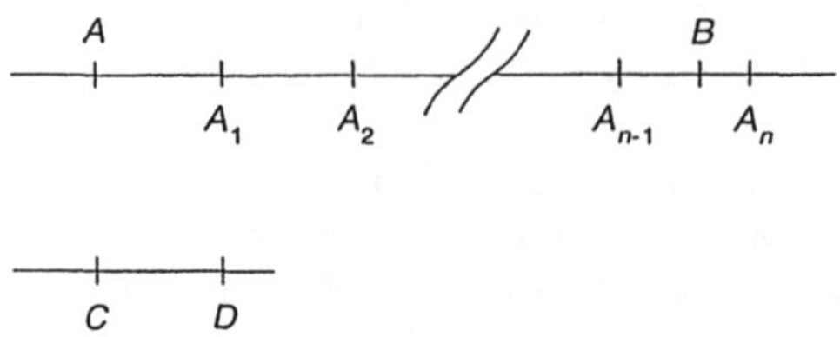

Fig. 5.5. Archimedes' axiom

into the stipulation of a concrete structure. Moreover, they play a crucial role in ensuring that the said structure is nothing but the usual analytic (Euclidean) geometry based on the field of real numbers. According to Hilbert himself [Hil], the reason for placing them at the end of the series of axioms is to put naturally into the foreground the problem of how and to what extent well-known theorems of elementary geometry can be established independently of the demand for continuity. This is an important consideration for us too, because reliance on domain-specific knowledge, while it will be practically inevitable, should be separated as clearly as possible from the purely logical nucleus of the framework.

A final remark we want to make on the paradigmatic significance of Hilbert's work is that we may take it as an exemplar either in the restricted dimension of geometric description or for a more comprehensive modeling of design objects. Axiomatic approaches to geometric reasoning (e.g. [Arb]) are considered as following Hilbert's paradigm strictly within its proper domain, and may be worthwhile in view of the importance of geometry in design. In our case, however, the scope is not limited to geometry or in any specific way but rather open-ended, which necessitates as well as justifies a more liberal adoption of the formal-axiomatic method in a much broader context.

References

[Arb] Arbab, F, Wing, J.M.: Geometric Reasoning: A New Paradigm for Processing Geometric Information. In: Yoshikawa, H., Warman, E.A. (eds.): Design Theory for CAD. North-Holland, 1987

[Aye] Ayer, A.J. (ed.): Logical Positivism. Free Press, 1959

[Bar] Barnes, J. (ed.): The Complete Works of Aristotle Vol. I, II. Princeton Univ. Press, 1984

[Bra] Brachman, R.J.: On the Epistemological Status of Semantic Networks. In: [Fin]

[Car] Carnap, R.: Meaning and Necessity. 2nd ed., Univ. of Chicago Press, 1956

[Che] Chen, P.P.-S.: The Entity-Relationship Model – Toward a Unified View of Data. ACM TODS, Vol. 1, No. 1, March 1976

[Cor] Cornford, F.M.: Plato's Theory of Knowledge. Routledge & Kegan Paul, 1935

[Fin] Findler, N.V. (ed.): Associative Networks: Representation and Use of Knowledge by Computers. Academic Press, 1979

[Ham] Hamilton, E., Cairns, H. (eds.): Plato: The Collected Dialogues. Princeton Univ. Press, 1961

[Hil] Hilbert, D.: Grundlagen der Geometrie. Teubner, 1930

[Kow] Kowalski, R.: Logic for Data Description. In: Gallaire, H., Minker, J. (eds.): Logic and Data Bases. Plenum Press, 1976

[Kuh(1)] Kuhn, T.S.: The Structure of Scientific Revolutions, 2nd ed., Univ. of Chicago Press, 1962

[Kuh(2)] Kuhn, T.S.: Second Thoughts on Paradigms. In: [Sup(1)]

[Mai] Maida, A.S., Shapiro, S.C.: Intensional Concepts on Propositional Semantic Networks. Cognitive Science 6(4), 1982

[McC] McCarthy, J.: First Order Theories of Individual Concepts and Propositions. In: Hayes, J.E. et al. (eds.): Machine Intelligence 9. Ellis Horwood, 1979

[Rei] Reid, C.: Hilbert. Springer, 1970

[Rus] Russell, B.: History of Western Philosophy. George Allen & Unwin, 1961

[Sch] Schubert, L.K. et al.: The Structure and Organization of a Semantic Net for Comprehension and Inference. In: [Fin]

[Smi] Smith, J.M., Smith, D.C.P.: Data Abstractions: Aggregation and Generalization. ACM TODS, Vol. 2, No. 2, June 1977

[Sup(1)] Suppe, F. (ed.): The Structure of Scientific Theories. 2nd ed., Univ. of Illinois Press, 1977

[Sup(2)] Suppe, F.: The Search for Philosophic Understanding of Scientific Theories. In: [Sup(1)]

[Web] Webster's Third New International Dictionary. G. & C. Merriam Company, 1969

[Wit] Wittgenstein, L.: Tractatus Logico-Philosophicus. Routledge & Kegan Paul, 1922

[Woo] Woods, W.A.: What's in a Link: Foundations for Semantic Networks. In: Bobrow, D.G., Collins, A.M. (eds.): Representation and Understanding: Studies in Cognitive Science. Academic Press, 1975

Chapter 6
The Fundamental Structure of the Design Object Model

We have outlined the model structure provisionally in Sect. 0.8 and expounded our position on the more crucial issues in constructing the conceptual framework in which to describe models in Sects. 4.2.2 (2), 4.3.2, 4.4, 4.5 and 4.6.4. Further, we have made clear the philosophical presumptions behind the technical decisions in Sects. 5.3.1 and 5.3.2. Our first task in this chapter is to put together these preliminary and fragmentary results, and present them in a more formal and comprehensive arrangement.

6.1 Physical Object, Theory and Interpretation

(1) *Formal Theory*
(1.1) *Sorts*
A countably infinite set of sorts $\{\sigma_1, \sigma_2, \ldots\}$ is provided. Sorts are special unary predicates which, when interpreted, divide the entire universe into subsets. For any two of them that overlap with each other, the intersection will form another sort.

(1.2) *Individual Constants and Variables*
For each positive integer n, a countably infinite set of distinct constants $\{a_1^n, a_2^n, \ldots\}$ of sort σ_n is provided. That is,

$$\sigma_n(a_i^n) \wedge (a_i^n \neq a_j^m) \quad (i \neq j \text{ or } n \neq m) \tag{6.1}$$

for each n, m, i and j. Another countably infinite set $\{x_1, x_2, \ldots\}$ is provided for representing variables.

(1.3) *Function and Predicate Symbols*
For each positive integer n, countably infinite sets of n-ary function symbols $\{f_1^n, f_2^n, \ldots\}$ and predicate symbols $\{p_1^n, p_2^n, \ldots\}$ are provided. Each argument of each function or predicate is of a predefined sort. The sort of the function value is also fixed for each function symbol.

(1.4) *Finite Theory*
The construction of a theory is done by selecting a subset from the reservoir of symbols of each category and giving a system of non-logical axioms. Only

such a theory is admitted that comprises a finite number of symbols and axioms, though the number may be arbitrarily large.

(1.4.1) There must be a particular sort 'ι' and its sub-sorts to denote 'physical object' and other entities the meanings of which are defined only by the current theory being constructed.

(1.4.2) Other sorts denote the classes of entities the structures of which are assumed to have been completely established and known, such as integers, real numbers, records, polygons, spline functions defined over a fixed interval, etc.

(1.4.3) It cannot be decided objectively but depends on designers' intention whether an abstract entity is to belong to ι or any 'known' sort. For example, a linear transformation in 3-dimensional Euclidean space can be conceived of either as a geometric entity or as a physical movement. If the former, it is simply an element of the domain of 4×4 matrices of real numbers with a fixed algebraic structure. If the latter, it allows the predication of an unrestricted number and variety of properties such as path, velocity, acceleration, constraints, optimality, etc., and yet its existence and identity are independent of any particular description.

(1.4.4) The semantics of a formula is a logical or functional relationship among physical objects, abstract entities and their properties. The system of axioms formally specifies the design requirements. The following should be noted, however:

- The requirement specification need not be complete or deterministic but may contain just those requirements that designers desire to have explicitly stated.
- Every axiom must contain at least one constant or variable of sort ι in order to express anything meaningful for design.

(2) *Interpretation*

(2.1) *The Universe of Interpretation*

A countably infinite reservoir of symbols $\tilde{I} = \{e_1, e_2, \ldots\}$ is provided to serve as the extension of sort ι. In an actual interpretation $\mathcal{I}$, a subset I of $\tilde{I}$, which may be arbitrarily large but remains finite, is taken. Another finite domain N is provided and contains character strings which are to be used as 'external names' of elements in I. If further the theory contains sorts $\sigma_{i_1}, \sigma_{i_2}, \ldots, \sigma_{i_s}$, the universe of $\mathcal{I}$ will be

$$U = I \cup A_{i_1} \cup A_{i_2} \cup \cdots \cup A_{i_s} \cup N \tag{6.2}$$

where A_{i_k} denotes the extension of σ_{i_k} ($k = 1, 2, \ldots, s$). The union of A_{i_k} denoted by A is to serve the domain of attributes.

(2.2) *Functions NAME and ID*

A built-in partial function NAME is defined over I with values in N. It is a bijection from the domain of definition, hence admits an inverse, here denoted

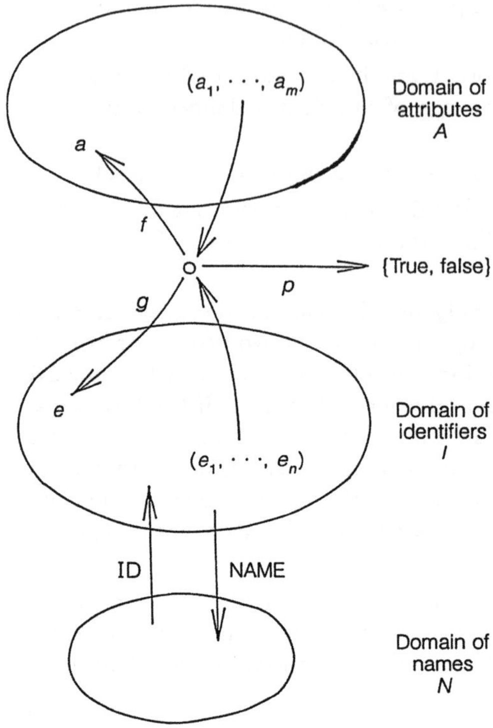

Fig. 6.1. The fundamental structure of an interpretation

by ID. That is,

$$\text{NAME} : I \to N \qquad \text{ID} : N \to I \qquad \text{ID} = \text{NAME}^{-1} \tag{6.3}$$

These enable human designers and application programs to access design objects by aliases which unlike identifiers (elements of I) are intelligible to them.

(2.3) *Other Functions and Predicates*
All the functions and predicates that are built-in over the domain of attributes A are included in $\mathcal{I}$. Those which contain arguments of sort ι may be defined in terms of more elementary ones. The chain of definitions continues until it ends up with primitive functions and/or predicates which are interpreted as either built-in or enumerative (table-like) type.

Figure 6.1 depicts the fundamental structure of an interpretation described above where f and g denote functions and p denotes a predicate.

Although natural sciences and design object modeling have much in common, as we have seen, they differ in one very important respect. Namely, it is not the case with the latter that theories are formed to explain and predict the behavior of what already exists. Instead, they are formed to stipulate

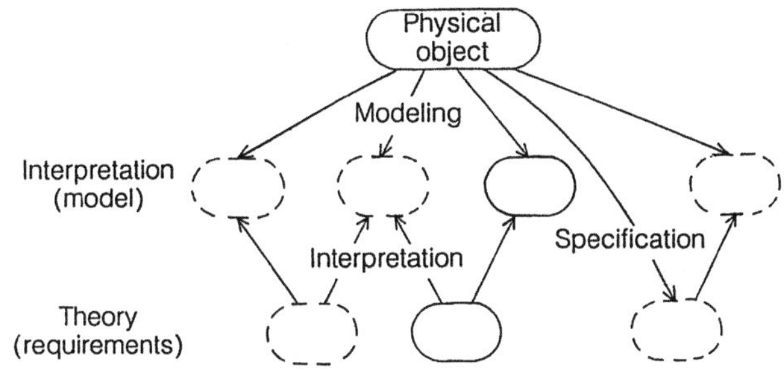

Fig. 6.2. Physical object, theory and interpretation

the characteristics of what is yet to come into existence. There will not exist any objective measure such as observation or experimentation in the light of which the correctness of a theory is to be established. It is possible to demonstrate the consistency of a theory only by creating a model of it. It is not possible to demonstrate physical implementability or manufacturability, as that depends on non-logical and non-conceptual factors such as technologies, materials or even the discoveries of new physical laws. Given the requirement specification as a consistent theory, there will generally be multiple interpretations of it, each of which might be physically realized in zero, one, more than one, or infinitely many ways. Thus, we obtain Fig. 6.2 corresponding to Fig. 5.4, where the islands encircled by solid lines indicate the actually selected versions and those drawn with dotted lines are alternatives.

Definition 6.1. A *design object model* $\mathcal{M}$ is the combination of a formal theory $\mathcal{T}$ and one of its models $\mathcal{I}$ (in the sense of first order predicate logic; see Definition 3.10) constructed in accordance with (1) and (2) above respectively.

Note that we require theoretical consistency of a design object model but leave open the question of manufacturability. Further qualifications will have to be imposed on the structure of model so that the requirements proper to CAD as presented in Chap. 1 may be naturally satisfied. This is our next theme.

6.2 Qualifications on Theories and Interpretations

With our primary interest being in the conceptual realm of modeling, due considerations will be given to the subsections

(1)–(10) of Sect. 1.1,
(1)–(3), (6), (7), (9) of Sect. 1.2, and

(1), (4) of Sect. 1.3,

which pertain directly to the said realm whereas the subsections

(4), (5), (8) of Sect. 1.2, and
(2), (3) of Sect. 1.3,

which may be dealt with more appropriately as requirements in the realms of physical data management and/or applications administration will be set aside.

(1) *Structures and Semantics*
That a design object model comprises both theory and interpretation is not meant to indicate the mere juxtaposition of these. Its true significance lies in the enrichment of theoretical structure it affords by establishing the exclusive universe of interpretation and adding to the basic vocabulary of $\mathcal{T}$ all the non-logical apparatus available in that universe.

(1.1) *Individual Constants and Variables*
All the constants and variables that occur in the theory $\mathcal{T}$ denote some elements of U. Conversely, legally expressed elements of A are counted among legitimate terms. Direct references to I's elements within $\mathcal{T}$ are not possible, however, because:

- first, the way to express elements in $\tilde{I}$ is not (and should not be) known to theory constructors, and
- second, the construction of a theory logically precedes the creation of an interpretation, hence that of I.

(1.2) *Quantifications and Finite Evaluations*
As each argument of a function or predicate is of a certain sort, the range of values that a bound variable may take within its scope of quantification is restricted to the extension of a particular sort. Methods proper to each structure can be exploited in exhausting the possible values, mathematical induction for the domain of natural numbers, for example. Particularly important is the case where the cardinality of the extension is known to be small enough to make the enumeration of members of any subsets and set-theoretic operations on them practical. To treat the matter more generally and precisely, we introduce the notion of finite evaluability of a formula. First, the universe of interpretation must be extended to contain the whole set of $\tilde{I}$ instead of I so that we may refer to a single, fixed universe and talk non-trivially about finiteness.

Definition 6.2. The *I-completion* $\tilde{\mathcal{I}}$ of an interpretation $\mathcal{I}$ is that which is obtained by replacing I by $\tilde{I}$ and leaving the rest unchanged. Hence its universe can be expressed as

$$\tilde{U} = \tilde{I} \cup A \cup N \tag{6.4}$$

corresponding to (6.2). An interpretation is called *I-completed* if it is the I-completion of a certain $\mathcal{I}$.

Definition 6.3. A formula, open or closed, of the theory $\mathcal{T}$ is ***positive-finitely evaluable*** if and only if there exists an algorithm that, for any I-completed interpretation $\tilde{\mathcal{I}}$, enumerates all the elements of the extension of the said formula and halts in a finite number of steps.

It is ***negative-finitely evaluable*** if its negation is positive-finitely evaluable. It is simply ***finitely evaluable*** if it is either positive- or negative-finitely evaluable.

Note that the finiteness of either the extension or its complement is implied. However, this is not a sufficient condition for establishing the finite evaluability of a formula. It is desired (and we feel it is a reasonable desire) that all the formulas that occur in a theory be finitely evaluable. To ensure this, we require that the following be accepted as basic presumptions:

(1.2.1) Every atomic formula of a theory $\mathcal{T}$ that contains at least one argument of sort ι is positive-finitely evaluable.

(1.2.2) Every formula of the form

$$f(x) = a$$

where f is an arbitrary function symbol, x is a tuple of free variables and a is an arbitrary value in its range is positive-finitely evaluable.

(1.2.3) Every atomic formula that contains no argument of sort ι may occur only as a conjunct in a conjunction in which another member is positive-finitely evaluable. If we write the conjunction as

$$w(x, y) \wedge p(x, z)$$

where $p(x, z)$ is the atomic formula in question, and x, y, and z are tuples of free variables, a further requirement is that for each tuple of values (x_1, y_1) in the extension of w, the atomic formula $p(x_1, z)$ be positive-finitely evaluable with respect to z, hence that the whole conjunction be positive-finitely evaluable.

The intent of stipulating this last condition is to allow the application of predicates over the domain of attributes strictly for further qualifying the (finite) statements concerning physical objects that have been obtained through finitely evaluating other formulas. We are now ready to present the following lemma.

Lemma 6.1. Under the conditions (1.2.1)–(1.2.3) above, every formula of a theory $\mathcal{T}$ is equivalent to a boolean combination of finitely evaluable formulas.

Proof. The proof is by induction on the depth of quantification. It is sufficient to show that if every formula of the outermost level within the scope of quantification is either a literal or finitely evaluable, then the result of quantification has the said property.

First, by conditions (1.2.1) and (1.2.3), every atomic formula that is not finitely evaluable can be absorbed into a finitely evaluable conjunction. Hence, we may consider every next outermost formula as finitely evaluable. We may further assume that the formula F in question is written as

$$\forall x(u_1(x, y^{(1)}) \vee \cdots \vee u_k(x, y^{(k)}) \vee \neg v(x, z))$$

where $u_1, \ldots, u_k$ and v are positive-finitely evaluable.

Let us take a value c for z such that

$$X = \{x : v(x, c)\} \quad \text{is not empty}$$

and define for each $i = 1, \ldots, k$

$$Y^{(i)} = \{y^{(i)} : \exists x(u_i(x, y^{(i)}))\}$$
$$w_i(y^{(i)}) \equiv \forall x(v(x, c) \rightarrow u_i(x, y^{(i)}))$$
$$Y^{(i)\prime} = Y^{(i)} - \text{extension of } w_i$$
$$\alpha : \text{defining predicate of } Y^{(1)\prime} \times \cdots \times Y^{(k)\prime}$$
$$w(y^{(1)}, \ldots, y^{(k)}) \equiv \forall x(v(x, c) \rightarrow (\vee_i u_i(x, y^{(i)}) \wedge \alpha(y^{(1)}, \ldots, y^{(k)})))$$

It is easy to see that w_i, α and w are all positive-finitely evaluable and that

$$F(y^{(1)}, \ldots, y^{(k)}, c) \equiv \vee_i w_i(y^{(i)}) \vee w(y^{(1)}, \ldots, y^{(k)})$$

Adding the case where X is empty, we obtain

$$F(y^{(1)}, \ldots, y^{(k)}, z)$$
$$\equiv \neg \exists x(v(x, z)) \vee \vee_i w_i(y^{(i)}) \vee w(y^{(1)}, \ldots, y^{(k)})$$ *End proof*

(1.3) *Functions and Predicates*
Not only the individual objects but the functions and predicates defined over the domain of attributes (A) are permitted to be freely used in constructing the theory $\mathcal{T}$. All the valid statements concerning these are automatically included in the axioms. In short, the logic that underlies $\mathcal{T}$ is not pure FOPL but A-logic, so to speak [Bar].

(1.4) *Sets, Set Operations and Functions*
References to the extensions of formulas within the theory are also possible. This is because:

- the universe of interpretation U is always taken from the fixed domain $\tilde{U}$,
- all the components of U are fixed and known except I, and
- no direct references to the elements of I (which is unknown to $\mathcal{T}$) are made.

The extensions are finitely expressible by Lemma 6.1. Use of the usual set-theoretic operations on these extensions raises no difficulty, though it produces nothing new either, since we can perform logical operations on component formulas and take the extension of the result to the same effect.

Frequently used 'standard' set functions should be available, too. These include among others such functions as:

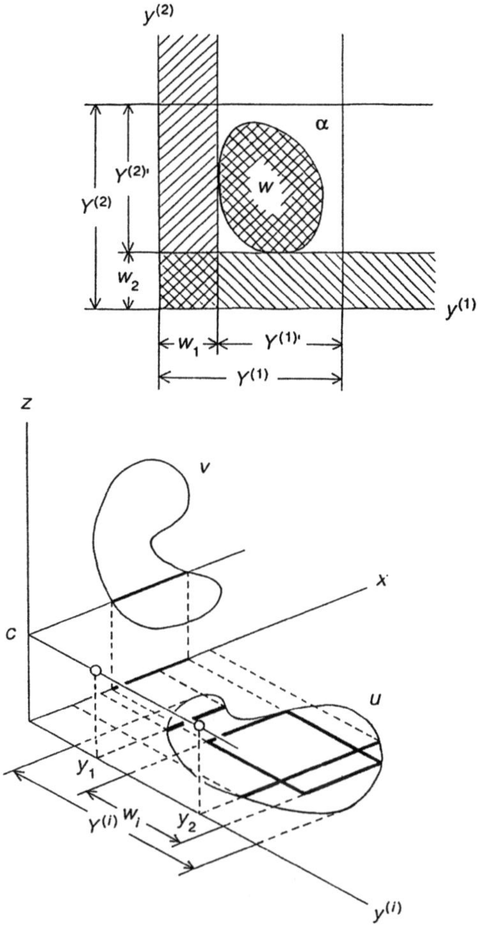

Fig. 6.3. Finite evaluability of formula

- the cardinality of a set,
- the maximum or minimum member for an ordered set,
- the sum, product or average of members for an arithmetically structured set, etc.

(1.5) *Recursive Functions and Predicates*
It should be noted that while recursions may occur only for postulating axioms in a formal theory, they can be exploited for defining functions and predicates in an interpretation. As previously noted (Subsect. 4.2.2 (1)), the transitive closure (r) of a binary predicate (p) gives a simple yet typical example to indicate the distinction between the two roles of recursive statements. That is, given an interpretation of p, there will generally exist multiple

interpretations of r that satisfy the axiom

$$\forall x \forall y(r(x,y) \leftrightarrow p(x,y) \vee \exists z(p(x,z) \wedge r(z,y))) \tag{6.5}$$

The 'canonical' one is characterized by having minimum extension, hence inherently presupposes the notion of set inclusion which has no representation in a first order formal theory. We write

$$r(x,y) \equiv p(x,y) \vee \exists z(p(x,z) \wedge r(z,y)) \quad \text{for any } x,y \tag{6.5$'$}$$

to denote r in this sense, or more generally

$$r(x,y) \equiv F(r(x,y)) \quad \text{for any } x,y \tag{6.6}$$

to define r as the least fixed point of a recursive formula scheme F. To ensure the existence of the least fixed point, we require that F be continuous as usual [Man].

An alternative way to express the intended meaning of r precisely is to state explicitly that the two arguments are connected to each other by a finite chain of arcs:

$$\begin{aligned} &r(x,y) \equiv p(x,y) \vee \exists n(r(x,y,n)) \\ &r(x,y,n) \equiv \exists z(p(x,z) \wedge r(z,y,n-1)) \\ &r(x,y,1) \equiv p(x,y) \qquad\qquad \text{for any } x,y \end{aligned} \tag{6.5$''$}$$

That the recursion will terminate in a finite number of steps depends of course on the structure of $\mathcal{N}$, and our knowledge of it is implicitly used in giving the above definition.

The recursive definition of a function is given in much the same way as that of a predicate.

(1.6) *Views, Theories and Interpretations*
It is possible to provide multiple theories rather independently and yet have them interpreted in the same universe U. So that these theories may constitute a coherent whole, we agree on the following convention:

- Levels of theories are distinguished: those of topmost level, or 'global' theories, which are directly interpreted and those of subordinate levels, or 'local' theories, which are not.
- A top level theory may be given multiple instances of interpretation, which can be compared to the placement of a variational part at different locations with different parametrizations. A distinct qualifier is supplied for each instance and affixed to the external names and the constant symbols of the original theory.
- Constants, function symbols, and predicate symbols of the top level theories are global in the sense that while they may occur in more than one theory, each occurrence is interpreted in the same way as others.

A local theory is to be included in other local or global theories with a proper renaming of non-logical symbols. Some of them may be given the final interpretation over the domain U at this stage. Rather as with interpretations,

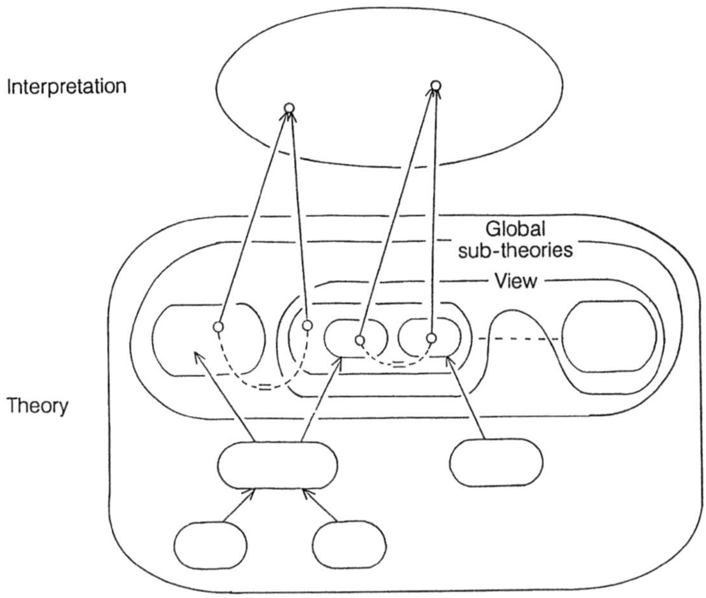

Fig. 6.4. Views, theories and interpretations

a theory may have multiple instances of inclusion in one or more theories. (It might be suggested that the introduction of a local-global distinction among non-logical symbols within a theory is appropriate to reduce the cost of achieving coherency. Though we feel this might be the case, we refrain from giving further consideration to the idea in order to keep theories as structurally simple as possible.)

A view (that is, an external schema) is defined as a collection of global theories together with local theories included therein.

Figure 6.4 presents an image of the complex of views, theories and interpretations described above.

(2) *Dynamic Aspect*

We envisage an agent which communicates with human designers, observes and acts on the design object model, makes use of it for various applications, and has intelligence concerning designers' intentions. The dynamic aspect of the model is the manner in which it interacts with this agent, and will be considered on two levels. The primitive, microscopic and non-intelligent level is treated in the following while the compound, macroscopic and intelligent level which constitutes the core of the design process model will be discussed in Chap. 8. Furthermore, accesses to the object model are divided into two categories, one for constructing theories and the other for manipulating interpretations.

(2.1) *Construction of Theory*

The construction of a theory will be done normally and mostly at a stage that precedes the initiation of its interpretation, which can be compared to the static description of a database schema with a DDL (data definition language). The nature of design should, however, allow extensions and modifications of a theory to be made dynamically, that is, in a situation where there exists a partial interpretation of it which should remain intact. Four basic kinds of actions on a theory need to be examined separately.

(2.1.1) *Addition of Non-Logical Symbols*

This can be done freely as there is no danger of interfering in the state of interpretation. Notice should be simply taken of the possibility that the symbols to be added exist already in other theories.

(2.1.2) *Deletion of Non-Logical Symbols*

In principle, this is forbidden. There are two exceptional cases: one is after all the instances of the symbols to be deleted are already deleted from the interpretation and the other is when the same symbols continue to exist in other theories. All the other symbols and formulas the definitions of which depend on the symbols to be deleted are subject to automatic deletion.

(2.1.3) *Addition of (Non-Logical) Axioms*

This is permitted, only the interpretation which may be consistent with the current theory or even fully satisfy it may get invalidated. It is fairly straightforward by Lemma 6.1 to see if the axioms being added hold true. If not, we will tentatively accept them, as future extensions of the interpretation may possibly satisfy them. Unfortunately, it is generally impossible to establish in advance whether this will be the case or not, or still worse, whether the new theory will itself be consistent or not. We shall not concern ourselves too seriously about this undecidability, however, because the system will sooner or later come to a dead end if an inconsistency exists, or in the meantime, designers might change the requirements.

(2.1.4) *Deletion of Axioms*

There is no restriction on the deletion of axioms, which is nothing but the relaxation of design requirements. The theoretical consistency or the satisfaction of axioms by the current interpretation will not be adversely affected.

(2.2) *Manipulation on Interpretation*

Manipulation on an interpretation is done by the combinations of four basic operations which constitute the functional core of most data manipulation languages (DMLs).

(2.2.1) *Evaluation of a Formula*

In response to the agent's inquiry presenting a formula

$$w(x_1, \ldots, x_n) \tag{6.7}$$

where $x_1, \ldots, x_n$ represent free variables, w's extension

$$C = \{(c_1^{(i)}, \ldots, c_n^{(i)}) : w(c_1^{(i)}, \ldots, c_n^{(i)})\} \tag{6.8}$$

is supplied. Lemma 6.1 guarantees the finite expressibility of C even when it is an infinite subset of $\tilde{U}^n$. In case $n = 0$ (that is, w is a closed formula), either TRUE or FALSE will be returned.

(2.2.2) *Assertion of a Closed Formula*
This directs the structure of interpretation to be modified so that the given formula may become true. A fundamental restriction to be imposed on the formation of the formula is that it must be logically equivalent to a conjunction of ground literals, each of which takes one of the following forms:

$$p(c_1, \ldots, c_n) \tag{6.9.1}$$

$$\neg p(c_1, \ldots, c_n) \tag{6.9.2}$$

$$f : c_1, \ldots, c_n \rightarrow c \tag{6.9.3}$$

where p and f are defined enumeratively. Furthermore, these are the only symbols that are yet to be interpreted (that is, the c_is and c are assumed to represent specific elements of U), and the assertion of the above literals should be understood as the addition or deletion of an instance to or from the interpretation of p or f. In particular, the assertion of

$$a : \ \rightarrow c$$

where a is a constant symbol (function symbol with zero argument) and c represents an element of the same sort will have a interpreted as c even if it has already been interpreted otherwise.

To slightly relax the restriction, we permit the assertion of a Horn clause

$$\forall x_1 \ldots \forall x_n (\wedge_i p_i(x_1, \ldots, x_n) \rightarrow q(x_1, \ldots, x_n)) \tag{6.10}$$

on the condition that those instances of the p_is that hold true currently are not to be falsified. In other words, for any $c_k \in U$ such that

$$\wedge_i p_i(c_1, \ldots, c_n) \wedge \neg q(c_1, \ldots, c_n) \tag{6.11}$$

is true, $q(c_1, \ldots, c_n)$ is to be asserted. A justification for this is given below:

- Each ground instance $p_i(c_1, \ldots, c_n)$ is understood to represent a positive design decision, hence should not be eliminated from the model unless so declared explicitly.
- The formula should be devoid of existential quantifiers so that its being true determines unequivocally the truth values of the conjuncts constituting it. Thus, we only have to consider the case

  $$\forall x_1 \ldots \forall x_n (\wedge_i p_i(x_1, \ldots, x_n) \rightarrow \vee_j q_j(x_1, \ldots, x_n))$$

- Furthermore, there must be only one positive literal within each conjunct in order for the change to be deterministic, hence we obtain (6.10).

Extending the class of permissible formulas to a more general one entails consideration of multiple choices and reliance on a macroscopic strategy to dissolve ambiguities, hence will be studied as a part of modeling the design process.

N.B. The assertion of (6.9.3) will establish the truth of

$$f(c_1, \ldots, c_n) = c \qquad (6.9.3)'$$

The reason for not writing this directly is to eliminate the possible ambiguity as to the side of the equality that is to be interpreted.

(2.2.3) *Creation of Objects*
Creating objects of sort ι is tantamount to bringing the entities representing them into the domain I from the reservoir $\tilde{I}$. As these entities are by themselves nothing but pure symbols, the simultaneous assertion of a formula expressing whatever meanings designer may want to give them is permitted and will be normal.

(2.2.4) *Deletion of Objects*
By being deleted from I, objects will be returned to $\tilde{I}$ to be available for future creation. The identification of objects to be deleted is done by specifying the condition they are to meet in a formula which contains just one free variable. All the elementary facts (ground atoms) involving them will be automatically deleted (falsified).

N.B. – When objects are created, it is completely under the control of the 'system' (model manager) to assign to them specific symbols of $\tilde{I}$ and should not concern the agent instructing the creation. The latter will be informed of the assignment, however.
– Formulas are evaluated under the closed world assumption. Namely, all the ground atoms other than those that have been explicitly asserted and remain so are assumed to be false.

(3) *Alternate Designs, Distributed Databases and Version Control*
(3.1) *Design Alternatives*
As pointed out previously (Sect. 1.3 (2)), producing versions is similar to creating alternate designs in its technical aspect on which we will focus our present attention. First, let us give the precise definition of design alternatives.

Definition 6.4. Two substructures $\mathcal{I}_1'$, and $\mathcal{I}_2'$ of $\mathcal{I}$ are said to be *design alternatives* to each other in the context of $\mathcal{I}$ *in the strong sense* when the following conditions are satisfied:
(1) Both $\mathcal{I}_1'$ and $\mathcal{I}_2'$ are the models of the same sub-theory $\mathcal{T}'$ of $\mathcal{T}$.
(2) The domains I_1' and I_2' of sort ι for $\mathcal{I}_1'$ and $\mathcal{I}_2'$ respectively are mutually disjoint.
(3) $\mathcal{I}_1'$ and $\mathcal{I}_2'$ have a common interface with the rest of $\mathcal{I}$ so that each may be substituted for the other without affecting the overall structure of $\mathcal{I}$. Pre-

cisely, this is stated as follows:
(3.1) The interface of $\mathcal{I}'_\nu$ denoted by $int(\mathcal{I}'_\nu)$ ($\nu = 1, 2$) is the set of all the ground instances of predicates and functions each of which contains both the elements of I'_ν and $I - I'_1 - I'_2$ as arguments and/or function value. That is, each statement in $int(\mathcal{I}'_\nu)$ takes the form of

$$p(e_1^{(\nu)}, \ldots, e_l^{(\nu)}, e_1, \ldots, e_m, a_1, \ldots, a_n),$$
$$f(e_1^{(\nu)}, \ldots, e_l^{(\nu)}, e_1, \ldots, e_m, a_1, \ldots, a_n) = c \quad \text{or}$$
$$f(c_1, \ldots, c_n) = e^{(\nu)}$$

with an appropriate reordering of arguments where $e^{(\nu)}$, e_j and a_k denote elements of I'_ν, $I - I'_1 - I'_2$ and the domain of attributes A respectively, l, $m \geq 1$, and c is of arbitrary sort.
(3.2) Let $int(I'_\nu)$ denote the set of $e^{(\nu)}$s occurring in the statements of $int(\mathcal{I}'_\nu)$. Then $int(I'_1)$ and $int(I'_2)$ are of the same cardinality.
(3.3) There exists a bijection α' from $int(I'_1)$ onto $int(I'_2)$ such that if we extend it to α by defining

$$\alpha(e) = \begin{cases} \alpha'(e) & \text{for } e \in int(I'_1) \\ e & \text{for } e \in I - I'_1 - I'_2 \end{cases}$$

α will provide an isomorphism between $\mathcal{I} - \mathcal{I}'_1 - (\mathcal{I}'_2 + int(\mathcal{I}'_2))$ and $\mathcal{I} - (\mathcal{I}'_1 + int(\mathcal{I}'_1)) - \mathcal{I}'_2$.

N.B. – All the elementary facts in $\mathcal{I}'_\nu$ are stated strictly in terms of I'_ν, A and N.
– $\mathcal{I}'_\nu + int(\mathcal{I}'_\nu)$ comprises all the elementary facts each of which involves at least one element of I'_ν.

Let us give another definition of design alternatives which is more theory-oriented as well as weaker than the above.

Definition 6.5. Two substructures $\mathcal{I}'_1$ and $\mathcal{I}'_2$ of $\mathcal{I}$ are said to be *design alternatives* to each other in the context of $\mathcal{I}$ *in the weak sense* when the following conditions are satisfied:
(1) Same as condition (1) of Def. 6.4.
(2) Same as condition (2) of Def. 6.4.
(3) Two substructures

$$\mathcal{I} - \mathcal{I}'_1 - (\mathcal{I}'_2 + int(\mathcal{I}'_2)) \quad \text{and} \quad \mathcal{I} - (\mathcal{I}'_1 + int(\mathcal{I}'_1)) - \mathcal{I}'_2$$

are elementarily equivalent. (See Def. 3.17 for the concept of elementary equivalence.)

Lemma 6.2. Let $\mathcal{I}'_1$ and $\mathcal{I}'_2$ be substructures of $\mathcal{I}$.
(1) If $\mathcal{I}'_1$ and $\mathcal{I}'_2$ are design alternatives to each other in the strong sense, they are so in the weak sense.
(2) Suppose $\mathcal{I}'_1$ and $\mathcal{I}'_2$ are design alternatives in the weak sense. For each

constant symbol of $\mathcal{T}'$, the elements $e^{(\nu)} \in int(I'_\nu)$ $(\nu = 1,2)$ corresponding to it have the same NAME. These are related by the isomorphism α if, further, the $\mathcal{I}'_\nu$ are alternatives in the strong sense.
(3) If $\mathcal{I}'_\nu$ are alternatives in the weak sense and every element of I that occurs in some statement of $int(\mathcal{I}'_\nu)$ corresponds to some constant symbol of $\mathcal{T}$, $\mathcal{I}'_\nu$ are alternatives in the strong sense.

The proof is straightforward. This lemma provides the theoretical basis for a practical means to initiate planning a design alternative, that is, structural copying and imbedding.

(3.2) *Copying and Imbedding*

Definition 6.6. Given an interpretation $\mathcal{I}$ on I and its substructure $\mathcal{I}'_1$ on I'_1, another interpretation $\mathcal{I}'_2$ is said to be *structurally imbedded* onto $\mathcal{I}'_1$ in the context of $\mathcal{I}$ when the following conditions are satisfied:
(1) The universe of interpretation I'_2 of $\mathcal{I}'_2$ and I are disjoint.
(2) A collection of elementary facts is asserted which relate elements of I'_2, $I - I'_1$, A and N, and is denoted by $int(\mathcal{I}'_2)$.
(3) There exists a bijection β' from I'_2 onto I'_1 such that if we extend it to β by defining

$$\beta(e) = \begin{cases} \beta'(e) & \text{for } e \in I'_2 \\ e & \text{for } e \in I - I'_1 \end{cases}$$

β will provide an isomorphism between $(\mathcal{I} \mid I - I'_1) + \mathcal{I}'_2 + int(\mathcal{I}'_2)$ and $\mathcal{I}$.

It must be possible for an application agent to specify $\mathcal{I}$ and $\mathcal{I}'_1$, and have $\mathcal{I}'_2$ and β constructed so that $\mathcal{I}'_2$ may be structurally imbedded onto $\mathcal{I}'_2$ in the context of $\mathcal{I}$. It should be noted that once I'_2 is created and β' is defined, the structures of $\mathcal{I}'_2$, $int(\mathcal{I}'_2)$ and the isomorphism β will be uniquely determined. There are a few technical matters which should concern the agent, however. First, after the creation of I'_2 and the definition of β' are done, the reproduction of instances of a specific predicate p or function f can be accomplished by asserting

$$\begin{aligned} &\forall x_1 \ldots \forall x_n (p(\beta(x_1), \ldots, \beta(x_n)) \rightarrow p(x_1, \ldots, x_n)) \\ \text{or}\quad &\forall x_1 \ldots \forall x_n \forall y ((f(\beta(x_1), \ldots, \beta(x_n)) = \beta(y)) \rightarrow f(x_1, \ldots, x_n) = y)) \end{aligned}$$

The reproduction of the entire structure, however, will generally require the universal quantification of predicate and function symbols which of course passes the limit of FOPL. One way to alleviate the difficulty is to make available the information on what predicates and functions are defined over the universe of interpretation. An interface will have to be provided through which the agent can inquire about the construction of each theory and its relations with other theories. Another and more direct way is to make it a primitive operation to create a copy and imbed it onto the original structure in the given context. We will implement both ways.

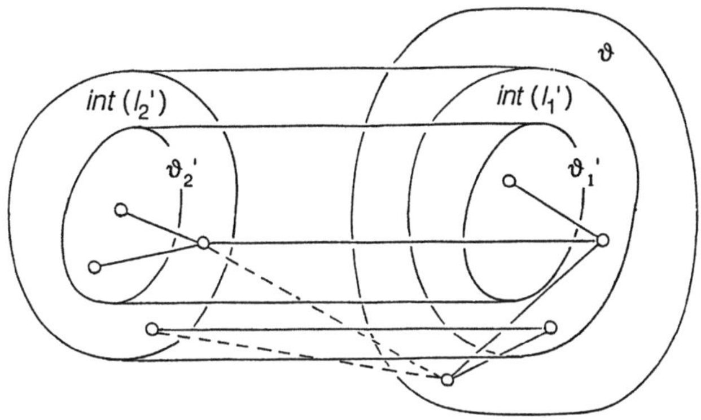

Fig. 6.5. Structural copying and imbedding

Second, $int(\mathcal{I}'_2)$ must be kept intact while the agent works on $\mathcal{I}'_2$ to design an alternative to $\mathcal{I}'_1$, though it may modify the 'internal structure' $\mathcal{I}'_2$ so long as the axioms of $\mathcal{T}'$ remain valid. In particular, elements of $int(I'_2)$ must not be deleted. A convenient way to ensure this is to require that for every interpretation of $\mathcal{T}'$ (say $\mathcal{I}'$), only those elements that correspond to constant symbols of $\mathcal{T}'$ be included in its interface with the outside ($int(\mathcal{I}')$). It is only necessary for the agent to recognize these special elements as such. Unless there is a need of information about $int(\mathcal{I}'_2)$ during the work of alternate designing, copying the structure of $int(\mathcal{I}'_1)$ can be left over until the final selection from the alternatives is made.

The decision on a specific alternative is made by simply deleting all the universes of other alternatives (I's). Deletion of structures ($\mathcal{I}$'s) will follow automatically.

(3.3) *Distributed Databases*
By distributed databases, we are considering the situation where a single central database and multiple and partial copies of it exist. Transferring data between different locations is somewhat similar to copying and imbedding in its abstract functionality. There are other aspects, however, which are peculiar to database distribution. That is:

- Each local agent may want to see and manipulate the data through his own view (external schema).
- Unlike the case discussed in the preceding paragraph, copies are placed under the control of separate systems. Transfer of them will inevitably be through intervening agents, hence pertinent pieces of information must at some time be put into a form intelligible for these agents.
- In general, the result of processing at any specific location will not wholly substitute the originals in the central database, but all the updates under different views must be cohered and incorporated into the originals.

Because of these, the central agent is enabled to do the following:

- To specify the desired view and retrieve non-logical symbols, their arities, sorts of domains and/or ranges, and axioms for each theory within the specified view.
- To define a transaction as a sequence of data update operations, that is, assertions of formulas, and creations and/or deletions of objects, and to have them replayed on the central database.

The confirmation of input and output conditions for each transaction is to be made by evaluating the formulas that express these conditions at the start and end of the transaction.

(3.4) *Archival Store*
Copying, discussed in (3.2) above, was actually the creation of a different object with the same structure as the original. Central and local databases in a distributed environment were placed in a hierarchical and complementary relationship. As previously noted (Sect. 1.2 (7)), neither is the case with conventional business data processing where all the copies, despite the varieties of media and locations, carry the same information content. The same can be said of a finished design and its archival store in that the latter, when restored to the active state, will behave in precisely the same manner as the former. Only the assignment of symbols of $\tilde{I}$ to entities of sort ι (that is, the creation of I) will be done completely afresh. There will be no connection between the old and new copies (such as the isomorphism β in the case of structural copying and imbedding) if the two come ever to coexist in the same database. Standard utilities are provided to perform saving and restoring between an active database and an archival store.

6.3 Illustrative Examples

We will present a few examples to illustrate concisely how conceptually important requirements of CAD databases can be fulfilled by the framework described in the preceding sections.

(1) *CSG and BREP Models of Polyhedron*
A miniature scale, yet typical example of a multifaceted description of an object is the consolidation of CSG (constructive solid geometry) and BREP (boundary representation) models of a polyhedron [Req]. For simplicity, we restrict the repertoire of primitives to cuboid and truncated cone, and as their specializations, cube, cone and cylinder. The prototype for each of these is represented as a lowest level sub-theory. We define another sub-theory to contain all the primitives together with the functions for placing them within the global space, which in turn is contained in the theory for the CSG model of a still higher level. It is also contained in the theory for BREP. Finally, a

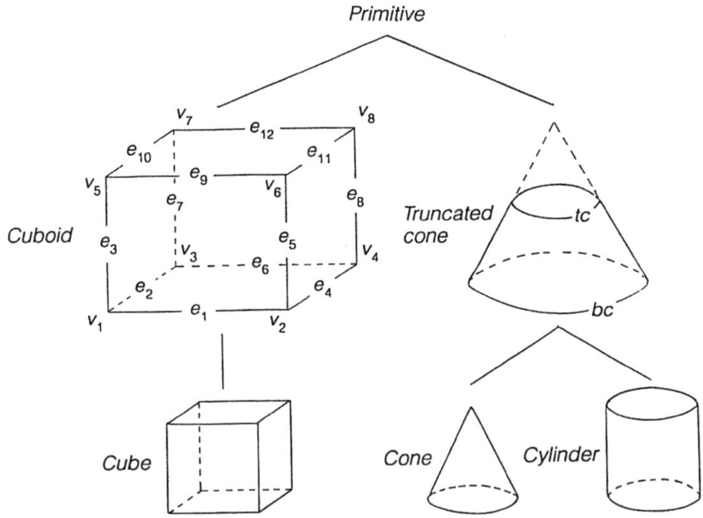

Fig. 6.6. Prototypical solid primitives

global theory consolidates both models by postulating the axioms that govern the updating of BREP in accordance with that of CSG.

(1.1) *Sub-Theories for Prototypical Primitives* (See Fig. 6.6)

(1.1.1) *Cuboid* ($\mathcal{T}p_1$)
- Sort $\quad cuboid(\iota)$
- Function $\quad l, w, h : cuboid \rightarrow \mathbf{real}$

  These denote the length, width and height of a cuboid.

(1.1.2) *Truncated Cone* ($\mathcal{T}p_2$)
- Sort $\quad tcone(\iota)$
- Function $\quad rt, rb, h : tcone \rightarrow \mathbf{real}$

  These denote the radiuses of top and bottom circles, and the height of a truncated cone.

(1.1.3) *Cube* ($\mathcal{T}p_{1.1}$)
- Sub-theory $\quad \mathcal{T}p_1$
- Sort $\quad cube(\iota)$
- Axiom 1) $\quad cube(\iota) \leftrightarrow (cuboid(\iota) \wedge (l(\iota) = w(\iota) = h(\iota)))$

  For convenience's sake, we write $F(\sigma)$ for $\forall x(\sigma(x) \rightarrow F(x))$ where σ is a sort and F is a formula.

(1.1.4) *Cone* ($\mathcal{T}p_{2.1}$)
- Sub-theory $\quad \mathcal{T}p_2$
- Sort $\quad cone(\iota)$
- Axiom 1) $\quad cone(\iota) \leftrightarrow (tcone(\iota) \wedge (rt(\iota) = 0))$

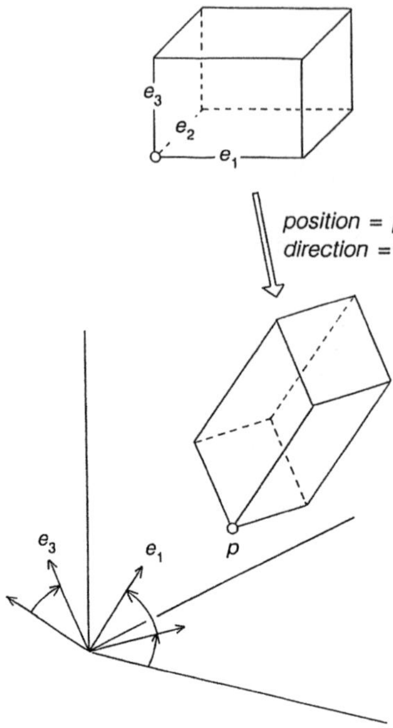

Fig. 6.7. Placing primitives

(1.1.5) *Cylinder* ($\mathcal{T}p_{2.2}$)
- Sub-theory $\mathcal{T}p_2$
- Sort *cylinder*(ι)
- Axiom 1) $cylinder(\iota) \leftrightarrow (tcone(\iota) \wedge (rt(\iota) = rb(\iota)))$

(1.2) *Sub-Theory for Placing Primitives* ($\mathcal{T}p$, see Fig. 6.7)
- Sub-theory $\mathcal{T}p_1, \mathcal{T}p_2, \mathcal{T}p_{1.1}, \mathcal{T}p_{2.1}, \mathcal{T}p_{2.2}$
- Sort *primitive*(ι) (abbreviated to p)
- Function $position : p \rightarrow \mathbf{real}^3$
  $direction: p \rightarrow \mathbf{real}^3$
- Axiom 1) $p(\iota) \leftrightarrow (cuboid(\iota) \vee tcone(\iota))$

(1.3) *Sub-Theory for CSG* ($\mathcal{T}c$)
- Sub-theory $\mathcal{T}p$
- Sort *object*(ι) (abbreviated to o)
- Constant *object-fin*, *object-semifin*
- Function $\cup, \cap, - : o \times o \rightarrow o$
- Predicate $constitute(o, o)$
- Axiom 1) $o(object\text{-}fin) \wedge o(object\text{-}semifin) \wedge o(p)$
  2) $object\text{-}semifin[\ \cup \mid \cap \mid -\ ]\ p = object\text{-}fin$
  3) $(o = object\text{-}fin) \vee constitute(o, object\text{-}fin)$

(1.4) *Sub-Theory for BREP of Primitives* ($\mathcal{T}bp$)

- Sub-theory $\mathcal{T}p$
- Sort *face*(ι), *loop*(ι), *edge*(ι), *vertex*(ι) (abbreviated to f, k, e, v respectively)
- Function $cf,\ ck,\ ce,\ cv : cuboid\ \times$ **positive integer** $\rightarrow f,\ k,\ e,\ v$
  ($cf(\ ,\ i), ck(\ ,\ j), ce(\ ,\ m)$ and $cv(\ ,\ n)$ are denoted by f_i, k_j, e_m and v_n respectively.)
  $tp,\ bm,\ sd : tcone \rightarrow f$
  $tc,\ bc : tcone \rightarrow e$

  These relate a cuboid to its faces, loops, edges and vertices (ordered lexicographically with respect to coordinate values, see Fig. 6.6), and a truncated cone to its top, bottom, side and two perimeters.
- Predicate $pf(p, f)$, $flout(f, k)$, $flin(f, k)$, $lee(k, e, e)$
- Function $ev0,\ ev1\colon e \rightarrow v$

  These relate the body of a primitive and each of its faces, a face and its unique outer perimeter, and each of its inner loops, a loop and each adjacent pair of its edges, an edge and its start and end vertices. Two edges in *lee* are placed in such a direction that the face lies at the left-hand side when viewed from outside of the body.
- Formula $str\text{-}cuboid(cuboid,\ f_i(i=1,6),\ k_j(j=1,6),\ e_m(m=1,12),\ v_n(n=1,8),\ l,\ w,\ h,\ position,\ direction)$

$$
\begin{aligned}
\equiv\ & \wedge_i(f_i : cuboid \rightarrow f_i) \wedge \wedge_j(k_j : cuboid \rightarrow k_j) \wedge \\
& \wedge_m(e_m : cuboid \rightarrow e_m) \wedge \wedge_n(v_n : cuboid \rightarrow v_n) \wedge \\
& (l : cuboid \rightarrow l) \wedge (w : cuboid \rightarrow w) \wedge (h : cuboid \rightarrow h) \wedge \\
& (position : cuboid \rightarrow position) \wedge (direction : cuboid \rightarrow direction) \wedge \\
& \wedge_i\, pf(cuboid, f_i) \wedge \wedge_i flout(f_i, k_i) \wedge \\
& lee(k_1, e_1, e_2) \wedge \cdots \wedge lee(k_1, e_4, e_1) \wedge \\
& \ldots \\
& lee(k_6, e_9, e_{11}) \wedge \cdots \wedge lee(k_6, e_{10}, e_9) \wedge \\
& (ev0 : e_1 \rightarrow v_1) \wedge \cdots \wedge (ev0 : e_{12} \rightarrow v_7) \wedge \\
& (ev1 : e_1 \rightarrow v_2) \wedge \cdots \wedge (ev1 : e_{12} \rightarrow v_8)
\end{aligned}
$$

$str\text{-}tcone(tcone, tp, bm, sd, tc, bc, rt, rb, h, position, direction)$

$$
\begin{aligned}
\equiv\ & (tp : tcone \rightarrow tp) \wedge (bm : tcone \rightarrow bm) \wedge (sd : tcone \rightarrow sd) \wedge \\
& (tc : tcone \rightarrow tc) \wedge (bc : tcone \rightarrow bc) \wedge \\
& (rt : tcone \rightarrow rt) \wedge (rb : tcone \rightarrow rb) \wedge (h : tcone \rightarrow h) \wedge \\
& (position : tcone \rightarrow position) \wedge (direction : tcone \rightarrow direction) \wedge \\
& pf(tcone, tp) \wedge pf(tcone, bm) \wedge pf(tcone, sd) \wedge \\
& flout(tp, tc) \wedge flout(bm, bc) \wedge flout(sd, bc) \wedge \\
& flin(sd, tc) \wedge lee(tc, tc, tc) \wedge lee(bc, bc, bc)
\end{aligned}
$$

(1.5) *Sub-Theory for BREP of Polyhedron* ($\mathcal{T}b$)

- Sub-theory $\mathcal{T}bp$
- Sort *object*(ι) (abbreviated to o)
- Constant *object-fin*
- Predicate $of(o, f)$

  This relates an object (body of a polyhedron) and each of its faces.

- Formula $fl(f,k) \equiv flin(f,k) \vee flout(f,k)$
  $le(k,e) \equiv \exists e'(lee(k,e,e') \vee lee(k,e',e))$
- Predicate $contain(k_1,k_2)$
  We presume the existence of a procedure to determine if a loop k_2 is contained by (lies inside of) a loop k_1 or not.
- Axiom 1) $o(object\text{-}fin)$
  2) $o(p)$
  3) $\exists 1k(flout(f,k))$
  4) $le(k,e) \rightarrow (\exists 1e_1 \exists 1e_2(lee(k,e_1,e) \wedge lee(k,e,e_2)))$
  5) $(flin(f,k) \wedge flout(f,k')) \rightarrow contain(k',k)$
  6) $\exists x((e = tc(x)) \vee (e = bc(x))) \vee \exists 2k(le(k,e))$
  7) $\exists x((e = tc(x)) \vee (e = bc(x))) \vee$
  $\exists v_0 \exists v_1((ev0(e) = v_0) \wedge (ev1(e) = v_1))$

  N.B. $\exists nx(formula(x))$ (as in Axioms 1) and 6)) where n is a positive integer stipulates that there exist precisely n elements x that satisfy $formula(x)$.

(1.6) *Theory for Consolidating CSG and BREP* ($\mathcal{T}$)
- Sub-theory $\mathcal{T}c$, $\mathcal{T}b$
- Function $on, onp : f \rightarrow f$

$$onp(f) = \begin{cases} on(f) & \text{if } p(\varepsilon o(of(o, on(f)))) \\ onp(on(f)) & \text{otherwise} \end{cases}$$

  on assigns a face to each of its sub-faces created by a boolean operation ($\cup$, $\cap$ or $-$) between an object and a primitive. *onp* maps each sub-face to the face of the primitive from which it was originally clipped.
- Predicate $intersect(e$ or f, f' or $o)$, $contain(o$, e or $f)$
  We presume the existence of a procedure to determine if an edge or a face intersects with an object or another face, or if it is wholly contained in an object.
- Sort $line(e)$, $arc(e)$, $free(e)$, $segment(\iota)$
- Predicate $fs(free$, $segment$, **positive integer**, **positive integer**$)$
  $fs(c,s,i,n)$ relates a free curve c and its i-th member of the total n segments.
- Function $eparam : line \rightarrow \{(a,b) : a,b \in \mathbf{real}^3\}$
  $: arc \rightarrow position \times direction \times radius \times \mathbf{real} \times \mathbf{real}$
  $: segment \rightarrow \mathbf{real}^t$
  The function *eparam* assigns to each edge a set of parameter values to decide its precise shape and position, which will be a pair of 3-dimensional coordinate values, a record comprising the attribute of an underlying circle and the terminal angles, or a t-tuple of real values depending on its type.
- Sort $plane(f)$, $cone(f)$
- Predicate $perpendicular(plane, cone)$
  A plane and a cone are related to each other by this predicate if the former is perpendicular to the axis of the latter. In particular, *tp* and *bm* of a *tcone* are perpendicular to *sd*.

- Predicate $on(edge, face)$
  This predicate relates an edge to each of the underlying faces. As asserted by the axiom 1) below, that an edge e is a boundary of a face f implies $on(e, f)$ but not vice versa.
- Formula $onp(e, f) \equiv \exists g(on(e, g) \wedge (f = onp(g)))$
  $onp(e, f_1, f_2) \equiv onp(e, f_1) \wedge onp(e, f_2) \wedge (f_1 \neq f_2)$
- Axiom 1) $on(e, f) \leftarrow \exists k(le(k, e) \wedge fl(f, k))$
  2) $line(e) \vee arc(e) \vee free(e)$
  3) $line(e) \leftrightarrow \exists f_1 \exists f_2(onp(e, f_1, f_2) \wedge plane(f_1) \wedge plane(f_2))$
  4) $arc(e) \leftrightarrow \exists f_1 \exists f_2(onp(e, f_1, f_2) \wedge plane(f_1) \wedge cone(f_2) \wedge perpendicular(f_1, f_2))$
  5) $plane(f) \leftrightarrow \exists p(pf(p, f) \wedge (cuboid(p) \vee (tcone(p) \wedge ((f = tp(p)) \vee (f = bm(p))))))$
  6) $cone(f) \leftrightarrow \exists p(pf(p, f) \wedge (tcone(p) \vee (f = sd(p))))$
  7) $intersect(f, o) \leftrightarrow (\exists f'(of(o, f') \wedge intersect(f, f')) \vee contain(o, f))$
  8) $((of(o, f) \wedge \neg intersect(f, p)) \vee (of(p, f) \wedge \neg intersect(f, o))) \rightarrow of(o \cup p, f)$
  9) $((of(o, f) \wedge intersect(f, p) \wedge \neg contain(p, f)) \vee (of(p, f) \wedge intersect(f, o) \wedge \neg contain(o, f))) \rightarrow ((of(o \cup p, f_1) \wedge (on(f_1) = f)) \leftrightarrow \exists e \exists k(fl(f_1, k) \wedge le(k, e) \wedge eligible(e)))$
  An edge e is 'eligible' if and only if
  9.1) it is an existing edge of f possibly clipped by one or two intersecting faces, or
  9.2) it is a newly created edge possibly clipped by one or two edges of f.
  (We do not further elaborate on this predicate as it would lead us into unduly detailed geometric analysis. We will give a complete definition of a simpler yet similar predicate later in connection with the clipping of a polygon (see Sect. 8.2.2., Example (2))).

Axioms similar to 8) and 9) above can be postulated for operators $\cap$ and $-$, but we omit them here. Also, functional relationships can be stipulated to determine the coordinate values of vertices, the characterizing points and other parameters for each type of edge.

(1.7) *Creating a Primitive Instance and Constructing a New Object by a Union Operation*

First, Tc is declared as the theory to be applied. The creation of a primitive instance is accomplished by

- specifying the object to be created be of one of the five primitive sorts,
- simultaneously giving a triplet of real values which is to be interpreted as either (l, w, h) or (rt, rb, h) according to the specified sort,
- also giving the pair (*position*, *direction*) to locate the object.

The creation of another object is instructed together with the assertion that it be the union of *object-fin* and the primitive, and that it be named *object-fin* anew.

(1.8) *Updating the BREP Structure to Reflect the Union Operation*
To begin with, theory $\mathcal{T}$ instead of $\mathcal{T}c$ must have been declared, all the axioms of which are presumed to be satisfied by the current *object-fin*. The application agent must do the following, in addition to (1.7) above to create a primitive instance:

- create entities of sorts f, k, e and v,
- assert the functional relationships between the body and these entities as defined in theory $\mathcal{T}bp$, and
- assert either one of two formulas of $\mathcal{T}bp$ depending on which sort this primitive belongs to, *cuboid* or *tcone*.

Next, an inquiry is made for all the pairs of intersecting faces (f, f') of *object-fin* and the primitive. The identifiers and geometric attribute values of loops, edges, vertices and adjacent faces are given to the geometric processor which in turn generates the new and clipped eligible edges, their vertices and other control points. It is fairly straightforward for the application agent to organize these into loops, identify sub-faces enclosed by them and assert the appropriate ground instances of *flout*, *flin*, *lee* and *evn* so that the axioms of $\mathcal{T}$ may all get satisfied. Figure 6.8 depicts the creation and clipping of edges, and the resulting division of faces into subfaces.

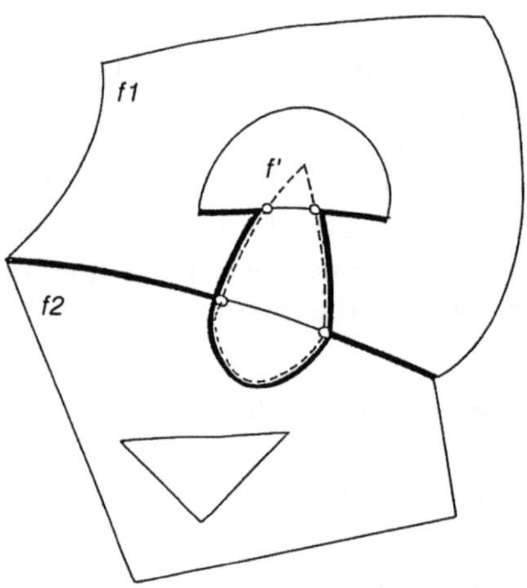

The geometric processor receives: pair of intersecting faces (*f1*, *f'*) and (*f2*, *f'*),
and returns: new and/or clipped edges shown in thick lines.

Fig. 6.8. Division of faces into sub-faces

It should be noted that while the geometric processing forms the most substantial part of the updating, the task is significantly eased by the appropriate management of the topological structure.

(2) *Form Features*
Many a function of mechanical products is embodied in a peculiar type of shape or so-called form feature [Pra] such as hole, groove, step, fillet, etc. Each can be viewed as a generic solid body structured with various boundary relations and measured in symbolic dimensions. We represent it as a small formal theory in much the same way as in the preceding paragraph. Two or more of them may be combined to define a more complex form. To precisely specify the relative positions of component forms, we introduce a ternary predicate *on* between two faces and either IN or OUT. The third argument indicates which side of the second face the first one is in contact with. When interpreted, $on(f_1, f_2, \text{IN/OUT})$ will cause:

- the shape and position of f_1 to be adjusted so that it may lie precisely on f_2, and
- the space bounded by f_1 to be taken from/added to the body to which f_2 belongs.

An example is a blind hole with its side bored by another hole (Fig. 6.9).

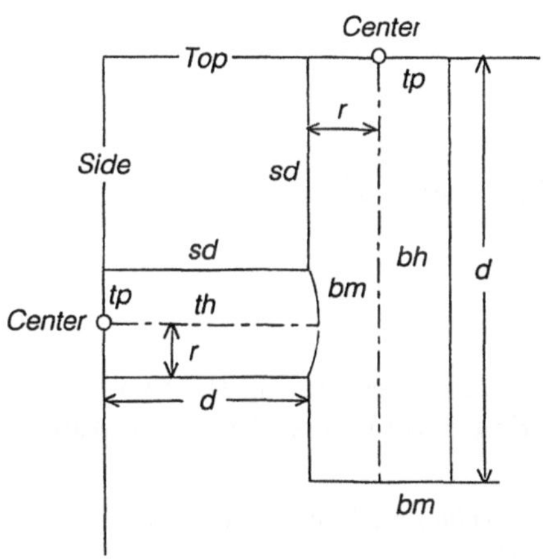

Fig. 6.9. A blind hole with its side bored by a through hole

(2.1) *Sub-Theory for Blind Hole* ($\mathcal{T}bh$)
- Sort *blind-hole*(ι), *face*(ι) (abbreviated to *bh*, *f* respectively)
- Function $tp, sd : bh \rightarrow f$

  $r, d : bh \rightarrow \mathbf{real}$

  $center : bh \rightarrow \mathbf{real}^3$

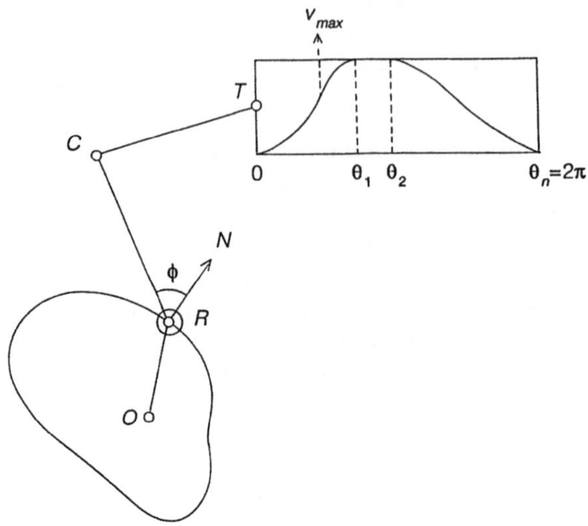

Fig. 6.10. Kinematics of a simple cam mechanism

(2.2) *Sub-Theory for Through Hole* ($\mathcal{T}th$)

- Sort *through-hole*(ι) (abbreviated to *th*), *face*(ι)
- Function $tp, bm, sd : th \rightarrow f$
  $r : th \rightarrow$ **real**
  $center : th \rightarrow$ **real**^{3}

(2.3) *Sub-Theory for Blind Hole Bored by Through Hole* ($\mathcal{T}bth$)

- Predicate $on(f, f, \{\text{IN, OUT}\})$
- Axiom $on(bm(th), sd(bh), \text{OUT})$

(2.4) *Interpretation*
Given an object together with its top and side faces, boring the two holes is done by declaring the theory $\mathcal{T}bth$, creating two objects of sorts *bh* and *th*, and making assertions to effect:

- the assignment of actual values to *center*(*bh*) and *center*(*th*), and
- $on(tp(bh), top, \text{IN}) \wedge on(tp(th), side, \text{IN})$

(3) *A Cam Mechanism*
Let us try to formulate a theory for the class of simple cam mechanisms such as shown in Fig. 1.5. We concentrate on the kinematic aspect which is abstracted and diagramatically represented in Fig. 6.10.

(3.1) *Sub-Theory for Lever* ($\mathcal{T}l$)

- Sort *point*(ι), *moving-point*(*point*) (abbreviated to *p* and *mp* respectively)
- Constant $center : p$

- Function *distance*, *displacement* : $mp \rightarrow$ **real**
  These denote the distance of an mp from the center and the displacement (at an arbitrarily given instant of time) from its original position, and are abbreviated to *dist* and *disp* respectively.
- Axiom 1) $dist(mp_1) \times disp(mp_2) = dist(mp_2) \times disp(mp_1)$

(3.2) *Theory for Cam Mechanism* ($\mathcal{T}$)

- Sort *point*(ι), *moving-point*(*point*), *contour-segment*(ι) (abbreviated to p, mp and cs respectively)
- Sort *curve* (defined as an ADT of continuous and monotonic functions mapping [0, 1] onto [0, 1])
  $c : curve\text{-}type(ct) \times list\text{-}of\text{-}parameters(par) \rightarrow curve$
  $v : curve \times [0,1] \rightarrow [0,1]$
- Constant $T, R : mp$ $C, O : p$ n : **positive integer**
- Function $position\text{-}initial : p \rightarrow \mathbf{real}^2$
  $angle\text{-}of\text{-}rotation : [0,n] \rightarrow [0,2\pi]$ $(i \rightarrow \theta_i)$
  $stroke : [1,n] \rightarrow ct \times param \times \mathbf{real}$
  $displacement : mp \times \{(i,[\theta_{i-1},\theta_i]) : i \in [1,n]\} \rightarrow \mathbf{real}$
  $cs : [1,n] \rightarrow cs$ $(i \rightarrow cs_i)$
  $trajectory : \{(cs_i,[\theta_{i-1},\theta_i]) : i \in [1,n]\} \rightarrow \mathbf{real}$
  $radius : \{R,O\} \rightarrow \mathbf{real}$
  $r\text{-}of\text{-}curvature : \{cs_i : i \in [1,n]\} \rightarrow \mathbf{real}$
  The cam will rotate from θ_{i-1} to θ_i to cause the tool end (T) to complete the i-th stroke. The value of the function *trajectory* is the displacement of point R from its initial position (that is, the position so that angle-of-rotation is equal to zero), and is easily determined by axiom 1) of $\mathcal{T}l$ from the displacement of point T. The latter is presumed to be calculated in a procedure by applying functions c and v to *stroke*. We also presume that another procedure is available to evaluate the minimum radius of curvature for each contour segment.
- Sub-theory $\mathcal{T}l$
  - Redefinition : $center \equiv C$
    $disp(mp) \equiv disp(mp,i,\theta)$ for any i, θ
- Axiom 1) $radius(O) = (v_m \cdot l/\theta_1 \cdot tan(\psi))$
  2) $radius(R) \leq r\text{-}of\text{-}curvature(cs_i)$ for any i
  In the first axiom, v_m is the maximum velocity of the rising stroke, l is the lift (that is, the difference between the maximum and minimum of distance of point R from O) and ψ is the maximum pressure angle of the cam. Note that these can be calculated from other functions defined above and that the pertinent procedures are assumed to be available (Fig. 6.11). The second axiom is postulated to ensure the smooth transmission of movement of the cam through its contact with the roller.

(3.3) *Interpretation*

Interpreting theory $\mathcal{T}$ is tantamount to designing a specific cam to meet the axioms and will proceed as follows:

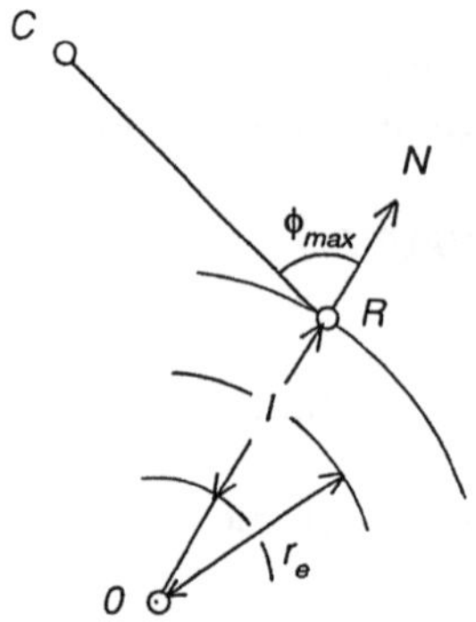

Fig. 6.11. Radius, lift and pressure angle of cam

(3.3.1) select the type and parameters of the curve for each stroke of movement of point T,

(3.3.2) assign the angle of rotation (θ_i) to each stroke,

(3.3.3) set the values of the lift (l) and the maximum pressure angle (ψ) of the cam, and obtain its radius by axiom 1),

(3.3.4) set the values of radius (R) and determine the position of C,

(3.3.5) give the direction of CT and place point T at the distance determined from *disp*(T), *disp*(R) and *dist*(R) by axiom 1) of sub-theory Tl,

(3.3.6) calculate the shape of each contour segment (*trajectory* of *cs*) and see if axiom 2) holds or not.

If axiom 2) fails to hold, one of the following remedial actions must be taken:

- decrease the radius of the roller R,
- increase the radius of the cam O, or
- expand the interval $[\theta_{i-1}, \theta_i]$ of the stroke corresponding to the contour segment in question,

and the appropriate steps must be repeated to reestablish axiom 1).

References

[Bar] Barwise, J.: An Introduction to First Order Logic. In: Barwise, J. (ed.): Handbook of Mathematical Logic. North-Holland, 1977

[Man] Manna, Z.: Mathematical Theory of Computation. McGraw-Hill, 1974

[Pra] Pratt, M.J., Wilson, P.R.: Requirements for Support of Form Features in a Solid Modelling System. Final Report, CAM-I Inc., 1985

[Req] Requicha, A.A.G.: Representations for Rigid Solids: Theory, Methods, and System. ACM Computing Surveys, Vol. 12, No. 4, Dec. 1980

Chapter 7
The Model Description Language (MDL)

In this chapter, we will expound the precise syntax and semantics of the language with which to model design objects. To facilitate understanding it, we will first describe the overall structure and behavior of an application agent, that is, a model constructor.

7.1 General

(1) *Mode of Operation*
An application agent operates in either of the two modes, theory construction or interpretation. Updates are made respectively to the formal theory (design specification) and interpretation (implementation), both of which are, in their normal states, only halfway completed. To keep the interaction between the two categories of update manageable, only one of the modes can be active at any given instant, and when an exit is made from it, the crucial messages concerning the change of state made during the current mode are written on a 'blackboard' to be transmitted to the succeeding mode.

(2) *Static Structure*
For either mode, the part of the application agent that operates in it is programmed in an amalgamation of the host language and data sub-language. The two languages, however, stand on equal terms, complementing each other, rather than in a master-and-servant relation where the sub-language is a mere addition to the host language. Specifically, the design object model is viewed as an ADT with the theory, interpretation and data manipulation commands forming the specification, set of instances and operators respectively, and the host language in turn as a supply of tools and materials for defining functions, predicates and data types of the model. Figure 7.1 grossly depicts the structure and mode of operation of an application agent.

We presume in the following the use of a PASCAL-like procedural language as the host language with the added capability of abstract data typing (for classes of entities pertinent to CAD such as geometric elements, graphic objects, etc.).

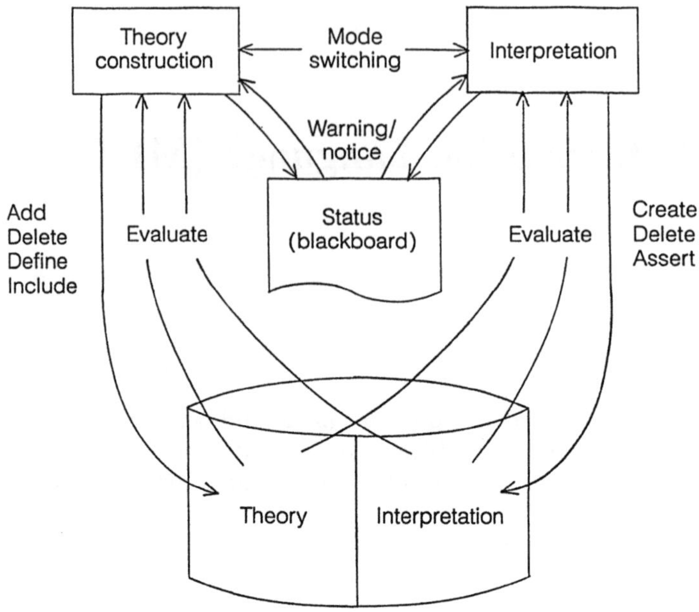

Fig. 7.1. **The structure and mode of operation of an application agent**

(3) *Alphabet*
The alphabet of MDL contains the following characters in addition to that of the host language:

logical symbols	:	$\wedge \quad \vee \quad \neg \quad - \quad \rightarrow \quad \forall \quad \exists \quad \equiv$
set-theoretic symbols	:	$\{ \quad : \quad \} \quad \cap \quad \cup \quad - \quad \times \quad \in \quad \subset$
prefixes of individuals	:	$\varepsilon \quad ?$

(4) *Meta-linguistic Notation*
The syntax is defined in a form which can be converted to, yet is more readable than the rigorous BNF. Reserved words (terminal symbols) appear in boldface. Non-terminal symbols are put in between ⟨ and ⟩ which are also used to enclose annotations in actual programs.

$[\alpha \mid \beta]$ and $[\frac{\alpha}{\beta}]$ indicate that either α or β, but not both of them, may occur at this place within a statement of the type being defined. A phrase of the form ⟨list of α⟩ appears frequently, which stands for an arbitrarily long series of symbols of type α separated by commas $(\alpha_1, \alpha_2, \ldots, \alpha_n)$.

7.2 Theory Construction

7.2.1 Declarations and Definitions

(1) ⟨sort declaration⟩ ::= **sort** ⟨list of id⟩ ::= **sort** ⟨id⟩, ... , ⟨id⟩

(2) ⟨sort id⟩ ::= ⟨type id⟩
Each sort is a sub-type of sort ι and its identifier may appear at any place where type identifiers can appear. Other sorts should be declared as host language data types.

(3) ⟨standard sort id⟩ ::= **identifier**
'identifier' is the identifier of sort ι.

(4) ⟨constant declaration⟩ ::= **const** ⟨list of id⟩ : ⟨type id⟩; … ; ⟨list of id⟩ : ⟨type id⟩

(5) ⟨variable declaration⟩ ::= **var** ⟨list of id⟩ : ⟨type id⟩; … ; ⟨list of id⟩ : ⟨type id⟩

(6) ⟨function definition⟩ ::= **function** ⟨id⟩

$$\left[\frac{(\langle\text{list of type id}\rangle) : \langle\text{type id}\rangle}{(\langle\text{list of var id : type id}\rangle) : \langle\text{type id}\rangle; \langle\text{block}\rangle}\right]$$

The first alternative which contains only the types of arguments and value specifies that the function is to be defined enumeratively.

(7) ⟨standard function id⟩ ::= **name** | **qualifier** | **qualifier.name** | **id**

(8) ⟨predicate definition⟩ ::= **predicate** ⟨id⟩ $\left[\frac{(\langle\text{list of type id}\rangle)}{(\langle\text{list of var id : type id}\rangle); \langle\text{block}\rangle}\right]$

(9) ⟨predicate id⟩ ::= ⟨boolean function id⟩

(10) ⟨formula definition⟩ ::= **formula** ⟨id⟩ (⟨list of free var id⟩) ≡ ⟨formula⟩

(11) ⟨formula⟩ ::= ⟨atomic formula⟩ | (⟨formula⟩) | ¬ ⟨formula⟩
| [⟨formula⟩ |] ⟨logical operator⟩ ⟨formula⟩
| ⟨quantifier⟩ ⟨var id⟩ … ⟨quantifier⟩ ⟨var id⟩ ⟨formula⟩

(12) ⟨atomic formula⟩ ::= ⟨predicate id⟩ (⟨list of term⟩) | ⟨term⟩ = ⟨term⟩
| ⟨function id⟩ : ⟨list of term⟩ → ⟨term⟩

(13) ⟨term⟩ ::= ⟨const id⟩ | [? |] ⟨var id⟩
| ⟨function id⟩ (⟨list of term⟩)
| ⟨set function id⟩ ⟨set-theoretic expression⟩
| ⟨ε-term⟩

A var id to which '?' is prefixed denotes a free variable. Formal arguments of functions and predicates are considered to be universally quantified. Other free occurrences of var id are not formal-theoretic symbols but represent the entities of the universe of interpretation that have been most recently assigned to these variables.

(14) ⟨standard set function id⟩ ::= **card** | **max** | **min** | **sum** | **ave** | **product**

(15) ⟨set-theoretic expression⟩ ::= **set**{(⟨list of var id⟩) : ⟨formula⟩}
| (⟨set-th. expression⟩) | [⟨set-th. expression⟩ |] ⟨set-th. op.⟩ ⟨set-th. expression⟩

(16) ⟨set-theoretic operator⟩ ::= $\cap \mid \cup \mid - \mid \times$

(17) ⟨ε-term⟩ ::= ε ⟨var id⟩ ⟨formula⟩

An ε-term denotes the uniquely existing entity that will make the formula true when substituted for the sole free variable of it.

(18) ⟨logical operator⟩ ::= $\wedge \mid \vee \mid - \mid \rightarrow$

(19) ⟨quantifier⟩ ::= $\forall \mid \exists \mid \exists$ ⟨integer⟩

(20) ⟨set-theoretic formula⟩ ::= ⟨term⟩ $\in$ ⟨set-th. expression⟩
| ⟨set-th. expression⟩ $\subset$ ⟨set-th. expression⟩

(21) ⟨extended boolean expression⟩ ::= ⟨formula⟩ | ⟨set-th. formula⟩

(22) ⟨postulation of axiom⟩ ::= **axiom** ⟨id⟩ : ⟨closed formula⟩

(23) ⟨declaration of theory⟩ ::= **theory** ⟨id⟩

(24) ⟨inclusion of sub-theory⟩ ::= **subtheory** ⟨id⟩, ⟨list of [non-logical symbol | const] **for** non-logical symbol⟩

'*B* for *A*' where *B* is a non-logical symbol or constant will cause *A* to be renamed or given an interpretation.

(25) ⟨theory addendum declaration⟩ ::= **theory-addendum** ⟨id⟩, ... , ⟨id⟩

(26) ⟨theory addendum⟩ ::= ⟨id⟩ : ⟨declaration of theory⟩;
[− |] ⟨theory constituent⟩; ... ;
[− |] ⟨theory constituent⟩; **end-theory**

Each addendum contains a collection of constituents to be added to or deleted from the current theory. A '−' sign indicates a deletion. The actual modification is done by an 'update theory' statement (see next section). The 'id' serves as the label of the whole addendum as well as the first entry (that is, the declaration of theory). Succeeding entries can be referenced by 'id+1', 'id+2', and so on.

(27) ⟨theory constituent⟩ ::= ⟨sort declaration⟩ | ⟨constant declaration⟩
| ⟨var declaration⟩ | ⟨function definition⟩
| ⟨predicate definition⟩ | ⟨formula definition⟩
| ⟨inclusion of sub-theory⟩ | ⟨postulation of axiom⟩

(28) ⟨symbol addendum declaration⟩ ::= **symbol-addendum** ⟨id⟩, ... , ⟨id⟩

(29) ⟨symbol addendum⟩ ::= ⟨id⟩: ⟨symbol category⟩ ⟨symbol id⟩;
[− |] **theory** ⟨theory id⟩; ... ;
[− |] **theory** ⟨theory id⟩; **end-symbol**

A symbol addendum lists the identifiers of theories from which the specified symbol has been deleted or in which it remains to be contained.

(30) ⟨symbol category⟩ ::= **sort** | **const** | **function** | **predicate**

(31) ⟨sub-theory addendum declaration⟩ ::= **subtheory-addendum** ⟨id⟩, ... , ⟨id⟩

(32) ⟨sub-theory addendum⟩ ::= ⟨id⟩: **subtheory** ⟨theory id⟩; **theory** ⟨theory id⟩; ... ; **theory** ⟨theory id⟩; **end-subtheory**

A sub-theory addendum lists the identifiers of theories that include the specified sub-theory.

7.2.2 Procedural Statements

(1) ⟨initialize theory addendum⟩ ::= **initialize** ⟨theory addendum id⟩

(2) ⟨finalize theory addendum⟩ ::= **finalize** ⟨theory addendum id⟩

For each symbol that is explicitly declared to be deleted, all the symbols, formulas and axioms whose definitions depend on it directly or indirectly will be automatically added to the list.

(3) ⟨write to theory addendum⟩ ::= **put** ⟨[var | const] id of string type⟩ **to** ⟨theory addendum id⟩

This statement causes a theory constituent expressed as a string of characters to be converted into a compiled form and included in the specified addendum. Writing is done to the current cursor position which will then be advanced to the next position.

(4) ⟨read from theory addendum⟩ ::= **get** ⟨var id of string type⟩ **from** ⟨theory addendum id⟩

(5) ⟨delete entry of theory addendum⟩ ::= **delete** ⟨theory addendum id⟩

The entry at the current cursor position is deleted. Succeeding constituents will be moved up in turn to fill the emptied position. The cursor indication will remain the same.

(6) ⟨set cursor position⟩ ::= **set** ⟨theory addendum id⟩ **at** [**start** | + ⟨integer⟩ | **end**] **position**

This statement sets the cursor to the specified position. The contents of the addendum will not be affected.

(7) ⟨update theory⟩ ::= **update-theory-by** ⟨theory addendum id⟩

The body of the addendum is applied to the theory specified in the first entry. The addendum must have been finalized prior to executing this instruction.

(8) ⟨retrieve theory⟩ ::= **retrieve-theory-at** ⟨theory addendum id⟩

The constituents of the theory identified by the first entry are retrieved and placed in succeeding entries of the specified addendum.

(9) ⟨find theories containing a given symbol⟩ ::= **get-theory-id-at** ⟨symbol addendum id⟩

The identifiers of theories that contain the given symbol are returned in the second and succeeding entries of the specified addendum, each separated by a semicolon (;) from its neighbors. In the first entry, the category and id of the symbol in question must be declared.

(10) ⟨find theories containing a given sub-theory⟩ ::= **get-theory-id-at** ⟨subtheory addendum id⟩

(11) ⟨evaluate formula⟩ ::= **evaluate** ⟨list of [⟨open formula⟩ |
formula ⟨label of open formula definition⟩ |
⟨var id of boolean type⟩ = [⟨closed formula⟩ |
formula ⟨label of closed formula definition⟩]]
separated by ;⟩

Suppose that the formula to be evaluated is an open formula $F(?x_1,\ldots,?x_n)$ where $?x_i$ denote free variables. All the tuples $(s_1^{(j)},\ldots,s_n^{(j)})$ $(j=1,2,\ldots)$ that satisfy F are calculated and assigned to $(x_1^{(j)},\ldots,x_n^{(j)})$. A variable length array of record (say r) of which each x_i is a field (say $r.f_i$) must be declared. The set of $(s_1^{(j)},\ldots,s_n^{(j)})$ may be considered as defining the extension of a temporary predicate for which the record name serves as the identifier. Moreover, as the extension of F will generally be the union $A\cup\bar{B}$ where both A and B are the results of positive-finite evaluations (see Def. 6.3 and Lemma 6.1), its complement (that is, $\bar{A}\cap B$) will actually be delivered if B is not empty, An implementation dependent control message will indicate if this is the case or not. If instead a closed formula is to be evaluated, either **true** or **false** is returned in the specified boolean variable. Formulas separated by ; are evaluated one after another in the specified order.

(12) ⟨send next theory addendum⟩ ::= **send** ⟨theory addendum id⟩

The contents of the identified addendum is placed on the 'blackboard' which is automatically prepared for receiving the first message when the current mode is entered. Multiple addenda may be written within a single duration of mode. They must have been finalized before being written.

(13) ⟨receive next theory addendum⟩ ::= **receive** ⟨theory addendum id⟩

Next of the messages sent by the immediately preceding duration of mode is received in the specified addendum. Following the reception of the last message, another execution of this statement will cause either 'final' or 'to-be continued' to appear in the first entry of the addendum which marks the end of message.

(14) ⟨entry to theory ::= **theory-construction-[initial | continued]**
construction mode⟩
This must be the first statement to be executed after the mode is switched from 'interpretation' to 'theory construction'. A 'hold' state will be automatically put in effect. '-initial' indicates the very first one after the system is logged in.

(15) ⟨exit from theory ::= **exit-[to-be-continued | final]**
construction mode⟩
The symbol addendum is automatically created on the 'blackboard' for each of the symbols that have been deleted. These addenda follow the theory addenda that have been sent by the application agent. Either 'to-be-continued' or 'final' is written at the end of message indicating whether a reentry to the current mode will be accepted or not. See Fig. 7.2 for the coarse organization of messages between two modes. A 'hold' state, if in effect, will be 'freed'.

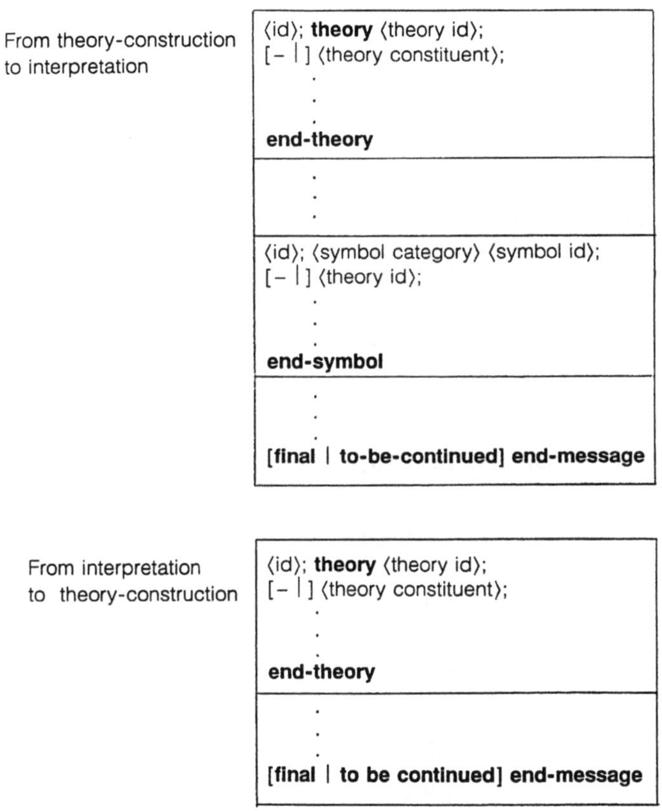

Fig. 7.2. Organization of messages transmitted between two modes

(16) ⟨mark the point of backtracking⟩ ::= [⟨id⟩ : |] **hold**

This marks the point to which a backtracking may possibly occur. The effect of the last issued 'hold' statement will in turn be 'freed'. Nesting states of holding will hence be impossible.

(17) ⟨release the state of holding⟩ ::= **free**

The state of holding effected by the last issued 'hold' statement is released, and the outstanding updates of theories and theory addenda, and blackboard messages are made permanent.

(18) ⟨backtrack to the marked point⟩ ::= **rollback** [**to** ⟨id⟩ |]

A simple 'rollback' which is normal nullifies all the updates of theories and theory addenda, and the sending of messages that have been made since the last execution of a 'hold' statement. The specification of an arbitrary point of backtracking will override the permanent updates to which the intervening 'free' statements have been committed. The use of this alternative is discouraged, however, hence should be considered exceptional in view of the cost incurred.

7.3 Interpretation

7.3.1 Declarations and Definitions

(1) ⟨sort declaration⟩ ::= same as 7.2.1 (1)

(2) ⟨sort id⟩ ::= same as 7.2.1 (2)

(3) ⟨standard sort id⟩ ::= same as 7.2.1 (3)

(4) ⟨constant declaration⟩ ::= same as 7.2.1 (4)

(5) ⟨variable declaration⟩ ::= same as 7.2.1 (5)

(6) ⟨function definition⟩ ::= same as 7.2.1 (6)

(7) ⟨standard function id⟩ ::= same as 7.2.1 (7)

(8) ⟨predicate definition⟩ ::= same as 7.2.1 (8)

(9) ⟨predicate id⟩ ::= same as 7.2.1 (9)

(10) ⟨formula definition⟩ ::= same as 7.2.1 (10)

(11) ⟨formula⟩ ::= same as 7.2.1 (11)

(12) ⟨atomic formula⟩ ::= same as 7.2.1 (12)

(13) ⟨term⟩ ::= same as 7.2.1 (13)

(14) ⟨standard set function id⟩ ::= same as 7.2.1 (14)

(15) ⟨set-theoretic expression⟩ ::= same as 7.2.1 (15)

(16) ⟨set-theoretic operator⟩ ::= same as 7.2.1 (16)

(17) ⟨ε-term⟩ ::= same as 7.2.1 (17)

(18) ⟨logical operator⟩ ::= same as 7.2.1 (18)

(19) ⟨quantifier⟩ ::= same as 7.2.1 (19)

(20) ⟨set-theoretic formula⟩ ::= same as 7.2.1 (20)

(21) ⟨extended boolean expression⟩ ::= same as 7.2.1 (21)

(22) ⟨postulation of axiom⟩ ::= same as 7.2.1 (22)

(23) ⟨declaration of theory⟩ ::= same as 7.2.1 (23)

(24) ⟨inclusion of sub-theory⟩ ::= same as 7.2.1 (24)

(25) ⟨view definition⟩ ::= **view** ⟨id⟩ : ⟨list of theory id⟩ [**standard-interpretation** |] [**with-qualifier** ⟨[const id | const] of string type⟩ |]

The optional 'standard-interpretation' causes separate objects to be automatically created, each corresponding to a constant symbol of sort ι and assigned to it as its value. The other option prefixes the specified qualifier to all the constants and external names. In turn, constants with or without qualifier serve to identify corresponding objects.

(26) ⟨theory addendum declaration⟩ ::= same as 7.2.1 (25)

(27) ⟨theory addendum⟩ ::= same as 7.2.1 (26)

(28) ⟨theory constituent⟩ ::= same as 7.2.1 (27)

(29) ⟨symbol addendum declaration⟩ ::= same as 7.2.1 (28)

(30) ⟨symbol addendum⟩ ::= same as 7.2.1 (29)

(31) ⟨symbol category⟩ ::= same as 7.2.1 (30)

(32) ⟨sub-theory addendum declaration⟩ ::= same as 7.2.1 (31)

(33) ⟨sub-theory addendum⟩ ::= same as 7.2.1 (32)

(34) ⟨transaction declaration⟩ ::= **transaction** ⟨list of id⟩

(35) ⟨transaction⟩ ::= ⟨id⟩ : ⟨data update operation⟩; ... ;
⟨data update operation⟩; **end-transaction**

(36) ⟨data update operation⟩ ::= ⟨create objects⟩ | ⟨delete objects⟩
| ⟨assert formula⟩ | ⟨copy and imbed⟩

7.3.2 Procedural Statements

(1) ⟨initialize theory addendum⟩ ::= same as 7.2.2 (1)

(2) ⟨finalize theory addendum⟩ ::= same as 7.2.2 (2)

(3) ⟨write to theory addendum⟩ ::= same as 7.2.2 (3)

(4) ⟨read from theory addendum⟩ ::= same as 7.2.2 (4)

(5) ⟨delete entry of theory addendum⟩ ::= same as 7.2.2 (5)

(6) ⟨set cursor position⟩ ::= same as 7.2.2 (6)

(7) ⟨retrieve theory⟩ ::= same as 7.2.2 (8)

(8) ⟨find theories containing a given symbol⟩ ::= same as 7.2.2 (9)

(9) ⟨find theories containing a given sub-theory⟩ ::= same as 7.2.2 (10)

(10) ⟨evaluate formula⟩ ::= same as 7.2.2 (11)

(11) ⟨assert formula⟩ ::= **assert** ⟨list of [⟨closed formula⟩ |
formula ⟨label of closed formula definition⟩]
separated by ;⟩

The structure of the interpretation is modified so that the specified formulas may become true. The restriction described in Sect. 6.2 (2.2.2) must be observed. Formulas separated by ; are asserted one after another in the specified order.

(12) ⟨create objects⟩ ::= **create** ⟨list of var id⟩ [⟨asserting formula⟩ |]
Suppose the actual statement is

create $x_1, \ldots, x_n$ **asserting** $formula(x_1, \ldots, x_n)$

then the following series of events will take place:
1) objects of sort ι (or its specific sub-sorts) $s_1, \ldots, s_n$ are created,
2) these are assigned to $x_1, \ldots, x_n$ respectively, and
3) assert $formula(x_1, \ldots, x_n)$ is performed.

(13) ⟨delete objects⟩ ::= **delete** ⟨list of [? |] id⟩ [⟨evaluating formula⟩ |]
For the actual statement

delete $?x, y_1, \ldots, y_n$ **evaluating** $formula(?x, y_1, \ldots, y_n)$

where $?x$ is the sole free variable that occurs in *formula*,
1) **evaluate** $formula(?x, y_1, \ldots, y_n)$ is performed assigning $s_1, \ldots, s_m$ to x, and
2) objects $s_1, \ldots, s_m$, $t_1, \ldots, t_n$ are deleted from the domain I where $t_1, \ldots, t_n$ are currently assigned to $y_1, \ldots, y_n$ respectively.

(14) ⟨copy and imbed⟩ ::= **copy** [⟨set id⟩ | ⟨set-theoretic expression⟩]
to ⟨set id⟩ **under-theory** ⟨theory id⟩
with-qualifier ⟨[const id | const] of string type⟩
[**imbedding-function** ⟨function id⟩ |]
interface ⟨formula id⟩ **to** ⟨formula id⟩

The specified set of objects of sort ι which is an interpretation of the specified theory is copied. All the constants and named objects will retain their names with the new qualifier. If the imbedding function id is specified, each pair of original and copied objects will be related by that function. Whatever relations objects may have in the specified interface will also be copied (see Sect. 6.2 (3.2)).

(15) ⟨definition of interface⟩ ::= **formula** ⟨id⟩ $\equiv$
interface(⟨set id⟩,
[**all** | ⟨list of constants⟩] **in** ⟨theory id⟩,
[**all** | ⟨list of functions and predicates⟩]
in ⟨view id⟩

The interface is defined as the conjunction of those ground instances of specified functions and predicates that relate the specified constants and elements outside the specified set.

(16) ⟨send next theory addendum⟩ ::= same as 7.2.2 (12)

Normally, the message created in the interpretation mode will contain those constituents that the application agent considers might better be eliminated from the theory. Specifically, these will include the symbols (sorts, constants, functions and predicates) of which there does not remain any instance and the axioms which are found difficult to establish. A '–' sign is prefixed to each of them.

(17) ⟨receive next [theory | symbol] addendum⟩ ::= **receive** ⟨[theory | symbol] addendum id⟩

This functions in much the same way as statement (13) of 7.2.2. Only a notice 'symbol-addenda-to-follow' will be given for the first execution following the reception of the last theory addendum if there is indeed any symbol addendum to follow. The type of message delivery area should be changed to 'symbol-addendum'.

(18) ⟨entry to interpretation mode⟩ ::= **interpretation-[initial | continued]**

The function of this is similar to that of the statement (14) of 7.2.2.

(19) ⟨exit from interpretation mode⟩ ::= same as 7.2.2 (15)

This serves the same purpose as statement (15) of 7.2.2 except the automatic creation of symbol addenda.

(20) ⟨mark the point of backtracking⟩ ::= same as 7.2.2 (16)

(21) ⟨release the state of holding⟩ ::= same as 7.2.2 (17)

The function is the same as that of statement (17) of 7.2.2 except that 'updates of interpretation' is substituted for 'updates of theories'.

(22) ⟨backtrack to the marked point⟩ ::= same as 7.2.2 (18)

The function is the same as that of statement (18) of 7.2.2 except that 'updates of interpretation' is substituted for 'updates of theories'.

(23) ⟨initialize transaction⟩ ::= **initialize** ⟨transaction id⟩

(24) ⟨finalize transaction⟩ ::= **finalize** ⟨transaction id⟩

(25) ⟨write to transaction⟩ ::= **put** ⟨[var | const] id of string type⟩ **to** ⟨transaction id⟩

(26) ⟨read from transaction⟩ ::= **get** ⟨var id of string type⟩ **from** ⟨transaction id⟩

(27) ⟨execute transaction⟩ ::= **execute-transaction** ⟨transaction id⟩

The above statements (23) through (27) have functions similar to those of statements (1) through (4) and (7) of Sect. 7.2.2.

7.4 Illustrative Examples

(1) *CSG and BREP Structures of a Polyhedron*
We skip specifying the theory part as it is fairly staightforward to have items (1.1) through (1.7) of Sect. 6.3 restated in a syntactically precise form.

(1.1) *The Overall Structure of the Program Working on the Interpretation*

view $\mathcal{T}$ ⟨and other theories as required⟩
interpretation-initial
⟨getting prepared for receiving designers/operators' instructions and starting or resuming the work of designing⟩
interpretation-continued
⟨main program : deciding on particular actions to be taken based on the dialogue with designers/operators and directing the flow of control⟩
...
⟨creating a primitive instance and constructing a new object by each of the permissible point set operations⟩
⟨updating the BREP structure of object to reflect the result of each set operation⟩
...
exit-to-be-continued
⟨terminating the current session of dialogue with designers/operators and adjourning or concluding the work of designing⟩
exit-final.

(1.2) *Creating a Primitive Instance and Constructing a New Object by a Set Operation*

⟨We again limit ourselves to programming the case of union (∪) operator. Furthermore, we will assume the availability of some of the procedures to carry out geometric calculations and will refrain from elaborating on them.⟩
var *sort* : **character-string**; x : **array**[1..3] **of real**; *position*, *direction* : $\mathbf{real}^3$;
  p : *primitive*; f : **array**[1..6] **of** *face*; k : **array**[1..6] **of** *loop*;
  e : **array**[1..12] **of** *edge*;
  v : **array**[1..8] **of** *vertex*; *tp*, *bm*, *sd* : *face*; *tc*, *bc* : *edge*; *object* : *object*;
function *sort*(*primitive*) : **character-string**;
formula *axiom-of-cuboid* ≡ *cuboid*(p) ∧ *str-cuboid*(p, $(i = 1..6)f[i], k[i]$,
  $(m = 1..12)e[m], (n = 1..8)v[n], (j = 1..3)x[j]$, *position*, *direction*);
  axiom-of-tcone ≡ *tcone*(p) ∧ *str-tcone*(p, *tp*, *bm*, *sd*, *tc*, *bc*,
  $(j = 1..3)x[j]$, *position*, *direction*);
  axiom-of-cube ≡ *axiom-of-cuboid* ∧ *cube*(p);
  axiom-of-cone ≡ *axiom-of-tcone* ∧ *cone*(p);
  axiom-of-cylinder ≡ *axiom-of-tcone* ∧ *cylinder*(p);

```
begin
  ⟨Read in the actual parameter values and assign them to program
  variables sort, x[1] through x[n] (n = 1, 2 or 3),
  position and direction.⟩;
  create case sort of
    'cuboid' : p, (i = 1..6)f[i], k[i], (m = 1..12)e[m], (n = 1..8)v[n]
      asserting axiom-of-cuboid;
    'tcone' : p, tp, bm, sd, tc, bc asserting axiom-of-tcone;
    'cube' : p, (i = 1..6)f[i], k[i], (m = 1..12)e[m], (n = 1..8)v[n]
      asserting axiom-of-cube;
    'cone' : p, tp, bm, sd, tc, bc asserting axiom-of-cone;
    'cylinder' : p, tp, bm, sd, tc, bc asserting axiom-of-cylinder
  end asserting sort : p →sort;
  create object asserting (∪ : object-fin, p →object)
  ∧ (object-semifin :→ object-fin); object-fin : →object
end.
```

(1.3) *Updating the BREP Structure of Object*

```
var fpair = array of record f : face ⟨of object-semifin⟩;
                g : face ⟨of primitive p⟩
              end;
  pf = array of record f : face ⟨of either one of two faces in fpair⟩;
          f0 : face ⟨of the primitive p on which f lies⟩;
          p : primitive ⟨of which f0 is a part⟩;
          position : real³ ⟨position of p⟩;
          direction : real³ ⟨direction of p⟩;
          sort : character-string ⟨indicating the sort of p⟩;
          param : real³ ⟨parameter of p⟩
        end;
  e = array of record f : face ⟨the same as pf.f⟩;
        k : loop ⟨lying on f⟩;
        e1 : edge ⟨lying on k⟩;
        v0 : vertex ⟨ = ev0(e1)⟩;
        v1 : vertex ⟨ = ev1(e1)⟩;
        e2 : edge ⟨following e1 in k⟩;
        f1 : face ⟨adjacent to f that shares e1⟩;
        esort : character-string ⟨indicating the sort of e1⟩;
        eparam : case esort of
          line : real³ × real³ ⟨coordinates of terminal points⟩;
          arc : real³ × real³ × real × real × real
          ⟨attribute of underlying circle and terminal angles⟩;
          free : nil
          end
      end;
```

$s =$ **array of record** *e* : *edge* ⟨of sort *free*⟩;
i : **integer** ⟨relative position of segment *s* within *e*⟩;
n : **integer** ⟨total number of segments of *e*⟩;
s : *segment* ⟨in question⟩;
sparam : $\textbf{real}^t$ ⟨parameter of *s*⟩
end;
$k =$ **array of record** *k* : *loop* ⟨same as e.*k*⟩;
first : **integer** ⟨pointing to the entry of *e* that contains the first edge of loop *k*⟩;
last : **integer** ⟨pointing to the entry of *e* that contains the last edge of loop *k*⟩
end;
kpair = **array of record** *out* : *loop* ⟨one of the newly created outer loops lying on *e.f*⟩;
in : *loop* ⟨one of the newly created inner loops contained in *out*⟩
end;
first : *edge*;

begin
evaluate $intersect(?fpair.f, ?fpair.g) \wedge of(object\text{-}semifin, ?fpair.f) \wedge of(p, ?fpair.g); \exists x(fpair(x) \wedge ((?pf.f = x.f) \vee (?pf.f = x.g)) \wedge (?pf.f0 = onp(?pf.f)) \wedge pf(?pf.p, pf.f0) \wedge (?pf.position = position(?pf.p)) \wedge (?pf.direction = direction(?pf.p)) \wedge (?pf.sort = sort(?pf.p)) \wedge ((?pf.param = (l, w, h)(?pf.p)) \vee (?pf.param = (rb, rt, h)(?pf.p)))); \exists x(pf(x) \wedge (?e.f = x.f) \wedge fl(?e \cdot f, ?e \cdot k) \wedge lee(?e.k, ?e.e1, ?e.e2) \wedge (?e.v0 = ev0(?e.e1)) \wedge (?e.v1 = ev1(?e.e1)) \wedge \exists k((k \neq ?e.k) \wedge fl(?e.f1, k) \wedge le(k, ?e.e1)) \wedge (?e.esort = esort(?e.e1)) \wedge (?e.eparam = eparam(?e.e1))); \exists x(e(x) \wedge (?s.e = x.e1) \wedge free(?s.e) \wedge fs(?s.e, ?s.s, ?s.i, ?s.n) \wedge (?s.sparam = eparam(?s.s)));$

⟨Based on the information provided by the above statement, the geometric processor is to determine the set of all eligible edges, and for each of them, to have the topological structures and geometric attribute values stored in the database and the working memory of the following format;

eligible = **array of record** *f* : *face* ⟨of either *p* or *object-semifin* on which *e1* lies⟩;
e1 : *edge* ⟨in question⟩;
v0 : *vertex* ⟨ = $ev0(e1)$⟩;
v1 : *vertex* ⟨ = $ev1(e1)$⟩;
e2 : *edge* ⟨following *e1* in such a direction that *f* lies at the left-hand side when viewed from outside of the body⟩;
f1 : *face* ⟨adjacent to *f* that shares *e1*⟩;
esort : **character-string** ⟨indicating the sort of *e1*⟩;

```
                    eparam : case esort of
                      line : real³ × real³ ⟨coordinates of terminal points⟩;
                      arc : real³ × real³ × real × real× real ⟨attribute of
                      underlying circle and terminal angles⟩;
                      free : nil
                      end
                  end;
    segment =  array of record e : edge ⟨of sort free⟩ ;
                    i : integer ⟨relative position of segment s within e⟩;
                    n : integer ⟨total number of segments of e⟩;
                    s : segment ⟨in question⟩;
                    sparam : realᵗ ⟨parameter of s⟩
                 end.
```

New entities are created to represent clipped edges and their vertices as well as new edges. The identifiers, attribute values and relations with other entities of those entities that remain intact after the set operation will appear as they currently are in the above list. The main body of application agent will resume the control to complete structuring the new *object-fin.*⟩;

```
foreach ∃x(e(x) ∧ (?f = x.f)) do
begin
  i := 1; create k[1].k; k[1].first := 1; first := eligible[1].e1;
  for j := 1 to number-of-eligible-edges do
  begin
    assert lee(k[i].k, eligible[j].e1, eligible[j].e2);
    if (eligible[j].e2 = first) ∧ (j ≠ number-of-eligible-edges) then
    begin
      k[i].last := j; i := i + 1; create k[i].k; k[i].first = j + 1;
      first := eligible[j + 1].e1
    end
  end;
  evaluate ∃x∃y(k(x) ∧ k(y) ∧ (?kpair.out = x.k) ∧ (?kpair.in = y.k)
  ∧ contain(x.k, y.k));
  foreach ∃x(kpair(x)∧?out = x.out) do
    create fnew asserting of(object-fin, fnew) ∧ (on : fnew → f) ∧
    flout(fnew, out) ∧ ∀x(kpair(out, x) → flin(fnew, x))
end.
```

Discussion

As can be seen from the above program, the net effect of an 'evaluate' operation is a transcription of the logical structure of a specified portion of database into an array of records comprising primary data types. This is more than a homomorphism between two structures in that it enables the exploitation of the well-order relation and arithmetic functions defined over the index domain. Suppose, for example, we want to sequentially process a set of vertices lying on an edge in order of increasing or decreasing parameter values. For efficiency's sake, we can enjoy the applicability of sophisticated sorting algorithms in a preparatory step, which will not be the case when manipulating a purely logical structure. On the other hand, we introduced the 'foreach' statement in order to facilitate random accesses to the transcribed data. This is not really essential, however, as it is restricted to dealing with an array structure, hence we can repeat the sorting and sequential processing with the conventional 'for' statement to the same effect.

In a more significant extension of the host language which provides a 'foreach' control mechanism for a direct traversal over the database, the need for 'evaluating' open formulas will be reduced except for such a purpose as further transmitting the data to output devices. We present in the following a version of the above program along these lines. It should be noted that database manipulation commands are more fully integrated into the host language and there is no separation of main program and geometric processor, the former being responsible for accessing the database and the latter acting upon the array of records. Further notes will be given later on those statements marked with asterisks.

type *vertex-on-edge* = **record** v : *vertex*; r : **real end**;

var i, s : **integer**; r : **real**; v, w : *vertex*; $e, e', e0, e1, e2$: *edge*;
  k : *loop*; f, g, h, f', fo, fp : *face*;
  vx : **set of** *vertex*; ef : **set of** *edge*; *fset*,*fgh* : **set of** *face*;
  vset : **set of** *vertex-on-edge*; *vseq* : **array of** *vertex-on-edge*;

predicate *eligible*(*edge*); *clipped*(*edge*); *temporary*(*vertex*);
  close(x, y : **real**^{3});
  ⟨body to determine if x and y coincide within a predetermined tolerance⟩;
  fore(x : *face*, y : *edge*);
  ⟨body to determine if face x viewed from outside of the body
  lies at the left-hand side of edge y⟩;
  back(x : *face*, y : *edge*) ≡ ¬*fore*(x : *face*, y : *edge*);

formula *fpair*(x, y : *face*) ≡ *of*(*object-semifin*, x) ∧ *of*(p, y) ∧ *intersect*(x, y);
  sort(x : *edge*, y, z : *face*)
  ≡ *free*(x); ((*plane*(y) ∧ *plane*(z)) → *line*(x)) ∧
  ((*plane*(y) ∧ *cone*(z) ∧ *perpendicular*(y, z)) → *arc*(x))
  ⟨an instance of the serialized conjunction of axioms 2), 3) and 4)
  of Sect. 6.3 (1.6)⟩;
  same-sort(x, y : *edge*) ≡ (*line*(x) → *line*(y))
  ∧ (*arc*(x) → *arc*(y)) ∧ (*free*(x) → *free*(y));

function *counter-body*(*edge*) : {p, *object-semifin*}; *faces*(*vertex*) : **set of** *face*;
  coord(*vertex*) : **real**^{3}; *no-of-v*(*edge*) : **integer**;
  ev(*edge*, *integer*) : *vertex-on-edge*; *vw*(*vertex*) : *vertex*;
  intersection(x : *edge*, y : *face*) : **real**;
  ⟨body to calculate the value of normalized parameter

```
    (independent variable ranging over the interval [0, 1]) for
    the point of intersection of x with y⟩;
    coordinate-of-intersection(x : edge, y : face) : real³;
    ⟨body to calculate the coordinate value of the point of intersection⟩;
    parameter-of-intersection(x, y : face) : ∪ ⟨data type for each edge sort⟩
    ⟨body to calculate the parameter for the intersection
    (common edge) of x and y⟩;
    parameter-of-clipped(x : edge, y, z : real) : ∪ ⟨data type for each edge
    sort⟩ ⟨body to calculate the parameter for the portion clipped from
    edge x at the points corresponding to y and z⟩
  procedure create-segment1(x : edge, y, z : face);
    ⟨body to create segments of edge x of sort free given two intersecting
    faces⟩;
    create-segment2(x : edge, y, z : real);
    ⟨body to create segments of a portion of edge x given two of points
    of clipping⟩;

  ⟨main program⟩
  begin
    ⟨create intersections (e(dges)) of faces of object-semifin and p⟩
*1) foreach fpair(?fo, ?fp) do
    begin
      create e asserting on(e, fo) ∧ on(e, fp); sort(e, onp(fo), onp(fp));
      eparam : e → param-of-intersection(onp(fo), onp(fp));
      if free(e) then create-segment1(e, onp(fo), onp(fp))
    end;
    ⟨assert those edges that remain intact as eligible and others as
    to-be-clipped, and for the latter, create intersections (v(ertices))
    with faces of the counter-body⟩
    foreach ∃x(fpair(?f, x) ∨ fpair(x, ?f)) do
    begin
      foreach on(?e, f) do
      begin
*2)     assert counter-body : e → object-semifin;
        (of(object-semifin, f) ∧ of(object-semifin, εg(on(e, g) ∧ (f ≠ g))))
        → (counter-body : e → p);
        if ¬intersect(e, counter-body(e)) then assert eligible(e)
        else if ¬contain(counter-body(e), e) then
        begin
          assert clipped(e); fset := set{g : on(e, g)};
          vset : = {(ev0(e), 0), (ev1(e), 1)}
        end;
        ⟨associate each v with intersecting faces, coord and normalized
        param(r) values⟩;
        foreach intersect(e, ?h) ∧ of(counter-body(e), ?h) ∧
        (?r = intersection(e, ?h)) do
```

```
          begin
            create v asserting temporary(v) ∧ (faces : v → fset ∪ {k}) ∧
            (coord : v → coord-of-intersection(e, h));
            vset := vset ∪ {(v, r)}
          end;
*3)       ⟨sort vset in the order of increasing magnitude of the
          2nd argument (r) and place the result in the array vseq[0..n]
          of record (v, r)⟩;
          assert (i = 0..card(vset))
                 ∧i(ev : e, i → vseq[i]) ∧ (no-of-v : e → card(vset))
        end
      end;
      ⟨divide the set of v into equivalent classes and reduce multiplicities⟩
      foreach ∃v(?fgh = faces(v)) do
      begin
        vx := set{v : fgh = faces(v)};
        repeat
*4)       create w asserting ∀v(((v ∈ vx) ∧ close(coord(v),
          coord(εu(u ∈ vx)))) → (vw : v → w));
          vx := vx − set{v : w = vw(v)}
        until vx = empty
      end;
      ⟨clip edges and create eligible edges⟩
      foreach clipped(?e) do
      begin
        if contain(counter-body(e), ev0(e)) then s := 1 else s := 0;
        for i := 0 to ((card(vset) − 1 − s) div 2) do
        begin
          create e' asserting eligible(e') ∧ ∀f(on(e, f) → on(e', f)) ∧
          same-sort(e, e') ∧ (eparam : e' →
          param-of-clipped(e, ev(e, s + 2 * i).r, ev(e, s + 2 * i + 1).r)) ∧
          (ev0 : e' → vw(ev(e, s + 2 * i).v)) ∧
          (ev1 : e' → vw(ev(e, s + 2 * i + 1).v));
          if free(e')
          then create-segment2(e, ev(e, s + 2 * i).r, ev(ev(e, s + 2 * i + 1).r)
        end
      end;
      delete temporary(?v);
      ⟨organize eligible edges into loops and loops into faces⟩
      foreach ∃x(fpair(?f, x) ∨ fpair(x, ?f) do
      begin
        ef := set{e : on(e, f) ∧ eligible(e)};
        repeat
          create k asserting fl(f, k); e0 := εe(e ∈ ef); e1 := e0;
```

repeat
$ef := ef - \{e1\}$; $e2 := \varepsilon e((e \in ef)$
$\wedge (fore(f, e1) \rightarrow ((ev0(e) = ev1(e1)) \vee (ev1(e) = ev1(e1))))$
$\wedge (back(f, e1) \rightarrow ((ev0(e) = ev0(e1)) \vee (ev1(e) = ev0(e1)))))$;
assert $lee(k, e1, e2)$; $e1 := e2$
until $e1 = e0$
until $ef = empty$;
foreach $fl(f, ?k) \wedge \forall k'(fl(f, k') \rightarrow \neg contain(k', k))$ **do**
create f' **asserting** $on(f', f) \wedge of(\varepsilon o(o, f), f') \wedge flout(f', k) \wedge$
$\forall k'((fl(f, k') \wedge contain(k, k')) \rightarrow flin(f, k'))$
end
end.

Discussion

$*1)$ The general form of a 'foreach' statement is

foreach $formula(?x, ?y, \ldots, ?z)$ **do**
begin $statement(x, y, \ldots, z); \ldots; statement(x, y, \ldots, z)$ **end**

where $?x, ?y, \ldots, ?z$ denote the free variables of an open formula. Once selected, each tuple of values of these variables is treated as a constant. An implication of this worthy of particular note is that the statement

foreach $formula1(?x, ?y, \ldots, ?z)$
assert $formula2(x, y, \ldots, z)$

will make the formula $\forall x \forall y \ldots \forall z(formula1(x, y, \ldots, z) \rightarrow formula2(x, y, \ldots, z))$ true while the seemingly equivalent statement

assert $\forall x \forall y \ldots \forall z(formula1(x, y, \ldots, z) \rightarrow formula2(x, y, \ldots, z))$

will not generally be allowed, as the formula to be asserted may not be a definite clause.

$*2)$ The effect of semicolons within an 'assert' statement is to serialize asserting formulas. For example, the execution of the statement

assert $q1; p \rightarrow q2$

will establish the truth of $\neg p \rightarrow q1$. The statement

assert $(\neg p \rightarrow q1) \wedge (p \rightarrow q2)$

is not legitimate, however.

$*3)$ The purpose of sorting *vset* is to assign to each vertex its relative position among those lying on the same edge with respect to the parameter value. Every other fragment of the underlying edge created by clipping it at those vertices is eligible for constituting the boundary of a new face (Fig. 7.3). As noted in the preceding discussion, any efficient ordering of elements will normally require a reevaluation of the characteristic formula for the set of elements in question. This is the reason why each vertex together with the corresponding parameter value is retained in *vset* when it is created, the sorted array of *vset* is produced and the order relation between vertices is reflected on the function *ev* before proceeding to clipping next edge.

∗4) As a vertex is defined by the intersection of an edge and a face, an edge in turn by that of two faces, a coarse identification of a vertex may be given by three intersecting faces (say, f, g and h). Ambiguities of two kinds will arise, however. First, the same vertex will be defined in three different ways, that is $(f \cap g) \cap h$, $(g \cap h) \cap f$ and $(h \cap f) \cap g$. Second, a single triplet of faces can intersect more than once (Fig. 7.3). Two vertices are regarded as equivalent if they lie on the same set of faces and are 'close' enough to each other. Ambiguities are dissolved by substituting a new vertex for each equivalent class of (temporary) vertices.

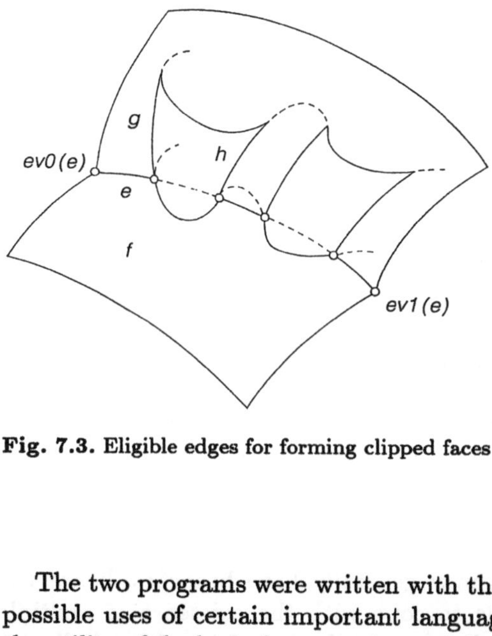

Fig. 7.3. Eligible edges for forming clipped faces

The two programs were written with the primary intent of illustrating the possible uses of certain important language constituents as well as showing the utility of the logical approach to solving an inherently geometric problem. Neither efficiency nor elegance is particularly claimed for them. It is claimed, however, that the language offers itself as one in which to allow describing varieties of solution in a unified and conceptually economical manner.

(2) *Intersections of Surfaces and a Plane*
Given a connected cluster of surfaces, find those that intersect with a given plane (pl) and arrange them in such an order that two successive faces are contiguous to (share a common boundary with) each other. The approach taken here may as well apply to the designing of the body of predicate *intersect* in the preceding example.

var pl : $plane$; $s0$: $surface$; $cont$: **array of record** $s1, s2$: $surface$, $e1$: $edge$ **end**;

function $enclosure(edge)$: **record** $d1, d2$: $\mathbf{real}^3 \times \mathbf{real}^3$ ⟨defines for each edge the smallest rectangle containing it with $d1$ and $d2$ denoting the diagonals⟩;

predicate $se(surface, edge)$ ⟨indicates the edge being a boundary of the surface⟩;
$cross(pl : plane, ls : line\text{-}segment)$;
⟨body to determine if pl and ls intersect with each other⟩;
$intersect(pl : plane, e : edge)$
$\equiv cross(pl, enclosure(e).d1) \vee cross(pl, enclosure(e).d2)$;
formula $wavefront(n : \mathbf{integer}, s1 : surface, e1 : edge)$
$\equiv ((n = 0) \rightarrow (s1 = s0)) \wedge ((n > 0) \rightarrow (\exists s \exists e(wavefront(n-1, s, e)$
$\wedge\, se(s1, e)) \wedge \forall i \forall s(i < n \rightarrow \neg wavefront(i, s, e1))) \wedge$
$se(s1, e1) \wedge intersect(pl, e1)$;
$contiguous(s1, s2, e1)$
$\equiv \exists n(wavefront(n, s1, e1) \wedge \forall e(\neg wavefront(n, s2, e))) \wedge se(s2, e1)$;
⟨See Figs. 7.4 and 7.5 for an intuitive understanding of the meanings of the above function and predicates⟩

⟨main program⟩
begin
⟨create the surfaces defining the enclosure for each of their edges, asserting the predicate se of each pair of a surface and its boundary, and create the plane pl⟩;
$s0 := \varepsilon s(\exists e(intersect(pl, e) \wedge se(s, e)))$;
evaluate $contiguous$ $(?cont.s1, ?cont.s2, ?cont.e1)$;
⟨sort the array $cont$ so that for any successive pair, the field $s2$ of the preceding record matches the field $s1$ of the succeeding record⟩;
end.

Discussion
The chain of contiguous surfaces may generally have branches, hence the application program must be able to cope with the existence of multiple segments in the array *cont*. Of each segment, the last entry either indicates the termination of *wavefront* propagation or points to an entry in the midst of another segment. A jump from a segment to the first or intermediate entry of another respectively starts a new thread of *wavefront* propagation or makes a connection to a parallel one (Fig. 7.5).

Looking at the construction of formula *contiguous*, one might be tempted to eliminate the ordinal n from the list of arguments for *wavefront* and to replace it with a simpler one such as

$wavefront(s1, e1)$
$\equiv ((s1 = s0) \vee (\exists s \exists e(wavefront(s, e) \wedge se(s1, e))$
$\wedge \forall s((s \neq s1) \rightarrow \neg wavefront(s1, e1)))) \wedge se(s1, e1) \wedge intersect(pl, e1)$

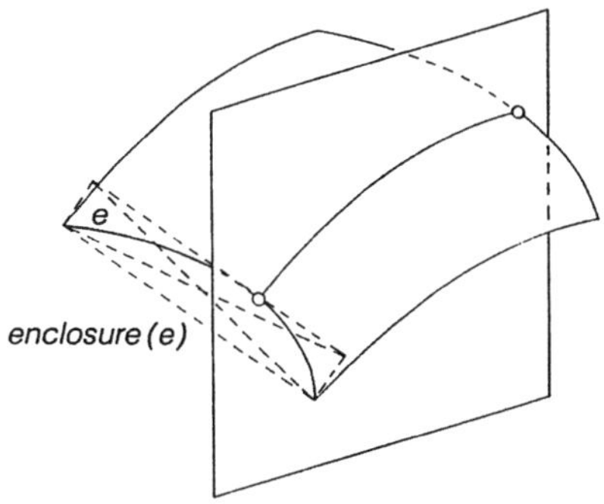

Fig. 7.4. Intersections of surfaces and a plane

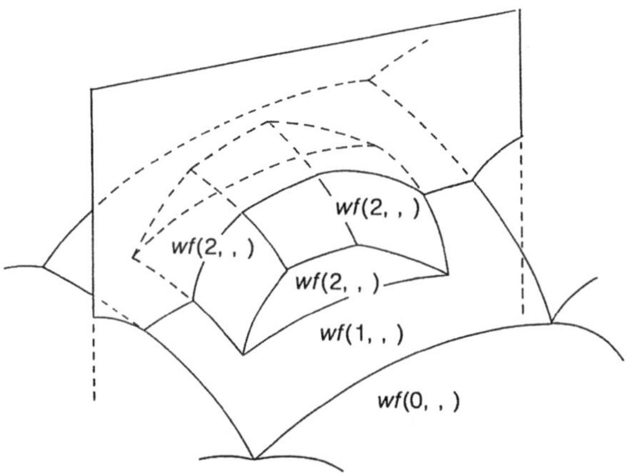

Fig. 7.5. Predicates *wavefront* and *contiguous*

This does not have the least fixed point, however, hence fails to define a predicate uniquely. Or, one might want to evaluate the primitive predicates *se* and *intersect* alone, and let an application program take responsibility for further organizing the retrieved fragments of information. Then, instead of constructing recursive formulas and having them evaluated, one must develop an algorithm which acts on the array of record structure to equivalent effect.

Again, it is possible to write such a program that traverses the database structure directly without relying on the evaluation of an open formula. For example :

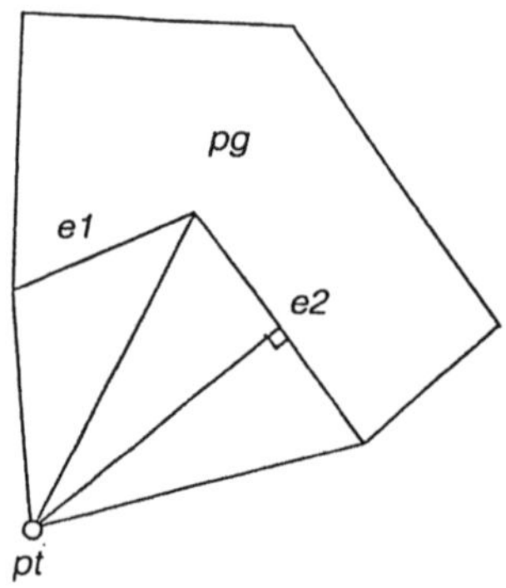

Fig. 7.6. Distance between a point and a polygon

begin
  assert $\forall s \forall e((se(s,e) \wedge intersect(pl,e)) \rightarrow target(s)); s1 := \varepsilon s(target(s));$
  repeat
    assert $head(s1);$
    repeat
      ⟨calculate the intersection *s1* and *pl*, and perform other units of processing⟩;
      assert $\neg target(s1); s1 := \varepsilon s(target(s) \wedge \exists e(se(s1,e) \wedge se(s,e)))$
    until $s1 =$ **nil**;
      $s1 := \varepsilon s(target(s))$
  until $s1 =$ **nil**
end.

This program works the same way as the previous version if there exists no branching in the chain of contiguous surfaces. Otherwise, it will follow different threads in exhausting the extension of *target*. Even if we choose to be guided by the predicates *wavefront* and *contiguous*, which of course is possible, we will have to live up to the many-to-many correspondence via *contiguous* and the resulting indeterminacy of the ε operator.

(3) *Distance between a Point and a Polygon*
Define the function that calculates the (minimum) distance between a polygon (*pg*) and a point (*pt*), and find the edge of *pg* that realizes the said distance. The reason for including this example, which involves nothing techically intricate, is that its straight-forwardness allows us to present a program in its completed form. For completeness' sake, we also have included the part of program that creates the polygon and the point. It should be noted that the function in question is formulated as an addition to the base theory of polygons and that if we desire, we may arbitrarily expand the theory by adding any other functions and predicates such as thickness, density or positional relations with other entities and by specifying constraints on these as axioms.

```
program theory-construction
theory polygon;
  sort point,line-seg,polygon;
  function coord(point) : record x,y : real end;
  predicate pge(polygon,line-seg),ev(line-seg,point);
  intersect(e1, e2 : line-seg);
    var a11, a12, a21, a22, b11, b12, b21, b22, : real;
     p11, p12, p21, p22 : record x, y : real end;
  begin
    p11 := coord(εp1(∃p2(ev(e1, p1, p2))));
    p12 := coord(εp2(∃p1(ev(e1, p1, p2))));
    p21 := coord(εp1(∃p2(ev(e2, p1, p2))));
    p22 := coord(εp2(∃p1(ev(e2, p1, p2))));
    a11 := p11.x; a12 := p12.x; a21 := p21.x; a22 := p22.x;
    b11 := p11.y; b12 := p12.y; b21 := p21.y; b22 := p22.y;
    intersect := (sgn((a11 − a12) * (b11 − b21) − (a11 − a21) *
    (b11 − b12)) * sgn((a11 − a12) * (b11 − b22) − (a11 − a22) *
    (b11 − b12)) = −1)
  end;
  formula non-intersecting(pg, e1, e2) ≡ ((pge(pg, e1) ∧ pge(pg, e2))
  → ¬intersect(e1, e2));
  axiom non-self-intersecting : ∀pg∀e1∀e2(non-intersecting(pg, e1, e2))
end-theory;
theory distance-pgpt;
  subtheory polygon;
  function d-lspt(ls : line-seg, pt : point) : real;
    var x1, x2, dx, dy, a, b, s : real; p1, p2 : record x, y : real end;
  begin
    p1 := coord(εp1(∃p2(ev(ls, p1, p2))));
    p2 := coord(εp2(∃p1(ev(ls, p1, p2))));
    x1 := p1.x; y1 := p1.y; x2 := p2.x; y2 := p2.y;
    dx := x2 − x1; dy := y2 − y1;
    a := coord(pt).x; b := coord(pt).y;
    s := ((a − x1) * dx + (b − y1) * dy)/(dx * *2 + dy * *2);
    if s ≤ 0 then s := 0 else if s ≥ 1 then s := 1;
    d-lspt := sqrt((a − ((1 − s) * x1 + s * x2)) * *2
    +(b − ((1 − s) * y1 + s * y2)) * *2)
  end;
  function d-pgpt(pg : polygon, pt : point)
  = min(set{d : ∃e(pge(pg, e) ∧ (d = d-lspt(e, pt)))});
end-theory

program create-pgpt(in1, in2, normal);
  sort polygon, line-seg, point;
  var in1, in2 : file of real; normal : boolean; i, n : integer;
  coord : array of record x, y : real end;
```

```
    pg : polygon; edge : array of line-seg; vtx : array of point; pt : point;
    view create-pgpt : polygon;
  begin
    interpretation-initial;
    reset(in1); n := 0;
    while ¬eof(in1) do
    begin
      n := n + 1; read(in1, coord[n].x, coord[n].y)
    end;
    create pg, (i = 1..n)edge[i], vtx[i]
      asserting (name : pg → 'pg') ∧ (i = 1..n) ∧ pge(pg, edge[i]) ∧
      (i = 1..n − 1) ∧ ev(edge[i], vtx[i], vtx[i + 1]) ∧
      ev(edge[n], vtx[n], vtx[1]) ∧ (i = 1..n) ∧ (coord : vtx[i] → coord[i]);
    evaluate ?normal = (i = 3..n)((j = 1..n − 2)
    ∧(¬intersect(edge[i], edge[j])));
    reset(in2); read(in2, coord[n + 1].x, coord[n + 1].y);
    create pt asserting (name : pt → 'pt') ∧ (coord : pt → coord[n + 1]);
    exit-final
  end

  program find-distance-pgpt(d, edge)
    var d : real; edge : edge;
    view find-distance-pgpt : distance-pgpt;
  begin
    interpretation-initial;
    evaluate (?d = d-pgpt(pg, pt)) ∧ pge(pg, ?edge) ∧ (?d = d-lspt(?edge, pt));
    exit-final
  end.
```

Discussion
The application agent in this example comprises one program for theory construction and two for interpretation. Each of the latter contains its own set of entry and exit statements. It is assumed that at each activation of the agent, a job control command will specify the program to be actually run. An alternative way of organizing the agent is to allow a single set of entries and exits to exist in a particular program, which serves as the monitor and transfers control to one of others that corresponds to the current view of interpretation (Fig. 7.7). The decision between these alternatives is left to the implementor as the question of which will be better can only be answered by considering the operating system and other executional environments.

(4) *Copying and Imbedding of a Complex Object*
We express in our language the method sketched in Sect. 6.2 (3.2) for creating a copy of a given complex object and imbedding it in a larger structure. It is assumed that the object is a model of a theory T with (a sole) axiom A and that its elements are hierarchically organized via the *part-of* relation with the element *object* lying at the topmost level.

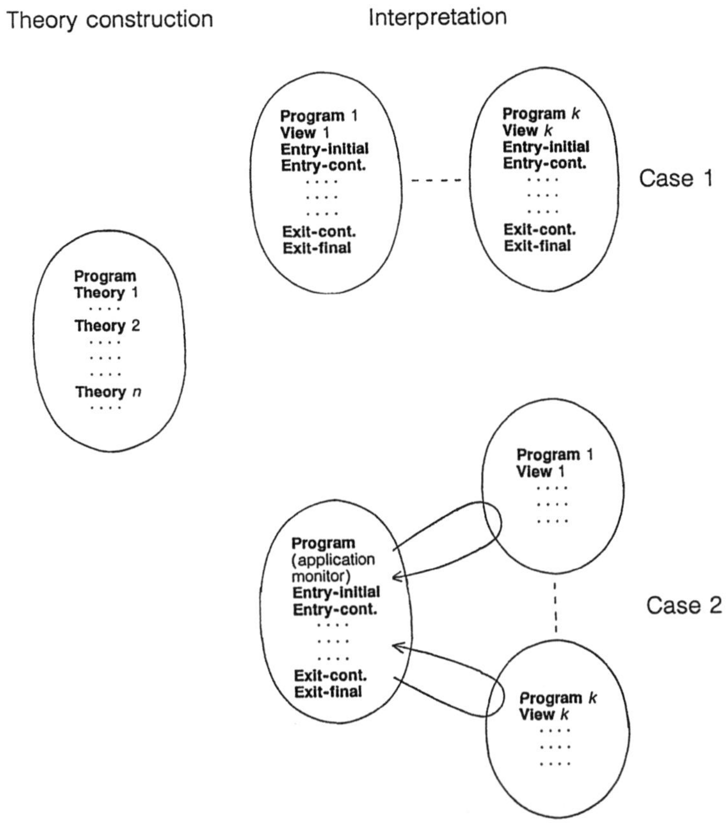

Fig. 7.7. Alternate organizations of an application agent

program *copy-and-imbed*$(t, t1, int1)$
  var $t, t1, int1$: **boolean**; $S, S1$: **set of identifier**;
  formula $P(element) \equiv (element = object) \vee$
    $\exists x(P(x) \wedge part\text{-}of(element, x))$;
    $\bar{A} \equiv \langle A$ relativized by P, that is, the version of A obtained by replacing each sub-formula of the form $\forall x(F(x))$ or $\exists x(F(x))$ by $\forall x(P(x) \rightarrow F(x))$ or $\exists x(P(x) \wedge F(x))$ respectively$\rangle$;
    $INT \equiv$ **interface**(S,[**all** | $\langle$list of constants$\rangle$ in $\mathcal{T}$,
    [**all** | $\langle$list of functions and predicates$\rangle$] in $\mathcal{V}$);
    $P1(element) \equiv (element =$ 'ver1'.$object)$
    $\vee \exists x(P1(x) \wedge part\text{-}of(element, x))$;
    $\overline{A1} \equiv \langle A$ relativized by $P1$, that is, the version of A obtained by prefixing the qualifier 'ver1' to each constant and replacing each sub-formula of the form $\forall x(F(x))$ or $\exists x(F(x))$ by
    $\forall x(P1(x) \rightarrow F(x))$ or $\exists x(P1(x) \wedge F(x))$ respectively$\rangle$;
  view $\mathcal{V}$;

begin
  interpretation-initial;
  S := **set**{$element : P(element)$};
  evaluate $t = \bar{A}$; **if** $\neg t$ **then exit-final**;
  copy S **to** $S1$ **under-theory** T **with-qualifier** 'ver1'
  imbedding-function $f1$ **interface** INT **to** $INT1$;
  ⟨$INT1$ is automatically defined as the interface of $S1$ with its complement in the universe of interpretation which is isomorphic to INT via $f1$⟩
  ⟨work on $S1$ maintaining the truth of $\overline{A1}$ and $INT1$⟩;
  evaluate $(?t1 = \overline{A1}) \wedge (?int1 = INT1)$; **if** $\neg(t1 \wedge int1)$ **then exit-final**;
  ⟨delete S or $S1$ in accordance with the designer's choice⟩;
  exit-final
end.

(5) *Clapper Bell and Its State Transitions*
Unlike the preceding examples, this last one involves the description of the dynamic characteristics of multi-state parts which determine the functions of the whole design object. The dynamic behavior shows a clear regularity which enables a static description of it. Namely, each part at any given instant will be in one of a finite number of distinct states, its behavior in a state is completely specified by a generic pattern together with a few parameter values, and only particular combinations of states are recognized as permissible with transitions from one to another being governed by a definite set of laws. Each series of transitions that forms a cycle is a realization of an elementary function. The modeling of such an object in our language can be generally done as follows:

- Each state is given the status of an entity. A function *state* is defined which maps a part to the set of states it may possibly take. That is,

$$state(part) : \textbf{set of } states$$

$$a_1 \rightarrow S_1 = \{s_1^{(1)}, \ldots, s_1^{(n_1)}\}$$

$$\ldots$$

$$a_m \rightarrow S_m = \{s_m^{(1)}, \ldots, s_m^{(n_m)}\}.$$

- An m-ary predicate *permissible* is declared, the extension of which should contain those and only those combinations of states that are permitted to exist.

$$permissible(s_1^{(i_1)}, \ldots, s_m^{(i_m)})$$

- Functions and predicates will generally be state-dependent, hence one or more of their arguments will take values from the domain $\cup_i S_i$.

$$f(e_1, \ldots, e_k, s_1^{(i_1)}, \ldots, s_m^{(i_m)}) = e$$

$$p(e_1, \ldots, e_k, s_1^{(i_1)}, \ldots, s_m^{(i_m)})$$

– A $2m$-ary predicate *trans* is declared, each instance of which has the form

$$trans(s_1^{(i_1)}, \ldots, s_m^{(i_m)}, s_1^{(j_1)}, \ldots, s_m^{(j_m)})$$

and stipulates that combined states $(s_1^{(i_1)}, \ldots, s_m^{(i_m)})$ change into $(s_1^{(i_1)}, \ldots, s_m^{(j_m)})$. If a condition ($\Gamma$) is further required for the transition to actually take place, an axiom

$$trans(s_1^{(i_1)}, \ldots, s_m^{(i_m)}, s_1^{(j_1)}, \ldots, s_m^{(j_m)}) \leftrightarrow \Gamma$$

should be postulated.
– Whether an intended elementary function is realized can be ascertained by evaluating the corresponding conjunction of instances of *trans*.

Figure 7.8 depicts the current example, which comprises three two-state parts (*coil*, *clapper* and *switch*), one state-independent and one state-dependent property (forces Fs and Fc), three permissible combinations of states and two elementary functions (to *clap* when the *switch* is *on*, remain *mute* otherwise). For simplicity, we will not include in our model the battery, its power (voltage) and the functional specification of Fc in terms of pertinent properties of the circuit.

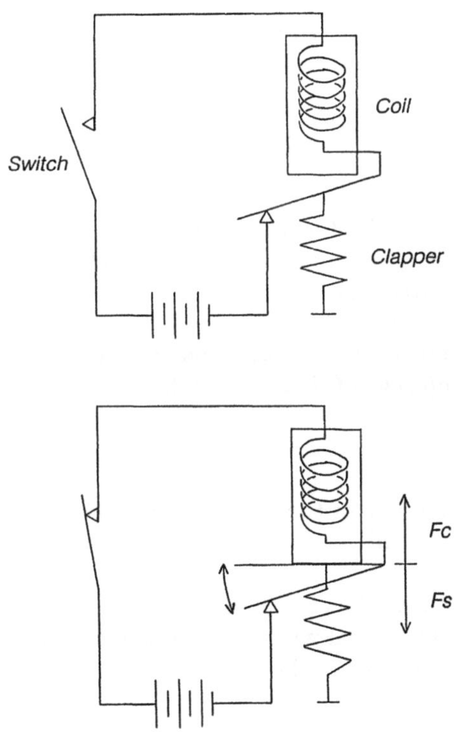

Fig. 7.8. Clapper bell and two elementary functions

program *theory-construction-for-clapper-bell*
theory *clapper-bell*
  sort *state*, *element*;
  const *clapper*, *coil*, *switch* : *element*; *Fs* : **real**;
  open, *closed*, *active*, *inactive*, *on*, *off* : *state*;
  function *state*(*element*) : **set of** *state*;
  Fc(*state-of-coil* : *state*(*coil*)) : **real**;
  predicate *permissible*(*state*, *state*, *state*);
  trans(*state*, *state*, *state*, *state*, *state*, *state*);
  axiom *state* : $\forall x \forall y((state : x \rightarrow y) \leftrightarrow$
    $(((x = clapper) \wedge (y = \{open, closed\})) \vee$
    $((x = coil) \wedge (y = \{active, inactive\})) \vee$
    $((x = switch) \wedge (y = \{on, off\}))))$;
   permissible : $\forall x \forall y \forall z(permissible(x, y, z) \leftrightarrow$
    $(((x, y, z) = (open, inactive, on)) \vee$
    $((x, y, z) = (closed, active, on)) \vee$
    $((x, y, z) = (closed, inactive, off))))$;
   trans : $\forall u \forall v \forall w \forall x \forall y \forall z(trans(u, v, w, x, y, z) \leftrightarrow$
    $(((u, v, w, x, y, z) = (open, inactive, on, closed, active, on)) \vee$
    $(((u, v, w, x, y, z) = (closed, active, on, open, inactive, on))$
    $\wedge (Fc(active) > Fs)) \vee$
    $(((w, z) = (on, off)) \wedge permissible(u, v, w) \wedge permissible(x, y, z)) \vee$
    $(((w, z) = (off, off)) \wedge permissible(u, v, w) \wedge permissible(x, y, z)) \vee$
    $((u, v, w, x, y, z) = (closed, inactive, off, closed, active, on))))$
end-theory;

program *interpretation*
  var *function1*, *function2* : **boolean**;
  const *force-of-spring* : **real**;
  view *clapper-bell* : *clapper-bell* **standard-interpretation**;
  formula *clapping* $\equiv trans(closed, inactive, off, closed, active, on) \wedge$
   $trans(closed, active, on, open, inactive, on) \wedge$
   $trans(open, inactive, on, closed, active, on) \wedge$
   $\forall x \forall y(permissible(x, y, on) \rightarrow trans(x, y, on, closed, inactive, off))$;
   mute $\equiv trans(closed, inactive, off, closed, inactive, off)$;
begin
  interpretation-initial;
  assert *Fs*: $\rightarrow$ *force-of-spring*;
  ⟨also have the values of other properties of the circuit specified so that the value of *Fc* may be calculated⟩;
  assert *state* $\wedge$ *permissible* $\wedge$ *trans*;
  evaluate (*function1* = *clapping*) $\wedge$ (*function2* = *mute*);
  ⟨Further processes are to follow, which deal with different combinations of truth values of *function1* and *function2*. The normal case is of course where both values are true.⟩
  exit-final
end.

Discussion

We have tried to capture within our framework the naive human understanding of a simple physical system. Unlike the purely qualitative approach [deK], we have included quantitative functions and predicates (that is, Fc, Fs and $Fc > Fs$) which play vital roles in determining the behavior of the system. It should be noted that our language serves for describing both the qualitative and quantitative aspects of a model equally well.

Another noticeable point is the division of the entire set of state combinations into mutually disjoint classes, each of which is characterized by a specific value of *state*(*switch*) and represents an elementary function (Fig. 7.9a).

Recognition of this fact can probably be generalized to the following statement of anticipation; there exists a certain 'crucial' set of elements (comprising only *switch* in the current example) which serves for organizing the space of states and state transitions into a collection of elementary functions. Each elementary function comprises a circular series of state transitions which is disjoint from others and has neither incoming nor outgoing branches (Fig. 7.9b).

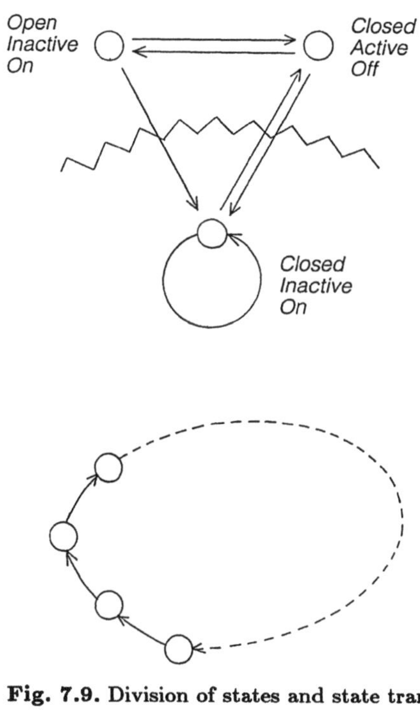

Fig. 7.9. Division of states and state transitions

References

[deK] de Kleer, J., Brown, J.S.: A Qualitative Physics Based on Confluences. In: Bobrow, D.G. (ed.): Qualitative Reasoning about Physical Systems. MIT Press, 1985

Chapter 8
Modeling the Design Process

As indicated earlier (Sect. 0.5), the process of designing is modeled as that of enlarging and enriching the database which represents the artifact being designed. Given the design requirements specified by a set of axioms $\alpha, \beta, \ldots, \gamma$, the statement

assert $\alpha \wedge \beta \wedge \cdots \wedge \gamma$

will be the simplest and most macroscopic form to express the process in its entirety, which, however, will not generally be susceptible to unequivocal interpretation or immediate execution. Expressions in logical grains of smaller and various sizes can be conceived of. At the opposite end of the spectrum is the class of statements

assert ⟨a conjunction of ground atoms⟩

(Sect. 6.2 (6.2.2)) which can be executed in a quite mechanical and deterministic way. If we recall that the modeling of the design process should be ultimately and inherently related to the elucidation of creative and intelligent works of human designers in a language precise enough to be interpretable by computers (Sects. 0.7 and 0.8), neither of these extremes can reasonably serve the purpose.

A natural alternative will be to

assert α

where α is an axiom, since each axiom is considered to embody a semantic unit of design requirement and to have associated with it an elementary body of designer's knowledge for interpreting the theory. Moreover, the control information to dissolve conflicts and indeterminacies, and direct the evolution of the database, can be specified appropriately in terms of relations among axioms such as precedence, dependence, independence, interaction, etc.

Having decided tentatively on the scale of the primary representational unit, what we must do first is:

- to precisely define the concept of a minimal extension of the database with respect to a given axiom,
- to develop the procedures to evaluate the current state of the database and to minimally extend it,

- to expand the model description language so that indeterminacies inherent in asserting a general closed formula may be explicated and properly handled, and
- to provide the rules and means to decrease the degree of freedom in choosing one alternative among others for advancing the database state.

Next, in order to facilitate guiding the design process on a more macroscopic level, we want:

- to introduce the notation with which to describe qualitatively the effect of the increase or decrease of extensions of constituent predicates on the validity of each axiom, and
- based on this, to provide the means to express the interrelationships between axioms.

8.1 The Concept of Minimal Extension

8.1.1 General

Suppose we are given a formal theory ($\mathcal{T}$), a particular axiom of it (α), and an extensional database, that is, a partial interpretation of it ($\mathcal{I}$) on which α is false, and we face the task of extending $\mathcal{I}$ so that α may get satisfied. Recall that such an extension of an interpretation is defined to be one with respect to α (Definition 3.16). There exists generally more than one extension with respect to the same axiom, among which we are primarily interested in those which are 'minimal'. Roughly speaking, minimality means that the axiom in question cannot be true with fewer entities or valid ground atoms. Its primacy derives from the economy of size and complexity of design to achieve the goal. We stand on the premise that designing in its most natural form will proceed in such a manner as to produce the smallest and simplest possible structure. The rigorous definition of a minimal extension is not immediately obtainable, however, as the simple example below illustrates.

Suppose we are given a polyline and an axiom which requires it to be a part of a polygon. Leaving aside worries about anomalies (that is, self-intersection, end edges being parallel to each other, etc.), there are many different ways to form a polygon out of a polyline, each of which may be claimed to be a minimal extension in its own right (Fig. 8.1).

Set membership and adjacency relations are maintained in all cases. Lengths are changed in cases 1 and 2. Lengths remain the same, but angles are changed in case 3. Cases 4 and 5 involve the creation of new entities while all the existing attributes and relations are unaffected. In trying to narrow down the possible choices, we face two issues to be resolved.

First, which parts or aspects of the current interpretation should be susceptible to change, and which should not? That lengths or angles (or more elementarily, coordinate values of end points) can be changed reflects the view that attributes which take values in the domain A (of integers, real numbers,

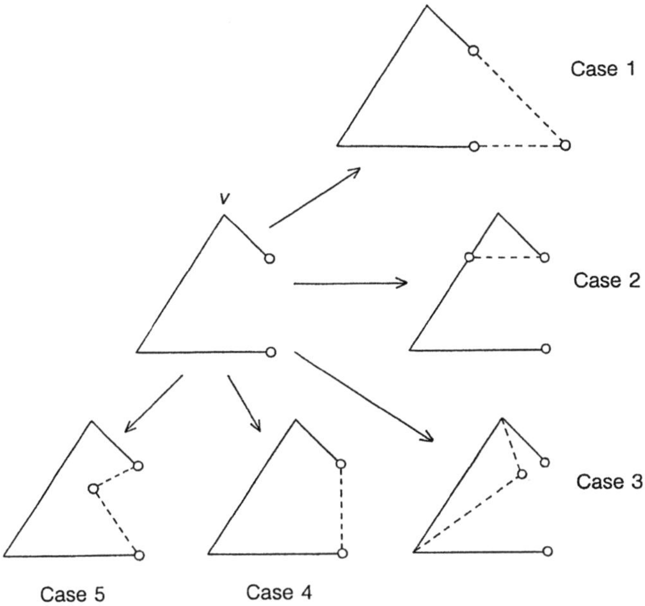

Fig. 8.1. Extending a polylinear structure into a polygon

etc.) can be changed without losing the identity or impairing the meaning of each entity. (See Sect. 5.3.1 (7)). On the other hand, relations predicable of physical entities (that is, entities of sort ι) must remain as they currently are. This will or will not be true of the predicate *adjacent* depending on whether it is defined explicitly or implicitly as the coincidence of the coordinate values of two end points. Here, explicity means enumerativity. More generally and precisely, we require the following rule be observed by our process model.

Rule P ('P' for physicality of objects):
Of each enumeratively defined predicate $p(x_1, \ldots, x_n : \iota, u_1, \ldots, u_k : A)$, the projection of its extension on the domain I^n, that is, the extension of the derived predicate

$$q(x_1, \ldots, x_n) \equiv \exists u_1 \ldots \exists u_k(p(x_1, \ldots, x_n, u_1, \ldots, u_k))$$

must not be decreased.

The idea behind this is that any explicit assertion of a relation among physical entities embodies the designer's intention, and as such, should not be retracted unless explicitly so instructed. We may as well say that with the current example, our prime interest is in contiguity, non-self-intersection and closure which constitute the 'topology' or abstract polygonarity of the object in question and any specific values of 'geometric' attributes are of no concern.

It might be the case, of course, that designers want to have certain attribute values remain fixed once they are given. In that case the attributes must have been so qualified by metatheoretic declarations. It should be noted that the qualifications cannot be made at theory level by postulating appropriate axioms, because what is involved is not the way to interpret the theory but the way to modify one interpretation to another. We will thus declare

$$\text{NOT CHANGEABLE } \mathit{coord}(v) \tag{8.1.1}$$

$$\text{NOT CHANGEABLE } \mathit{angle} \tag{8.1.2}$$

or

$$\text{NOT CHANGEABLE } \mathit{length} \tag{8.1.3}$$

to stipulate that the coordinate value of a specific vertex and all the lengths of edges, or all the angles, be insusceptible to change.

Second, what should be the order among different extensions of the same interpretation? Take two such extensions, $\mathcal{E}_1$ and $\mathcal{E}_2$. We may properly define that $\mathcal{E}_1$ is *prior to $\mathcal{E}_2$ in the strong sense*, denoted by

$$\mathcal{E}_1 \prec_s \mathcal{E}_2,$$

if there exists a monomorphism μ from $\mathcal{E}_1$ into $\mathcal{E}_2$ which preserves the structure of the former except for the functions and predicates that have not been declared as not-changeable. All the cases 1 through 5 in the current example are minimal according to the order of this sense. On the other hand, some may feel it intuitively more natural to put them into such an order as

$$\{\text{case 1, case 2, case 3}\} \prec \text{case 4} \prec \text{case 5} \tag{8.2.1}$$

where those within the braces are incomparable to each other and are all minimal. Clustering cases 1 through 3 is permitted only under (8.1.1), however. The addition of (8.1.2) or (8.1.3) should result respectively in

$$\{\text{case 1, case 2}\} \prec \text{case 4} \prec \text{case 5} \tag{8.2.2}$$

or

$$\text{case 3} \prec \text{case 4} \prec \text{case 5} \tag{8.2.3}$$

with the missing cases not being permitted to exist. We will only have

$$\text{case 4} \prec \text{case 5} \tag{8.2.4}$$

if all of the three declarations are made.

Two questions will be raised immediately. First, what is the basis for the above orderings of extensions besides the comparison of cardinalities of underlying sets, and second, should those within the braces be considered as equivalent rather than incomparable? Indeed, case 4 with the single edge connecting the terminal points of the original polyline is not only of smaller size but structurally simpler than the two-edged version of case 5, while both serve the same function for the whole object. Unlike in the strong order, however, the former will not be imbedded elementwise in but substituted as a whole for the latter. These are somewhat reminiscent of the commonality and distinction between design alternatives in the strong and weak senses (Definitions 6.4 and 6.5).

Generalizing the above analysis, we are now ready to present the following definitions of a precedence relation among extensions and the concept of equivalence as well as of minimality based on it.

Definition 8.1. Given a formal theory $\mathcal{T}$, an axiom α and a partial interpretation $\mathcal{I}$ on which α is false, one extension $\mathcal{E}_1$ of $\mathcal{I}$ is said to *precede* another $\mathcal{E}_2$ (or, $\mathcal{E}_2$ *succeed* $\mathcal{E}_1$) *with respect to* α *in the weak sense*, denoted by

$$\mathcal{E}_1 \prec_{\mathrm{w}} \mathcal{E}_2,$$

if the following conditions are satisfied;
(1) α is true on both $\mathcal{E}_1$ and $\mathcal{E}_2$.
(2) There exists a monomorphism μ from $I(\mathcal{E}_1)$ (the sub-universe of interpretation of sort ι of $\mathcal{E}_1$) into $I(\mathcal{E}_2)$ which is an identity mapping on $I(\mathcal{I})$.
(3) For each predicate p (or formula of the form $f(x) = y$) of enumerative type which involves two or more arguments of sort ι, there exists a formula q such that:
(3.1) q is not logically valid,
(3.2) $\forall x(p(x) \rightarrow q(x))$ is a theorem of $\mathcal{T}$.
(3.3) If $q(e_1, \ldots, e_n, a_1, \ldots, a_r)$ holds true in $\mathcal{E}_1$, so does $q(\mu(e_1), \ldots, \mu(e_n), a'_1, \ldots, a'_r)$ in $\mathcal{E}_2$ where
(3.3.1) $e_1, \ldots, e_n \in I(\mathcal{E}_1)$ and $\mu(e_1), \ldots, \mu(e_n) \in I(\mathcal{E}_2)$,
(3.3.2) $a_1, \ldots, a_r, a'_1, \ldots, a'_r \in A$ and a'_i may or may not differ from a_i depending on whether the value of i-th attribute is allowed to change or not.
(3.4) If $q(x_1, \ldots, x_n, a_1, \ldots, a_r)$ holds true in $\mathcal{E}_2$, so does p for the same tuple of arguments or others which contain both some objects x_i and objects from $I(\mathcal{E}_2) - \mu(I(\mathcal{E}_1))$. (Note q will be reduced to p if μ maps $I(\mathcal{E}_1)$ onto $I(\mathcal{E}_2)$.)

Definition 8.2. With the same notation as above, $\mathcal{E}_1$ is said to be *equivalent to* $\mathcal{E}_2$ *with respect to* α, denoted by

$$\mathcal{E}_1 \equiv_{\mathrm{w}} \mathcal{E}_2,$$

if $\mathcal{E}_1 \prec_{\mathrm{w}} \mathcal{E}_2$ and $\mathcal{E}_2 \prec_{\mathrm{w}} \mathcal{E}_1$ with respect to α.

For the predicate *adjacent* in the above example, we may give a recursive definition

$$q(x_1, x_2) \equiv adjacent(x_1, x_2) \vee \exists y(adjacent(x_1, y) \wedge q(y, x_2))$$

to establish any of (8.2.1) through (8.2.4).

If we focus our attention on the differences of extensions $\mathcal{E}_i$ from $\mathcal{I}$ rather than the structures of the extensions themselves, and let d_i and $d_{p,i}$ denote the cardinalities of

$$I(\mathcal{E}_i) - I(\mathcal{I})$$

and

$$\text{extension of } p \text{ in } \mathcal{E}_i - \text{extension of } p \text{ in } \mathcal{I}$$

respectively, with p ranging over the above-specified class of predicates, it is apparent that the condition

$$(\mathrm{WW}) : d_1 \leq d_2 \quad \text{and} \quad d_{p,1} \leq d_{p,2} \quad \text{for all } p$$

results from $\mathcal{E}_1 \prec_{\mathrm{w}} \mathcal{E}_2$. We define by the satisfaction of this condition a yet weaker order '$\prec_{\mathrm{ww}}$' (order *in the weakest sense*).

The following lemmas immediately follow from these definitions.

Lemma 8.1 With the notation again the same as above:
(1) With respect to any given α, $\mathcal{E}_1 \prec_{\mathrm{s}} \mathcal{E}_2$ implies $\mathcal{E}_1 \prec_{\mathrm{w}} \mathcal{E}_2$.
(2) Conversely, if μ maps $I(\mathcal{E}_1)$ onto $I(\mathcal{E}_2)$, $\mathcal{E}_1 \prec_{\mathrm{w}} \mathcal{E}_2$ implies $\mathcal{E}_1 \prec_{\mathrm{s}} \mathcal{E}_2$.
(3) $\mathcal{E}$ will be minimal in the strong sense if it is so in the weak sense.
(4) $\mathcal{E}$ will be minimum in the weak sense if it is so in the strong sense.
(5) $\mathcal{E}_1 \equiv_{\mathrm{w}} \mathcal{E}_2$ may as well be defined as $\mathcal{E}_1 \prec_{\mathrm{s}} \mathcal{E}_2$ and $\mathcal{E}_2 \prec_{\mathrm{s}} \mathcal{E}_1$.
(6) The relation '$\equiv_{\mathrm{w}}$' establishes an equivalence relation in the usual sense in the set of all the extensions of $\mathcal{I}$ with respect to α.
(7) It preserves the precedence structures in both the strong and weak senses. That is,

$$\text{if } \mathcal{E}_1 \equiv_{\mathrm{w}} \mathcal{E}_1',\ \mathcal{E}_2 \equiv_{\mathrm{w}} \mathcal{E}_2' \text{ and } \mathcal{E}_1 \prec \mathcal{E}_2 \text{ then } \mathcal{E}_1' \prec \mathcal{E}_2'$$

where $\prec$ may be whichever of $\prec_{\mathrm{s}}$ or $\prec_{\mathrm{w}}$.

Lemma 8.2. With respect to any given α:
(1) $\mathcal{E}_1 \prec_{\mathrm{w}} \mathcal{E}_2$ implies $\mathcal{E}_1 \prec_{\mathrm{ww}} \mathcal{E}_2$.
(2) Conversely, $\prec_{\mathrm{ww}}$ implies $\prec_{\mathrm{w}}$ for any pair of extensions which are comparable in the order of weak sense.
(3) $\mathcal{E}$ will be minimal in the weak sense if it is so in the weakest sense.
(4) $\mathcal{E}$ will be minimum in the weakest sense if it is so in the weak sense.
(5) If $\mathcal{E}_1 \equiv_{\mathrm{ww}} \mathcal{E}_2$ (that is, $\mathcal{E}_1 \prec_{\mathrm{ww}} \mathcal{E}_2$ and $\mathcal{E}_2 \prec_{\mathrm{ww}} \mathcal{E}_1$), either $\mathcal{E}_1 \equiv_{\mathrm{w}} \mathcal{E}_2$ or these extensions are incomparable in the order of weak sense.
(6) The relation '$\equiv_{\mathrm{ww}}$' establishes an equivalence relation in the usual sense in the set of all the extensions of $\mathcal{I}$ with respect to α.
(7) It preserves the precedence structures in the weakest sense. That is,

$$\text{if } \mathcal{E}_1 \equiv_{\mathrm{ww}} \mathcal{E}_1', \mathcal{E}_2 \equiv_{\mathrm{ww}} \mathcal{E}_2' \text{ and } \mathcal{E}_1 \prec_{\mathrm{ww}} \mathcal{E}_2 \text{ then } \mathcal{E}_1' \prec_{\mathrm{ww}} \mathcal{E}_2'.$$

Figure 8.2 illustrates how the entire class of extensions of $\mathcal{I}$ is organized by means of multiple levels of precedence and equivalence relations.

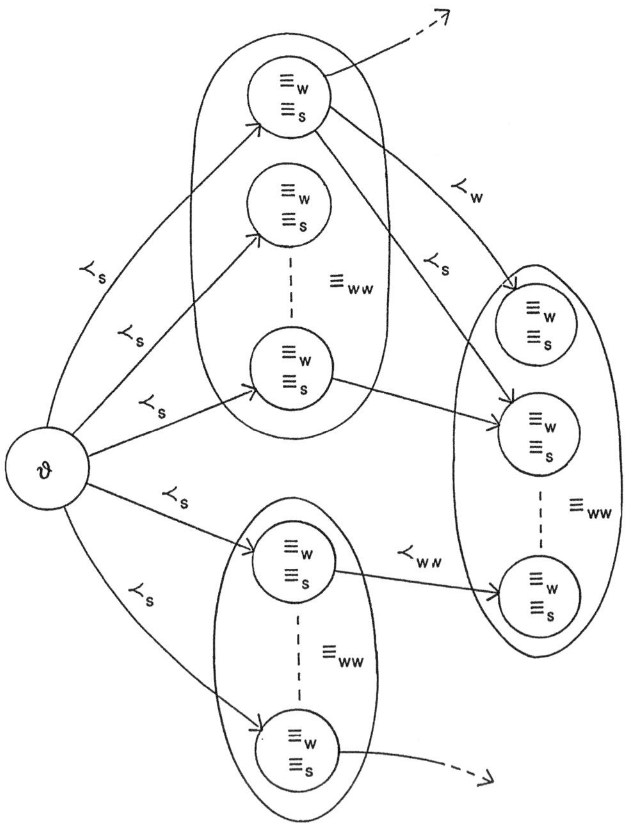

Fig. 8.2. Structured class of extensions

A few remarks will be due as to the relationship between the concepts of design alternative (Sect. 6.2 (3.1)) and equivalence of extensions in the strong or weak sense. On one hand, these are similar in that both allow a design to be partially substituted for by another. More precisely, two equivalent interpretations are considered as design alternatives to each other in the strong sense when changes of attributes are prohibited. Conversely, if sub-interpretations $\mathcal{I}_1'$ and $\mathcal{I}_2'$ are design alternatives in the context of $\mathcal{I}$ (Definition 6.4) and $\mathcal{I}'$ denotes $\mathcal{I} - (\mathcal{I}_1' + int(\mathcal{I}_1')) - (\mathcal{I}_2' + int(\mathcal{I}_2')), \mathcal{I}' + int(\mathcal{I}_1')$ and $\mathcal{I}' + int(\mathcal{I}_2')$ are equivalent extensions of $\mathcal{I}'$. On the other hand, the equivalence relation pertains to the internal structure of a part of model which the design alternative ignores. In other words, each equivalence class is considered to incarnate design alternatives at a finer level. The combination of these two concepts together with that of minimal extension provides a general and concrete scheme in which to plan for alternate designs as depicted as in Fig. 8.3.

Before going further into general discussion, let us examine another very simple example of an interpretation which exhibits various possibilities of

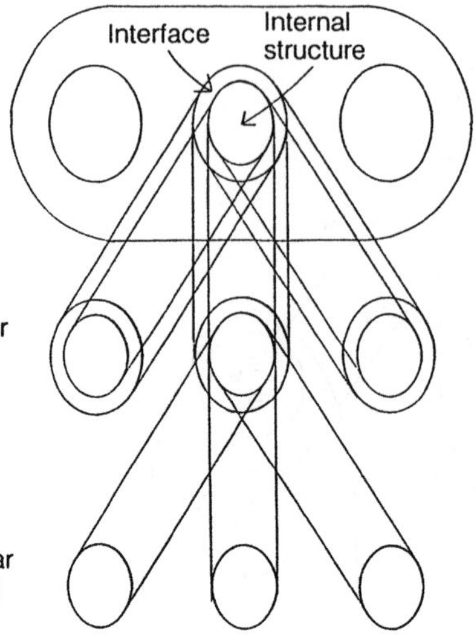

Fig. 8.3. Three levels of alternate design

extension. Suppose we have

predicate : $p(\iota), q(\iota), r(\iota), s(\iota, \iota)$
formula : $\tilde{s}(x, y) \equiv s(x, y) \vee \exists z(s(x, z) \wedge \tilde{s}(z, y))$
axiom : $\forall x \forall y(s(x, y) \rightarrow \neg s(y, x))$

as a part of the theory and

$I(\mathcal{I}) = \{a, b, c\}$
$G(\mathcal{I})$ (the set of ground atoms) $= \{p(a), p(b), q(c), s(a, b), s(b, c)\}$

as the current interpretation. Alternate cases of axioms are given and corresponding minimal extensions obtained as follows, with $\delta(I(\mathcal{E}))$ and $\delta(G(\mathcal{E}))$ denoting $I(\mathcal{E}) - I(\mathcal{I})$ and $G(\mathcal{E}) - G(\mathcal{I})$ respectively.

Case (1) **axiom** $\alpha \equiv \forall x(p(x) \rightarrow q(x))$
(1.1) extension $\mathcal{E}_1$, minimum in the order $\prec_s$ (hence in $\prec_w$ and $\prec_{ww}$)

$\delta(I(\mathcal{E}_1)) =$ empty
$\delta(G(\mathcal{E}_1)) = \{q(a), q(b)\}$, logical consequence of $\mathcal{I}$ and α

Case (2) **axiom** $\alpha \equiv \exists x(p(x) \wedge r(x))$
(2.1) extension $\mathcal{E}_1$, minimum in the order $\prec_{ww}$ (hence minimal in $\prec_w$ and $\prec_s$)

$\delta(I(\mathcal{E}_1)) =$ empty
$\delta(G(\mathcal{E}_1)) = \{r(a)\}$

(2.2) extension $\mathcal{E}_2$, minimum in the order $\prec_{ww}$ (hence minimal in $\prec_w$ and $\prec_s$)

$$\delta(I(\mathcal{E}_2)) = \text{empty}$$
$$\delta(G(\mathcal{E}_2)) = \{r(b)\}$$

(2.3) extension $\mathcal{E}_3$, minimal in the order $\prec_w$ (hence in $\prec_s$)

$$\delta(I(\mathcal{E}_3)) = \{d\}$$
$$\delta(G(\mathcal{E}_3)) = \{p(d), r(d)\}$$

(2.4) relations among $\mathcal{E}_1$, $\mathcal{E}_2$ and $\mathcal{E}_3$

$$\mathcal{E}_1 \equiv_{ww} \mathcal{E}_2 \prec_{ww} \mathcal{E}_3$$

$\mathcal{E}_1$ and $\mathcal{E}_2$ incomparable with $\mathcal{E}_3$ in the order $\prec_w$ and $\prec_s$

Case (3) **axiom** $\alpha \equiv \exists x(\neg p(x) \wedge \neg q(x) \wedge r(x))$

(3.1) extension $\mathcal{E}_1$, minimum in the order $\prec_s$ (hence in $\prec_w$ and $\prec_{ww}$)

$$\delta(I(\mathcal{E}_1)) = \{d\}$$
$$\delta(G(\mathcal{E}_1)) = \{r(d)\}$$

Case (4) **axiom** $\alpha \equiv \forall x \forall y(\tilde{s}(x, y))$

(4.1) extension $\mathcal{E}_1$, minimum in the order $\prec_w$ (hence in $\prec_{ww}$ and minimal in $\prec_s$)

$$\delta(I(\mathcal{E}_1)) = \text{empty}$$
$$\delta(G(\mathcal{E}_1)) = \{s(c, a)\}$$

(4.2) extension $\mathcal{E}_2$, minimal in the order $\prec_s$

$$\delta(I(\mathcal{E}_2)) = \{d\}$$
$$\delta(G(\mathcal{E}_2)) = \{s(c, d), s(d, a)\}$$

(4.3) relation between $\mathcal{E}_1$ and $\mathcal{E}_2$

$$\mathcal{E}_1 \prec_w \mathcal{E}_2$$

$\mathcal{E}_1$ and $\mathcal{E}_2$ incomparable with each other in the order $\prec_s$

Case (5) **axiom** $\alpha \equiv \forall x(\neg p(x) \wedge \neg q(x))$ unsatisfiable by any extension of $\mathcal{I}$.

It is noted that in the simplest cases where the strongly minimum extensions exist, these extensions are logical consequences of $\mathcal{I}$ and α, which makes us suspect this might be always the case. To see if it is indeed the case, let us define a theory $\mathcal{T}(\mathcal{I})$ by considering each element of $I(\mathcal{I})$ as a constant symbol and each ground atom in $G(\mathcal{I})$ as an axiom. Let us likewise consider $\delta(I(\mathcal{E}))$ and $\delta(G(\mathcal{E}))$ as interpreting the formula

$$F(\mathcal{E}) \equiv \wedge_i(\bar{e}_i \notin I(\mathcal{I})) \wedge \wedge_{i,j}(\bar{e}_i \neq \bar{e}_j) \wedge \wedge_\rho \rho[(e_i)_i \mid (\bar{e}_i)_i]$$

where $\bar{e}_i$ is the new constant symbol corresponding to an element e_i of $\delta(I(\mathcal{E}))$, ρ ranges over $\delta(G(\mathcal{E}))$ and $[(e_i)_i \mid (\bar{e}_i)_i]$ indicates the substitution of each occurrence of e_i in ρ by that of $\bar{e}_i$. This is the result of Skolemizing the existentially quantified variables in the formula

$$\tilde{F}(\mathcal{E}) \equiv (\exists x_i)_i((\wedge_i(x_i \notin I(\mathcal{I})) \wedge \wedge_{i,j}(x_i \neq x_j) \wedge \wedge_\rho \rho[(e_i)_i \mid (x_i)_i])$$

which is a legitimate formula of $\mathcal{T}(\mathcal{I})$. Assume

$$\mathcal{T}(\mathcal{I}) \wedge \alpha \vdash \tilde{F}(\mathcal{E})$$

and take any extension $\mathcal{E}'$ of $\mathcal{I}$ which is a model of α, hence of $\tilde{F}(\mathcal{E})$. Since $\mathcal{E}'$ is also a model of $F(\mathcal{E})$ (Lemma 3.16), we can define a monomorphic mapping μ from $\mathcal{E}$ into $\mathcal{E}'$ such that for each e_i, $\mu(e_i)$ interprets the same $\bar{e}_i$ as e_i does and establish the relation $\mathcal{E} \prec_s \mathcal{E}'$. Conversely, if $\mathcal{E}$ is the strongly minimum extension, it will be structurally imbedded in any extension $\mathcal{E}'$ of $\mathcal{I}$ with respect to α. If we additionally postulate the unique name axiom

$$\nu \equiv \wedge_{i,j}(\bar{c}_i \neq \bar{c}_j) \quad \text{for any pair of elements } c_i \text{ and } c_j \text{ of } I(\mathcal{I})$$

every model of $\mathcal{T}(\mathcal{I}) \wedge \nu \wedge \alpha$ is isomorphic to a certain $\mathcal{E}'$ in which F, hence also $\tilde{F}$, is true. By the completeness of FOPL (Theorem 3.2), $\tilde{F}$ must be a logical consequence of α in the theory $\mathcal{T}(\mathcal{I}) \wedge \nu$. We have thus proved:

Theorem 8.1. Given an interpretation $\mathcal{I}$ and axiom α, an extension $\mathcal{E}$ of $\mathcal{I}$ with respect to α is a model of the theory $\mathcal{T}(\mathcal{I}) \wedge \nu \wedge \alpha$. Conversely, any model of $\mathcal{T}(\mathcal{I}) \wedge \alpha \wedge \nu$ is isomorphic to some extension of $\mathcal{I}$. $\mathcal{E}$ is the minimum in the strong sense if and only if

$$\mathcal{T}(\mathcal{I}) \wedge \nu \wedge \alpha \vdash \tilde{F}(\mathcal{E})$$

where $\mathcal{T}(\mathcal{I}), \nu$ and $\tilde{F}(\mathcal{E})$ are as defined above.

Corollary. Given an interpretation $\mathcal{I}$, an axiom α allows for the minimum extension of $\mathcal{I}$ with respect to α in the strong sense if and only if it can be satisfied by the fewest possible additions of new objects to $I(\mathcal{I})$ and additions of its logical consequences in the augmented universe of interpretation to $G(\mathcal{I})$.

In the above case (3) for instance, we recognize the need for adding at least one object (d) since none of the existing objects can satisfy the matrix of α, and find immediately that $r(d)$ derives from and satisfies α.

This corollary characterizes the strongly minimum extension as the result of carrying the design one step forward in an analytical manner, much the same as in logic programming, and provides a means to actually enumerate the said extension when it does exist. However, more complex cases (cases (2) and (4)) show that on one hand, minimumness in the strong sense is too strict a condition to be required of an extension, and on the other, minimality in the same sense is too coarse a sieve to screen a proper one. The same can be said of the weak order (case (2)). We are therefore inclined to take minimality (or minimumness if attainable) in the weakest sense as the standard criterion for distinguishing a particular extension among alternatives. Hereafter, precedence, equivalence, minimality and minimumness are assumed to be in the weakest sense unless explicitly stated otherwise. $\prec_{ww}$ will be written simply as $\prec$ and defines a partial order in the set of equivalence classes under $\equiv$ (meaning $\equiv_{ww}$).

8.1.2 Fundamental Procedure

In this section, we will develop the fundamental procedure to evaluate how far the current interpretation $\mathcal{I}$ has come to satisfy the axiom α designated as the immediate subgoal and to extend $\mathcal{I}$ to the full satisfaction of α. The adjective 'fundamental' signifies that the said procedure does not depend on any domain specificities but on the purely logical properties of $\mathcal{I}$ and α, and that performance issues are not considered. Let us assume that α takes the form of either

$$\alpha_\forall \equiv \forall x(\rho(x) \to \beta(x)) \quad \text{or} \quad \alpha_\exists \equiv \exists x(\rho(x) \land \beta(x)) \tag{8.3}$$

where

$$\rho(x) \equiv (p(x) \land \cdots \land \neg p'(x) \land \cdots) \lor \cdots \lor (q(x) \land \cdots \land \neg q'(x) \land \cdots) \tag{8.4}$$

where p's and q's are unary predicate symbols (sorts, typically) with a single argument of sort ι and where at least one positive literal exists in each disjunct. A justification for this assumption lies in the recognition that examining only a positive-finitely evaluable part of $I(\mathcal{I})$ should suffice to establish the truth of α if it is ever possible. More precisely, if we let α be written as

$$\forall x(\alpha_1(x)) \quad \text{or} \quad \exists x(\alpha_1(x)) \tag{8.5}$$

and ρ be as above, either

$$\forall x(\neg\alpha_1(x) \to \rho(x)) \quad \text{or} \quad \forall x(\alpha_1(x) \to \rho(x)) \tag{8.6}$$

must hold true respectively. We will then have:

Lemma 8.3. α takes one of the forms of (8.3) if and only if (8.6) holds true of α_1.

Proof. The 'only if' part is clear. To prove the 'if' part, assume first

$$\forall x(\neg\alpha_1(x) \to \rho(x))$$

and define $\beta(x)$ as $\rho(x) \land \alpha_1(x)$. Then,

$$\begin{aligned}\forall x(\rho(x) \to \beta(x)) &\equiv \forall x(\neg\rho(x) \lor (\rho(x) \land \alpha_1(x))) \equiv \forall x(\neg\rho(x) \lor \alpha_1(x)) \\ &\equiv \forall x((\neg\rho(x) \lor \alpha_1(x)) \land (\neg\alpha_1(x) \to \rho(x))) \\ &\equiv \forall x((\neg\rho(x) \lor \alpha_1(x)) \land (\rho(x) \lor \alpha_1(x))) \\ &\equiv \forall x((\neg\rho(x) \land \rho(x)) \lor \alpha_1(x)) \equiv \forall x(\alpha_1(x)) \equiv \alpha_\forall.\end{aligned}$$

Likewise, if

$$\forall x(\alpha_1(x) \to \rho(x)),$$

then, by defining $\beta(x)$ as $\alpha_1(x)$,

$$\begin{aligned}\exists x(\rho(x) \wedge \beta(x)) &\equiv \exists x(\rho(x) \wedge \alpha_1(x)) \\ &\equiv \exists x(\rho(x) \wedge \alpha_1(x)) \vee \neg\forall x(\alpha_1(x) \to \rho(x)) \\ &\equiv \exists x(\rho(x) \wedge \alpha_1(x)) \vee \exists x(\neg\rho(x) \wedge \alpha_1(x))) \\ &\equiv \exists x((\rho(x) \vee \neg\rho(x)) \wedge \alpha_1(x)) \equiv \exists x(\alpha_1(x)) \equiv \alpha_\exists.\end{aligned}$$

End proof

The procedure consists of two stages. The first stage is to evaluate the extensions of formulas ρ and β, and the second is to assert $\beta(x)$ of each or some element x of ρ's extension. By virtue of the finite evaluability of formulas (Lemma 6.1), there is nothing difficult about the first stage. It will not be always easy or even possible, however, to assert β, not to speak of doing so minimally, because of the potential inconsistency of the theory $\mathcal{T}(\mathcal{I}) \wedge \alpha$ and the multiplicity of its interpretations. The following theorem, which should be apparent without proof, provides a basis for and sets a practical limitation on the development of the second stage of the procedure.

Theorem 8.2. Let $\alpha_\forall, \alpha_\exists, \beta$ and ρ be as in (8.3) and (8.4), and $\mathcal{E}$ be an extension of $\mathcal{I}$. Let $U_\alpha, \alpha'_\forall$ and $\alpha'_\exists$ be defined by

$$U_\alpha \equiv \{x : \rho(x) \text{ in } \mathcal{E}\} \tag{8.7}$$

$$\alpha'_\forall \equiv \textstyle\bigwedge_{x \in U'} \beta(x) \tag{8.8}$$

$$\alpha'_\exists \equiv \textstyle\bigvee_{x \in U'} \beta(x) \tag{8.9}$$

where U' is a subset of U_α. Assume α is false in the current interpretation $\mathcal{I}$. Then:

(1) If $\alpha_\forall$ is true in $\mathcal{E}$, so is $\alpha'_\forall$. Conversely, if $U' = U_\alpha$ and if $\alpha'_\forall$ is true in $\mathcal{E}$, so is $\alpha_\forall$.

(2) If $\alpha'_\exists$ is true in $\mathcal{E}$, so is $\alpha_\exists$. Conversely, if $U' = U_\alpha$ and if $\alpha_\exists$ is true in $\mathcal{E}$, so is $\alpha'_\exists$.

(3) The set U°_α defined by

$$U^\circ_\alpha \equiv \{x : \rho(x) \wedge \neg\beta(x) \text{ in } \mathcal{I}\} \tag{8.10}$$

is a subset of U_α for any $\mathcal{E}$, hence the corresponding $\alpha'_\forall$ or $\alpha'_\exists$ denoted by $\alpha^\circ_\forall$ or $\alpha^\circ_\exists$ is a necessary or sufficient condition for establishing $\alpha_\forall$ or $\alpha_\exists$ respectively. Moreover,

(3.1) $\alpha^\circ_\forall$ is sufficient if its assertion is possible without the addition of any element to U°_α.

(3.2) $\alpha^\circ_\exists$ is necessary if the addition of any element to U°_α is impossible.

(4) Let $\alpha_\vee$ and $\alpha_\wedge$ be defined by

$$\alpha_\vee \equiv \alpha^{(1)} \vee \alpha^{(2)} \quad \alpha_\wedge \equiv \alpha^{(1)} \wedge \alpha^{(2)} \tag{8.11}$$

where $\alpha^{(1)}$ and $\alpha^{(2)}$ are in either form of (8.3). Then:
(4.1) Every extension with respect to $\alpha^{(1)}$ or $\alpha^{(2)}$ is an extension with respect to $\alpha_\vee$, and vice versa.
(4.2) Every extension with respect to $\alpha^{(1)}$ and $\alpha^{(2)}$ is an extension with respect to $\alpha_\wedge$, and vice versa.

Note that the third proposition, which is the only part of the theorem that might not be immediately clear, calls upon the Rule P stipulated in the preceding section. By moving from $\mathcal{I}$ to $\mathcal{E}$, the extension of ρ which comprises only positive ground atoms may only be increased, while that of β may be increased or decreased. (Fig. 8.4)

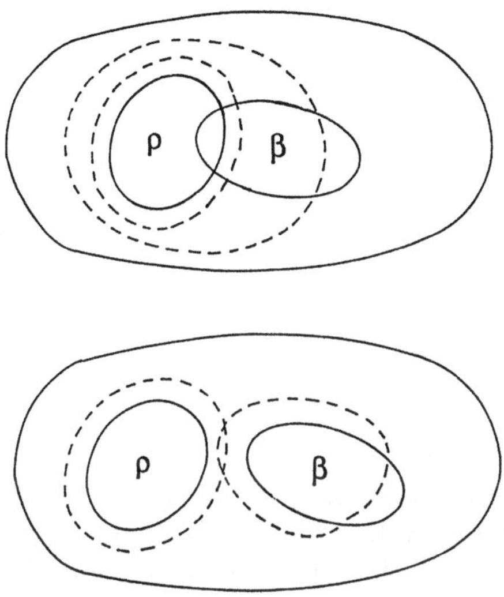

Fig. 8.4. Extensions with respect to $\alpha_\forall$ and $\alpha_\exists$

Corollary. (1) For each extension $\mathcal{E}$ of $\mathcal{I}$ with respect $\alpha_\forall$, there exists some extension $\mathcal{E}'$ with respect to $\alpha^\circ_\forall$ such that $\mathcal{E}' \prec \mathcal{E}$. If the assertion of β of any element of U°_α neither falsifies any other instances of β, nor entails the assertion of ρ of any other elements, then each extension with respect to $\alpha^\circ_\forall$ is also one with respect to $\alpha_\forall$.
(2) For each extension $\mathcal{E}'$ of $\mathcal{I}$ with respect to $\alpha^\circ_\exists$, there exists some extension $\mathcal{E}$ with respect to $\alpha_\exists$ such that $\mathcal{E} \prec \mathcal{E}'$. If the assertion of ρ and β of any existing element outside of U_α is impossible, each extension with respect to $\alpha_\exists$ is either one also respect to $\alpha^\circ_\exists$ or it entails the creation of some new element.
(3) Suppose $\alpha^\circ(\alpha^\circ_\forall$ or $\alpha^\circ_\exists)$ is put into a disjunctive form

$$(\alpha_1^{(1)} \wedge \cdots \wedge \alpha_{n_1}^{(1)}) \vee \cdots \vee (\alpha_1^{(r)} \wedge \cdots \wedge \alpha_{n_r}^{(r)}) \tag{8.12}$$

where each $\alpha_j^{(i)}$ is itself either a disjunction of closed sub-formulas or in one form of (8.3). Then, each extension with respect to α^o is one with respect to each of the conjuncts of some $\alpha_1^{(i)} \wedge \cdots \wedge \alpha_{n_i}^{(i)}$, and vice versa.
(4) Suppose instead α^o is in a conjunctive form

$$(\alpha_1^{(1)} \vee \cdots \vee \alpha_{n_1}^{(1)}) \wedge \cdots \wedge (\alpha_1^{(r)} \vee \cdots \vee \alpha_{n_r}^{(r)}). \tag{8.13}$$

Then, each extension with respect to α^o is one with respect to some of the disjuncts of each conjunct.

The second stage starts with unfolding α into the conjunction or disjunction of sub-formulas $\beta(x)$ according to the kind of the outermost quantification, with x ranging over the set U_α^o. The assertion of α is construed as that of every or some of the $\beta(x)$. The latter are in turn represented by the conjunctive or disjunctive assertion of sub-formulas. The same is repeated until the next level of quantification or atomicity is reached at which the step of unfolding is reentered or instances of predicates are simply added. The entire process forms a tree as illustrated in Fig. 8.6 where nodes, leaves and branches represent the assertion of sub-formulas, ground atoms and logical connectives respectively. An instance of it or a path is a collection of nodes and branches such that all the AND branches or precisely one OR branch emanating from its nodes are contained in it. It will come to a successful completion if all the assertions at leaves are actually made.

The sub-tree under each node embodies the sub-process of asserting the corresponding sub-formula. While we shun discussing the details of implementation issues, let us assume for simplicity that these sub-processes are carried out sequentially in the depth-first and left-to-right manner. Thus, the truth of a sub-formula at a node will be established when all or one of its immediate descendants taken up in turn are found to be true. Various anomalies may arise during the process, hindering the attainment of minimality or even the realization of extension itself. As examined in the following, some may be remedied while others may not.

(1) *Failure to Assert a Sub-Formula*
This is the situation where any alternate path originating at the current node (ν) is inconsistent with $\mathcal{I}$ augmented by those ground atoms that have been already asserted. Let $\tilde{\nu}, \nu_-$ and ν_+ be the immediate ascendant (root node), predecessor and successor of ν respectively within the conjunctive or disjunctive set of nodes to which ν belongs. Let μ be the last disjunctive node that has been successfully asserted by selecting a member node which is not the last one and does not allow itself any alternate path. If μ is nil (non-existent), α is inconsistent with $\mathcal{I}$. (See Fig. 8.5.) Several sub-cases are distinguished.

(1.1) *ν Is a Conjunct*

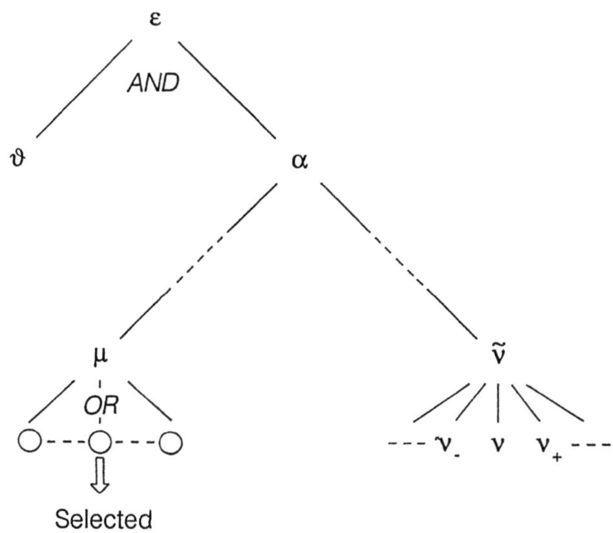

Fig. 8.5. A node's immediate neighbors and point of backtracking

(1.1.1) ν *Is the First (Leftmost) Member of a Conjunctive Set of Nodes*
The failure of $\tilde{\nu}$ is reported and a backtracking is made to μ in search of an alternate path.

(1.1.2) ν *Is Preceded by (Lying at the Right of) Another Node*
A backtracking is made to μ which might be ν_-.

(1.2) ν *Is a Disjunct of an Original Sub-Formula*

(1.2.1) ν *Is the Last (Rightmost) Member of a Disjunctive Set of Nodes*
The failure of $\tilde{\nu}$ is reported and a backtracking is made to μ.

(1.2.2) ν *Is Succeeded by (Lying at the Left of) Another Node*
No failure is reported but the process moves on to take up ν_+. An indication is given so that no backtracking may be made to this node from any (succeeding) node.

(1.3) ν *Is a Disjunct of an Unfolded (Existentially Quantified) Sub-Formula*
This is similar to the case (1.2) except for the possible creation of new objects, nodes and underlying sub-trees to assert the sub-formula of these objects. Such will occur if and only if the failure to assert ν inevitably implies that of α under the rule (1.2). More precisely, it will occur if and only if

- ν is the last member of the disjunction under $\tilde{\nu}$, hence its failure implies that of $\tilde{\nu}$ according to (1.2), and
- μ is nil.

(2) *Failure to Keep Intact the Extension of* α_1 *(as per* (8.5)*) in* $\mathcal{I}$ *for* $\alpha = \alpha_\forall$
Carrying out a successful path of extension may possibly falsify instances

of α_1 that are valid in the current interpretation. The possibility can be decreased by requiring that the truth values of ρ once given in an interpretation do not change in any extension of it. Instances of β may still be necessarily deleted, however, as the result of adding those for the elements of U°_α. The whole process will then have to be repeated for a new U°_α. This may eventually lead to a success or complete failure (inconsistency), or could continue ad infinitum if the assertion of β involves the creation of new elements.

(3) *Failure to Attain the Minimality of Extension for* $\alpha = \alpha_\exists$
β might be asserted of an element outside of U°_α more economically than of any element of U°_α. Such an extension will be prohibited, not to speak of a minimal one, by making the same requirement as stated above of ρ.

In concluding this section, let us return to the example of closing a polyline, put the requirement into a set of axioms and form the tree to represent the process of asserting it.

- Theory $\mathcal{T}$
  - Sort $edge(\iota)$
  - Function $coord_-, coord_+ : edge \rightarrow \mathbf{real}^2$
  - Predicate $adj(edge, edge)$
  - Axiom

$$A_- \equiv \forall e(edge(e) \rightarrow \exists e_-(edge(e_-) \wedge (e_- \neq e) \wedge adj(e_-, e) \wedge \forall e'((edge(e') \wedge adj(e_-, e')) \rightarrow (e = e'))))$$

$$A_+ \equiv \forall e(edge(e) \rightarrow \exists e_+(edge(e_+) \wedge (e_+ \neq e) \wedge adj(e, e_+) \wedge \forall e'((edge(e') \wedge adj(e', e_+)) \rightarrow (e = e'))))$$

$$A_c \equiv \forall e \forall e_+((edge(e) \wedge edge(e_+) \wedge adj(e, e_+)) \rightarrow (coord_+(e) = coord_-(e_+)))$$

$$\alpha \equiv A_- \wedge A_+ \wedge A_c$$

- Current interpretation $\mathcal{I}$
  - $I(\mathcal{I}) = \{e_i (i = 0 - n)\}$
  - $G(\mathcal{I}) = \{edge(e_i), c_{-i} = coord_-(e_i), c_{+i} = coord_+(e_i), adj(e_i, e_{i+1}) \; (i = 0 - n{-}1)\}$

8.2 The Expansion of Model Description Language (MDL)

Ideally, the application agent will only have to issue the statement

assert α

to extend the current interpretation with respect to α, letting all the inconsistencies and indeterminacies be automatically taken care of, which will require a high level of intelligence in the system. The most unintelligent way of achieving the purpose is for the application agent to determine U°_α by

evaluate $\rho(?x) \wedge \neg\beta(?x)$

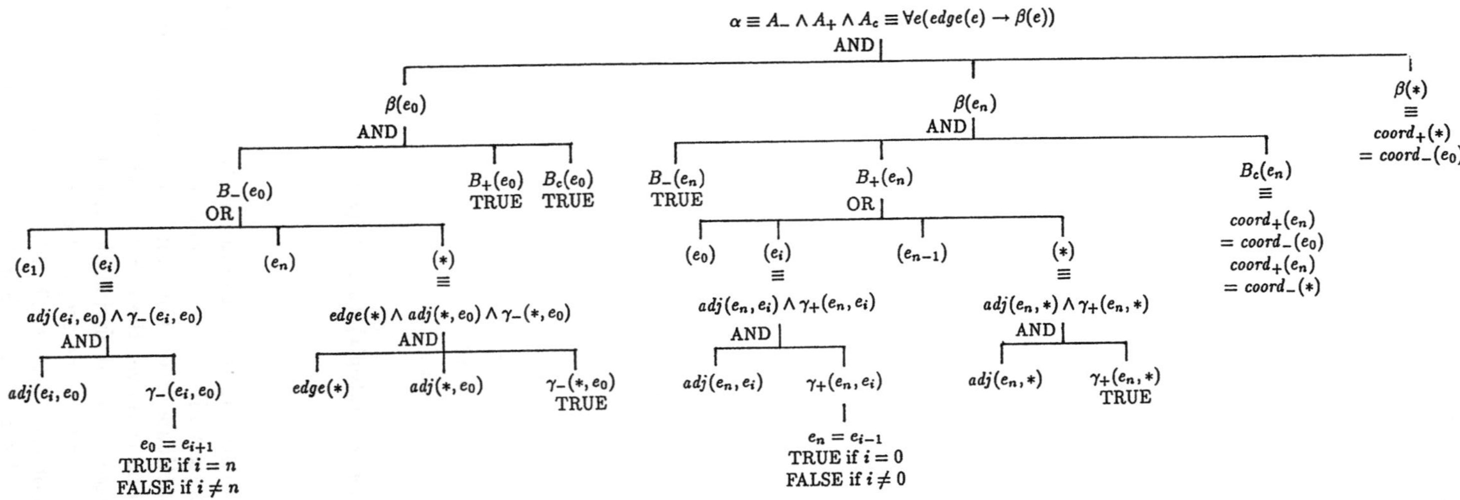

$$\gamma_-(x, y) \equiv \forall z((edge(z) \wedge adj(x, z)) \rightarrow (y = z))$$

$$\gamma_+(x, y) \equiv \forall z((edge(z) \wedge adj(z, y)) \rightarrow (x = z))$$

Fig. 8.6. The process tree for extending $\mathcal{I}$ with respect to α

where ρ and β are as per (8.3), to create new objects if necessary and to

assert ⟨a conjunction of ground atoms⟩

such that β will be valid in the thus extended interpretation. The present section aims somewhere in between by providing language constructs with which to describe the preferred process of extension as against others. To be more specific, let us consider the case of α taking the form

$$\alpha \equiv \forall x(\neg p(x) \vee q(x) \vee \beta(x)) \tag{8.14}$$

which allows two (or possibly more) processes toward satisfaction as represented by two trees of Fig. 8.7.

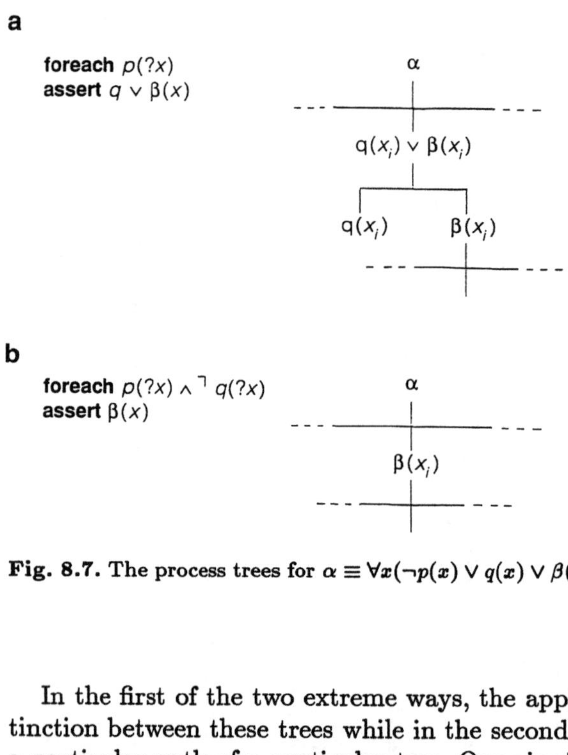

Fig. 8.7. The process trees for $\alpha \equiv \forall x(\neg p(x) \vee q(x) \vee \beta(x))$

In the first of the two extreme ways, the application agent makes no distinction between these trees while in the second, it concerns itself only with a particular path of a particular tree. Our aim is to enable it to specify explicitly a preference for one tree over the other, leaving the path selection to the system. The specification should obviously have its core in the treatment of the outermost formulas within the scope of each quantifier. It will be reduced to the convention of Sect. 6.2 (2.2.2) when the axiom is a conjunction of ground literals and/or Horn clauses. For the above case of (8.14), those corresponding to the two trees are to be given as follows:

- Tree 8.7(a) **assert** $\forall x(\neg p(x) \vee q(x) \vee \beta(x))$

begin
  foreach $p(?x)$ **do**
  begin
    assert $q(x)$ **or assert** $\beta(x)$
  end
end.

- Tree 8.7(b) **assert** $\forall x(\neg p(x) \vee q(x) \vee \beta(x))$

begin
  foreach $p(?x) \wedge \neg q(?x)$ **do**
  begin
    assert $\beta(x)$
  end
end.

The predicate q is treated as an assertion or a constraint. The function of **foreach** is of extra-logical level in that it serves to distinguish these meanings of q which are indistinguishable from the purely logical or set-theoretic standpoint. Likewise, **assert** $q(x)$ and **assert** $\beta(x)$ have different meanings than the truth values of $q(x)$ and $\beta(x)$ respectively, and the connective **or** should be understood as being applied to the sub-processes under these nodes rather than to the formulas themselves. The introduction of such extra-logical elements is necessitated and justified in view of the inherently extra-logical nature of the task of constructing models.

8.2.1 Syntax and Semantics

(1) ⟨process⟩ ::= **process** ⟨identifier⟩ [**to assert** ⟨open formula⟩ |] ⟨block⟩

(2) ⟨new statement⟩ ::= **assert** ⟨closed formula⟩ | ⟨enter process⟩ | [[**foreach** | **forsome**] ⟨open formula⟩ | **fornew** ⟨list of free var id⟩] ⟨statement⟩ | ⟨statement⟩ [**or** ⟨statement⟩ **or** ... **or** ⟨statement⟩ | **and** ⟨statement⟩ **and** ... **and** ⟨statement⟩]

A process is defined to assert a class of (closed) formulas which is to be instantiated and activated by an ⟨enter process⟩ statement specified below. Those free variables occurring in the open formula in the process header will be given specific values when 'entered', hence what will be actually accomplished is the assertion of a closed formula.

The (closed) formula in a simple 'assert' statement should be of the restricted form as in Sect. 6.2 (6.2.2). A process may enter other processes. This can be repeated an arbitrary number of times, yet no recursion is allowed. The control words **foreach** and **forsome** specify the way of asserting a formula with respect to a universally or existentially quantified variable it contains. Free variables denoted by ?⟨var id⟩ occurring in the open formula are those bound in the original formula and treated as if they represent certain instances in the formulas to be asserted. Suppose, for example, we write

foreach $p(?x)$ (**assert** $(q(x))$ or **forsome** $(r(x, ?y) \wedge \neg s(x, ?y))$
assert $(\beta(x, y))$),

then

$$\forall x(\neg p(x) \vee q(x) \vee \exists y(r(x, y) \wedge \neg s(x, y) \wedge \beta(x, y)))$$

will be asserted, for which alternatives exist such as

foreach $(p(?x) \wedge \neg q(?x))$ **forsome** $r(x, ?y)$ **assert** $(\neg s(x, y) \wedge \beta(x, y))$.

The difference between these is that the latter does not allow an increase of the extension of q while the former does not allow a selection from the extension of s.

The control word **fornew** indicates that new objects denoted by the following list of free variable id are to be created. These objects can be retrieved later by evaluating the predicate **new**.

(3) ⟨enter process⟩ ::= **enter** ⟨process id⟩ [**to assert** ⟨closed formula⟩ |]

(4) ⟨built-in function⟩ ::= **path**(⟨process id⟩)

(5) ⟨built-in predicate⟩ ::= **succeed**(⟨process id⟩) | **new**([? |] ⟨var id⟩)

(6) ⟨constrained evaluation⟩ ::= ⟨evaluate formula⟩ **in extension**(⟨path id⟩)

(7) ⟨reenter process⟩ ::= **reenter** ⟨process id⟩

(8) ⟨commit path⟩ ::= **commit** ⟨path id⟩

When the process terminates in success or in failure, control is returned to the statement following the 'enter' statement. If in success, the predicate **succeed** is made true and the successful path is given an identifier as the value of function **path**. This identifier can be used to constrain the search space of an 'evaluate' to the collection of those elementary facts (ground instances of functions and predicates) added to or changed from the current interpretation along the path. Such a constrained evaluation enables the application agent to present this particular solution to the designer for his/her judgement.

An alternate path will be searched for each time the process is reentered until all the possible paths are exhausted and the predicate **succeed** turns false. A 'commit path' statement will commit the selected extension to the permanent store after which 'reenter process' may no longer be executed.

8.2.2 Illustrative Examples

(1) *Closing a Polyline*
Let us paraphrase the process tree for the example of closing a polyline of the preceding section in the language just defined.

process *close-a-polyline* **to assert** $\forall e(edge(e) \rightarrow (B_- \wedge B_+ \wedge B_c))$;
begin
  foreach $edge(?e)$ (**enter** B_- **and enter** B_+ **and enter** B_c)
end;
process B_- **to assert** $\exists e_-(edge(e_-) \wedge (e_- \neq e) \wedge adj(e_-, e) \wedge \forall e'((edge(e') \wedge adj(e_-, e')) \rightarrow (e = e')))$;
begin
  (**forsome** $edge(?e_-) \wedge (?e_- \neq e) \wedge \neg\exists e'(adj(?e_-, e') \wedge (e' \neq e))$
  assert $adj(e_-, e)$) **or**
  (**fornew** $?e_-$ **assert** $edge(e_-) \wedge adj(e_-, e)$)
end;
process B_+ **to assert** $\exists e_+(edge(e_+) \wedge (e_+ \neq e) \wedge adj(e, e_+) \wedge \forall e'((edge(e') \wedge adj(e', e_+)) \rightarrow (e = e')))$;
begin
  (**forsome** $edge(?e_+) \wedge (?e_+ \neq e) \wedge \neg\exists e'(adj(e', ?e_+) \wedge (e' \neq e))$
  assert $adj(e, e_+)$) **or**
  (**fornew** $?e_+$ **assert** $edge(e_+) \wedge adj(e, e_+)$)
end;
process B_c **to assert** $\forall e_+((edge(e) \wedge edge(e_+) \wedge adj(e, e_+)) \rightarrow (coord_+(e) = coord_-(e_+)))$;
begin
  foreach $adj(?e, ?e_+)$ **assert** $(coord_+(e) = coord_-(e_+))$
end;

Actually, the assertion of $coord_+(e) = coord_-(e_+)$ cannot be made as simply as that of an enumerative predicate, and requires the execution of a process such as exemplified below. It may involve geometric calculations depending on which attributes are changeable and which are not.

process *coord* **to assert** $coord_+(e) = coord_-(e_+)$;
begin
  if $coord_+(e) \neq coord_-(e_+)$ **then**
    if new(e) **then assert** $coord_+ : e \rightarrow coord_-(e_+)$
    else if new(e_+) **then assert** $coord_- : e_+ \rightarrow coord_+(e)$
      else if changeable(*angle*) **then** ⟨solve the equation $coord_+(e) = coord_-(e_+)$ with $coord_-(e), coord_+(e_-)$ and lengths of e and e_+ fixed as they are⟩
        else if changeable(*length*) **then** ⟨solve the same equation with angles instead of lengths fixed as they are⟩
end;

(2) *Clipping a Polygon*
We want to reformulate in the expanded MDL the problem of clipping a given polygon with a given line which we presented in Sect. 4.2.1.3.

(2.1) *Theory*

sort *polygon*, *edge*, *vertex*, *line*;
function $pee_+(p : polygon, pe(p)) : pe(p)$; $ev_+(edge), ev_-(edge) : vertex$;
$coord(vertex) : \mathbf{real}^2$; $intersection(l : line, e : edge) : \mathbf{real}^2$;
⟨body to calculate the coordinate values of intersection of l and e, and to assign the pair of these as the value of function *intersection*⟩;
predicate $pe(polygon, edge)$;
$intersect(l{:}line, e{:}edge)$;
⟨body to determine if l and e intersect or not⟩;
$sameside(l : line, a, b : edge \cup vertex)$;
⟨body to determine if a and b lie on the same side of l⟩;
$internal(p : polygon, e : edge \cup (\mathbf{real}^2, \mathbf{real}^2))$;
⟨body to determine if e lies inside of p or not⟩;

Readers are referred to Fig. 8.8 for the intuitive meanings of these functions and predicates.

axiom $A1 : \forall p \forall e((polygon(p) \wedge edge(e) \wedge pe(p, e)) \rightarrow (ev_+(e) = ev_-(pee_+(p, e))))$;
$A2 : \forall p(polygon(p) \rightarrow \exists e(edge(e) \wedge pe(p, e)))$;
$A3 : \forall p \forall e(pe(p, e) \rightarrow \exists e^-(pee_+(p, e^-) = e))$;
$A4 : \forall p \forall l((polygon(p) \wedge line(l)) \rightarrow \exists \tilde{p}(polygon(\tilde{p}) \wedge clip\text{-}p(p, l, \tilde{p})))$;
formula $clip\text{-}p(p, l, \tilde{p}) \equiv \forall \tilde{e}(pe(\tilde{p}, \tilde{e}) \rightarrow (eligible(p, l, \tilde{e}) \wedge eligible(p, l, pee_+(\tilde{p}, \tilde{e})) \wedge sameside(l, \tilde{e}, pee_+(\tilde{p}, \tilde{e}))))$;
$eligible(p, l, \tilde{e}) \equiv (pe(p, \tilde{e}) \wedge \neg intersect(l, \tilde{e})) \vee \exists e(pe(p, e) \wedge clip\text{-}e(e, l, \tilde{e})) \vee clip\text{-}l(l, p, \tilde{e})$;
$clip\text{-}e(e, l, \tilde{e}) \equiv intersect(l, e) \wedge (((ev_-(e) = ev_-(\tilde{e})) \wedge (intersection(l, e) = coord(ev_+(\tilde{e})))) \vee ((ev_+(e) = ev_+(\tilde{e})) \wedge (intersection(l, e) = coord(ev_-(\tilde{e})))))$;
$clip\text{-}l(l, p, \tilde{e}) \equiv \exists e1 \exists e2((e1 \neq e2) \wedge pe(p, e1) \wedge pe(p, e2) \wedge (intersection(l, e1) = coord(ev_-(\tilde{e}))) \wedge (intersection(l, e2) = coord(ev_+(\tilde{e}))) \wedge internal(p, \tilde{e}))$;

(2.2) *Current Interpretation*
It is assumed that a polygon p exists in the current interpretation together with its edges and vertices, and the axioms *A1* through *A3* hold of these. Also assumed is that a line l has been given and an object $\tilde{p}$ of sort *polygon* has been created.

(2.3) *Completing a Model*
What is to be accomplished is the assertion of the formula

$$\forall \tilde{e}(pe(\tilde{p}, \tilde{e}) \rightarrow (ev_+(\tilde{e}) = ev_-(pee_+(\tilde{p}, \tilde{e})))) \wedge \exists \tilde{e}(edge(\tilde{e}) \wedge pe(\tilde{p}, \tilde{e})) \wedge clip\text{-}p(p, l, \tilde{p})$$

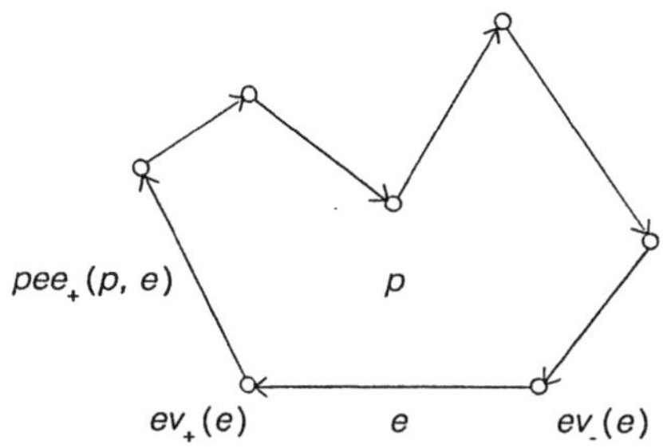

Fig. 8.8. The topology of a polygon and its constituents

which is an instance of $A1 \wedge A2 \wedge A4$. The process for this is defined in a stepwise refinement as follows and depicted as in Fig. 8.9.

process *clip-a-polygon* **to assert** $\widetilde{A1} \wedge \widetilde{A2} \wedge \widetilde{A4}$;
begin
((**forsome** $pe(p, ?\tilde{e}) \wedge \neg intersect(l, ?\tilde{e})$ **assert** $pe(\tilde{p}, \tilde{e})$) **or**
(**forsome** $pe(p, ?e) \wedge intersect(l, ?e)$
fornew $?\tilde{e}, ?v$ **assert** $pe(\tilde{p}, \tilde{e}) \wedge (ev_- : \tilde{e} \rightarrow ev_-(e)) \wedge (ev_+ : \tilde{e} \rightarrow v)$
$\wedge (coord : v \rightarrow intersection(l, e))$)) **and**
(**enter** $\widetilde{B4}$ **to assert** $clip\text{-}p(p, l, \tilde{p})$)
end;
process $\widetilde{B4}$ **to assert** $clip\text{-}p(p, l, \tilde{p})$;
begin
foreach $pe(\tilde{p}, ?\tilde{e}) \wedge \neg\exists \tilde{e}^+(pee_+(\tilde{p}, ?\tilde{e}) = \tilde{e}^+)$
(**forsome** $eligible(p, l, ?\tilde{e}^+) \wedge sameside(l, \tilde{e}, ?\tilde{e}^+) \wedge (ev_+(\tilde{e}) = ev_-(?\tilde{e}^+))$
assert $pe(\tilde{p}, \tilde{e}^+) \wedge (pee_+ : \tilde{p}, \tilde{e} \rightarrow \tilde{e}^+)$) **or**
(**fornew** $?\tilde{e}^+$ **enter** $\widetilde{B4.1}$ **to assert** $eligible(p, l, \tilde{e}^+) \wedge$
$sameside(l, \tilde{e}, \tilde{e}^+)$ **and**
assert $pe(\tilde{p}, \tilde{e}^+) \wedge (pee_+ : \tilde{p}, \tilde{e} \rightarrow \tilde{e}^+) \wedge (ev_- : \tilde{e}^+ \rightarrow ev_+(\tilde{e}))$)
end;
process $\widetilde{B4.1}$ **to assert** $eligible(p, l, \tilde{e}^+) \wedge sameside(l, \tilde{e}, \tilde{e}^+)$
begin
(**forsome** $pe(p, ?e) \wedge intersect(l, ?e) \wedge (ev_+(\tilde{e}) = ev_-(?e))$
fornew $?v$ **assert** $(ev_+ : \tilde{e}^+ \rightarrow v) \wedge (coord : v \rightarrow intersection(l, e))$ **or**
(**forsome** $pe(p, ?e) \wedge intersect(l, ?e) \wedge$
$internal(p, (coord(ev_+(\tilde{e})), intersection(l, ?e)))$
fornew $?v$ **assert** $(ev_+ : \tilde{e}^+ \rightarrow v) \wedge (coord : v \rightarrow intersection(l, e))$ **or**
(**forsome** $pe(p, ?e) \wedge intersect(l, ?e) \wedge (coord(ev_+(\tilde{e})) = intersection(l, ?e))$
assert$(ev_+ : \tilde{e}^+ \rightarrow ev_+(e))$
end.

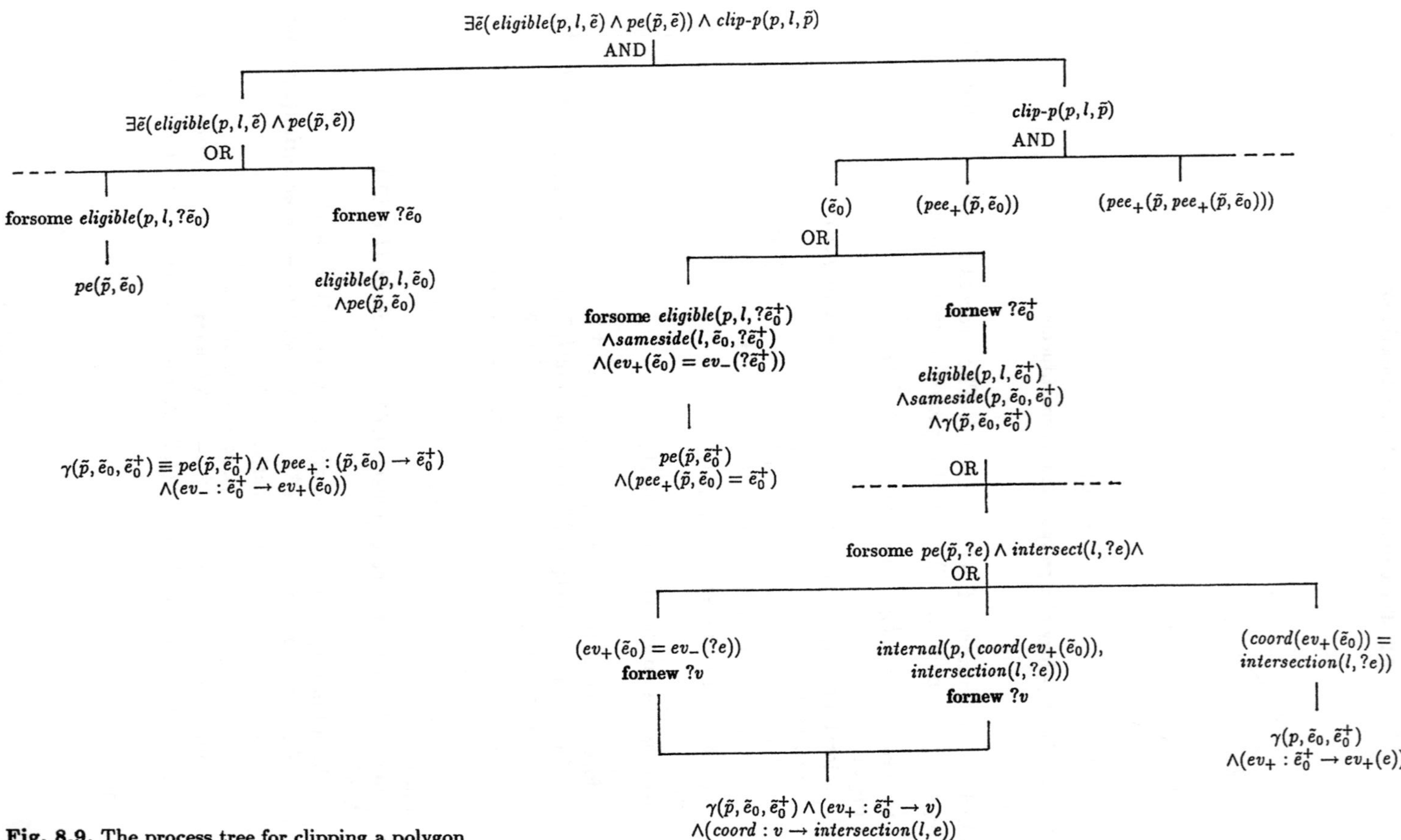

Fig. 8.9. The process tree for clipping a polygon

Discussion
We did not include *A3* among the axioms to be explicitly asserted because it would follow automatically from other axioms together with the properties of function pee_+ and predicate *sameside* in an finite interpretation of the theory. That is, the facts that

- if a vertex is shared by more than two eligible edges, it must lie at the intersection of an edge of p and the line l, hence there must be four pairs of incoming and outgoing edges, and that
- one of the pairs lies on different sides of l while another forms physically the same line segment, hence both incoming edges cannot be connected to the same outgoing edge via pee_+

establish a one-to-one, hence onto correspondence by pee_+. It should be clearly noted, however, that the above is not a logical truth but a piece of geometric knowledge and that the equivalence of one-to-one and onto relies vitally on the finiteness of the set $\{\tilde{e} : pe(\tilde{p}, \tilde{e})\}$ for any $\tilde{p}$. For the sake of theoretical completeness and derivability of other important consequences, *A3* cannot be dispensed with.

The same piece of knowledge was used in defining the process $\widetilde{B4.1}$ to the effect that direct references to the predicate *sameside* were saved. The primary motivation of this is not the executional efficiency but the structural simplicity of the process.

We also depend on our knowledge of geometry in providing the bodies of such non-enumerative function and predicates as *intersection*, *intersect*, *sameside* and *internal*. Domain specificities are normal and usual in formulating a theory or selecting the axioms to be asserted at each moment. It is emphasized, however, that the interpretation and execution of processes, once they are given, will be done independently of any specific domains.

We may take quite a different approach, of which the first step is to create all the eligible edges, the second to select an arbitrary starting edge and the third to make all the possible connections. The first step is mostly geometry-oriented while the succeeding steps are purely 'topological', hence domain independent. To facilitate a clear separation between domain-specific and independent aspects of the problem, we introduce two enumerative predicates *eligible* and *differentsides*, and postulate the defining formula for *eligible* as an axiom. What is to be asserted will then be:

$$C \equiv C1 \wedge C2 \wedge C3$$
$$C1 \equiv C1.1 \wedge C1.2$$
$$C1.1 \equiv \forall e((pe(p,e) \wedge \neg intersect(l,e)) \rightarrow eligible(e))$$
$$C1.2 \equiv \forall e1 \forall e2((pe(p,e1) \wedge pe(p,e2) \wedge (e1 \neq e2) \wedge intersect(l,e1) \wedge intersect(l,e2) \wedge internal(p,(intersection(l,e1), intersection(l,e2)))) \rightarrow \exists e1^- \exists e1^+ \exists e2^- \exists e2^+ \exists l1 \exists l2 \exists v1 \exists v2(D1.2))$$

$$D1.2 \equiv \textit{eligible}(e1^-) \wedge \textit{eligible}(e1^+) \wedge \cdots \wedge \textit{eligible}(l2) \wedge$$
$$(ev_-(e1^-) = ev_-(e1)) \wedge (ev_+(e1^-) = ev_-(e1^+) = v1) \wedge$$
$$(ev_+(e1^+) = ev^+(e1)) \wedge (ev_-(e2^-) = ev_-(e2)) \wedge$$
$$(ev_+(e2^-) = ev_-(e2^+) = v2) \wedge (ev_+(e2^+) = ev^+(e2)) \wedge$$
$$(ev_-(l1) = v1) \wedge (ev_+(l1) = v2) \wedge (ev_-(l2) = v2) \wedge (ev_+(l2) = v1) \wedge$$
$$(\textit{coord}(v1) = \textit{intersection}(l, e1)) \wedge (\textit{coord}(v2) = \textit{intersection}(l, e2)) \wedge$$
$$\textit{differentsides}(e1^-, e1^+) \wedge \textit{differentsides}(e2^-, e2^+) \wedge$$
$$\textit{differentsides}(l1, l2)$$
$$C2 \equiv \exists\tilde{e}(\textit{eligible}(\tilde{e}) \wedge pe(\tilde{p}, \tilde{e}))$$
$$C3 \equiv \forall\tilde{e}^+((\textit{eligible}(\tilde{e}^+) \wedge \exists\tilde{e}(pe(\tilde{p}, \tilde{e}) \wedge (ev_-(\tilde{e}^+) = ev_+(\tilde{e})) \wedge$$
$$\neg\textit{differentsides}(\tilde{e}, \tilde{e}^+)) \rightarrow (pe(\tilde{p}, \tilde{e}^+) \wedge (\tilde{e}^+ = pee_+(\tilde{p}, \tilde{e}))))$$

The process to assert C is defined as follows:

process *clip-a-polygon* **to assert** C;
begin
  foreach $pe(p, ?e) \wedge \neg\textit{intersect}(l, ?e)$ **assert** $\textit{eligible}(e)$;
  foreach $pe(p, ?e1) \wedge pe(p, ?e2) \wedge (?e1 \neq ?e2) \wedge \textit{intersect}(l, ?e1) \wedge$
  $\textit{intersect}(l, ?e2) \wedge \textit{internal}(p, (\textit{intersection}(l, ?e1), \textit{intersection}(l, ?e2)))$
    fornew $?e1^-, ?e1^+, ?e2^-, ?e2^+, ?l1, ?l2, ?v1, ?v2$
      assert $\textit{eligible}(e1^-) \wedge \textit{eligible}(e1^+) \wedge \cdots \wedge \textit{eligible}(l2) \wedge$
      $(ev_- : e1^- \rightarrow ev_-(e1)) \wedge (ev_+ : e1^- \rightarrow v1) \wedge \ldots \wedge (ev_+ : l2 \rightarrow v1) \wedge$
      $(\textit{coord} : v1 \rightarrow \textit{intersection}(l, e1); v2 \rightarrow \textit{intersection}(l, e2)) \wedge$
      $\textit{differentsides}(e1^-, e1^+) \wedge \textit{differentsides}(e2^-, e2^+) \wedge$
      $\textit{differentsides}(l1, l2)$;
  forsome $\textit{eligible}(?\tilde{e})$ **assert** $pe(\tilde{p}, \tilde{e})$;
  foreach $\textit{eligible}(?\tilde{e}^+) \wedge \exists\tilde{e}(pe(\tilde{p}, \tilde{e}) \wedge$
  $(ev_-(\tilde{e}^+) = ev_+(\tilde{e})) \wedge \neg\textit{differentsides}(\tilde{e}, \tilde{e}^+)$
    assert $pe(\tilde{p}, \tilde{e}^+) \wedge (pee_+ : \tilde{p}, \tilde{e} \rightarrow \tilde{e}^+)$
end.

Statements are marked off by semicolons (;) rather than logical connectives to indicate that these are to be executed sequentially in the written order. The programming of a process to assert the axiom looks quite straightforward in this case, which may well make one wonder if a system provided with an adequate level of intelligence could automatically develop the details out of the simple **assert** C. We will return to this question in succeeding sections.

Interested readers are encouraged to draw the trees to represent both of the above-defined processes and compare them.

(3) *Designing a Cam Contour*
In designing the contour of a cam to realize the specified motion of the tool end, one of the critical checkpoints was the verification of the constraint imposed on the radius of curvature of each segment. Specifically, we required that the axiom

$$\alpha : \forall i(\textit{r-of-curvature}(cs(i)) \geq \textit{radius}(R))$$

hold true where $cs(i)$ and R denote the i-th contour segment and the roller respectively. (See Sects. 1.1 (10) and 6.3 (3).) We refer to the minimum radius of curvature of $cs(i)$ also by

$$r\text{-}of\text{-}curvature(i, \theta(i), \theta(i-1), radius(O))$$

to explicate its dependence on the angle of rotation for the i-th stroke and the average radius of cam. The latter in turn depends on $\theta(1)$ and the maximum pressure angle ψ. A scenario for designing a cam mechanism was presented in Sect. 6.3 (3.3) which started by specifying the type of the curve representing the tool end's motion, proceeded to assigning the values of $\theta(i)$ and ψ, and after a few more steps, ended by verifying the axiom α. The current question is how we can enable the system to offer alternate plans for the second and succeeding steps when it finds α false. There are three basic means to answer this, that is, decreasing $radius(R)$, increasing $radius(O)$ and increasing $\theta(i) - \theta(i-1)$ for all the i for which the verification fails. The general solution which follows is a combination of these and contains two uncertainty factors ('forsome' phrases).

process *constraining r-of-c* **to assert** α;
begin
  forsome $0 \leq ?d \leq radius(R) - \mathbf{min}_i(r\text{-}of\text{-}curvature(cs(i))$
    assert $radius : R \rightarrow radius(R) - d$ **and**
    (**forsome** $0 \leq ?p \leq radius(R)/\mathbf{min}_i(r\text{-}of\text{-}curvature(cs(i))) - 1$
      enter *decreasing* ψ **to assert** $radius : O \rightarrow (1+p) * radius(O)$ **and**
      foreach $r\text{-}of\text{-}curvature(cs(?i)) < radius(R)$
        enter *increasing* θ **to assert**
        $r\text{-}of\text{-}curvature(i, \theta(i), \theta(i-1), radius(O)) = radius(R))$
end;
process *decreasing* ψ **to assert** $radius : O \rightarrow (1+p) * radius(O)$;
begin
  assert $\psi : O \rightarrow \mathbf{tan}^{-1}(\mathbf{tan}(\psi)/(1+p))$
end;
process *increasing* θ **to assert**
$r\text{-}of\text{-}curvature(i, \theta(i), \theta(i-1), radius(O)) = radius(R)$;
begin
  if $1 < i$ **then assert** $\theta : i-1 \rightarrow \theta(i) - angle(i, r\text{-}of\text{-}curvature, radius(O))$
    else assert $\theta : i \rightarrow \theta(i-1) + angle(i, r\text{-}of\text{-}curvature, radius(O))$
end.

The choices of d and p differ from those of 'forsome' variables in the preceding examples in that these are made from the domain of attributes, moreover from intervals of continuous data type. Unless further constrained by any other conditions, the system is free to use default values (say, the center values), and presents the results to the designer and requests his/her judgement and instructions. Changing the maximum value of the pressure angle (ψ) may likewise be subject to engineering constraints which may be stipulated as additional axioms if known beforehand.

The function *angle* is to determine the value of the angular span of a stroke given those of *r-of-curvature* and other arguments, the body of which is assumed to have been defined elsewhere. It will generally be the case that the assertion of a non-enumerative relation, typically one that involves arithmetic functions and/or predicates among arguments from the domain of attributes, requires the evaluation of inverse functions.

8.3 Controlling the Model Development

One might feel the examples given in the preceding section are still too microscopic, procedural and ad hoc. In this section, we will explore the methods and provide the facilities with which to develop process models in a more systematic and less procedural manner, and hopefully at a more macroscopic and intelligent level. First, we will focus our attention on some of the techniques we have already used with the intent of generalizing and classifying these into a set of working principles to guide programming processes. Next, we will try to expand the system's capability in such a direction as to allow it to treat entities of larger grain size and accept designers' instructions in making selections among alternatives.

8.3.1 Working Principles for Programming Processes

Issues and questions have been pointed out, the decisions on which will largely characterize the process being considered. These include:

- the distinction between the sub-formula to be asserted and that functioning as a constraint within each given axiom,
- the definition of the process body to assert instances of each non-enumerative predicate,
- the identification of the primary axioms to be asserted at a given stage of the design process as against those serving as auxiliaries or constraints, and
- the recognition of logical relationships between formulas, particularly those which could possibly contradict each other,

each of which will be addressed in turn in the following.

(1) *Assertion Versus Constraint*
As indicated in the last paragraph of Sect. 8.2, there is no distinction between assertion and constraint in the purely logical and set-theoretic senses. In fact, the requirement specification as a formal theory pertains only to the final states or completed interpretations where all the axioms are satisfied. On the contrary, the subject matter of process modeling is by definition incomplete interpretations and their extension into more complete ones, which the formal theory has nothing to do with. Distinguishing the constraint from the assertion within each axiom is tantamount to recognizing the part of any

given interpretation that should be kept invariant by any possible extension of it. Take an arbitrary axiom α and let it be in the form either of

$$\forall x(\neg\beta_1(x) \vee \cdots \vee \gamma_1(x) \vee \cdots) \tag{8.15.1}$$

or of

$$\exists x(\beta_1(x) \wedge \cdots \wedge \neg\gamma_1(x) \wedge \cdots). \tag{8.15.2}$$

The general form of the process of asserting either one of these is

$$\begin{array}{l}\textbf{foreach } (\beta_1(?x) \wedge \cdots \wedge \neg\gamma_1(?x) \wedge \cdots) \\ \quad \textbf{assert } (\neg\beta_{i_1}(x) \vee \cdots \vee \gamma_{j_1}(x) \vee \cdots)\end{array} \tag{8.16.1}$$

or

$$\begin{array}{l}\textbf{forsome } (\beta_{r_1}(?x) \wedge \cdots \wedge \neg\gamma_{s_1}(?x) \wedge \cdots) \\ \quad \textbf{assert } (\beta_1(x) \wedge \cdots \wedge \neg\gamma_1(x) \wedge \cdots)\end{array} \tag{8.16.2}$$

respectively. The rationale for demanding the existence of the 'foreach' or 'forsome' phrase in front of the main part is the realization that the act of advancing the state of design will necessarily occur under a certain constraining condition which must remain intact. An important by-product is a decrease in ambiguities and indeterminacies. While application programmers are responsible for explicitly stating (8.16.1) or (8.16.2), it will be quite possible and desirable to expand the MDL slightly so that they may write in a less procedural manner, such as:

$$\begin{array}{l}\textbf{axiom } \alpha : \forall x(\neg\beta_1(x) \vee \cdots \vee \gamma_1(x) \vee \cdots) \\ \quad \textbf{version } \alpha_1 : \forall x(\neg\beta_{i_1}(x) \vee \cdots \vee \gamma_{j_1}(x) \vee \cdots \\ \quad\quad \textbf{under } \beta_{r_1}(x) \wedge \cdots \wedge \neg\gamma_{s_1}(x) \wedge \cdots) \\ \quad \textbf{version } \alpha_2 : \cdots \\ \quad \cdots\end{array} \tag{8.17.1}$$

or

$$\begin{array}{l}\textbf{axiom } \alpha : \exists x(\beta_1(x) \wedge \cdots \wedge \neg\gamma_1(x) \wedge \cdots) \\ \quad \textbf{version } \alpha_1 : \exists x(\beta_{i_1}(x) \wedge \cdots \wedge \neg\gamma_{j_1}(x) \wedge \cdots \\ \quad\quad \textbf{under } \beta_{r_1}(x) \wedge \cdots \wedge \neg\gamma_{s_1}(x) \wedge \cdots) \\ \quad \textbf{version } \alpha_2 : \cdots \\ \quad \cdots\end{array} \tag{8.17.2}$$

where $\{i_1, \ldots, j_1, \ldots\}$ and $\{r_1, \ldots, s_1, \ldots\}$ complement each other within the set of indices for β and γ. The constraint following **under** must be positive-finitely evaluable. Each of the sub-formulas in the main part may in turn be divided into assertion and constraint in a similar manner. If a new object needs to be created in order to make true an existentially quantified formula, the predicate **new** should be specified as the whole constraint. Thus, (8.17.2) would contain:

$$\textbf{version } \alpha_3 : \exists x(\beta_1(x) \wedge \cdots \wedge \neg\gamma_1(x) \wedge \cdots \textbf{ under new}(x)). \tag{8.17.3}$$

The connectives **and** and **or** may occur in the main part as before. Moreover, the use of **or** is extended to combine two or more versions into a single version.

Some classes of formulas may serve as strictly the one or the other, either as assertions or constraints. For example, sorts of objects once given cannot be changed freely, hence will normally belong to the constraining part. Formulas that contain non-changeable attributes (e.g., *eligible* in the first version of the process *clip-a-polygon)* cannot be asserted either. On the other hand, negative literals and their boolean combinations which are not positive-finitely evaluable must not be used as constraints, though they might allow the assertion of individual instances. Many formulas may take both roles, however, depending on the contexts in which they occur as exemplified by $pe(\tilde{p}, ?\tilde{e})$ in the first version of *clip-a-polygon* and $eligible(?e)$ in the second version. A question is in what ways different contexts containing the same formula will interact each other within the same and more global process, which will be addressed later.

(2) *Asserting Predicates and Functional Relationships*
At some place on every path of the repeated refinement of an axiom's main part, an instance of a predicate

$$p(u(x, y, \ldots), v(x, y, \ldots), \ldots) \tag{8.18}$$

will come to be asserted. The variables $x, y, \ldots$ which are bound and instantiated by the accompanying constraint part should be considered as virtually denoting definite entities. One or more of the function values $u(x, y, \ldots)$, $v(x, y, \ldots)$ could remain undefined in the current interpretation ($\mathcal{I}$), however. Depending on whether it is indeed the case and on the kind of predicate p (that is, enumerative, non-enumerative or the equality sign '='), asserting (8.18) will be accomplished in different ways as discussed below, though these will all end up asserting ground instances of enumerative predicates and/or functions. We assume that (8.18) is currently false, since otherwise, no action is needed.

(2.1) *All the Arguments (Outermost Terms)* $u, v, \ldots$ *of* (8.18) *Are Defined*

(2.1.1) p *Is Enumerative*
The formula (8.18) is simply asserted, adding one entry to $G(\mathcal{I})$.

(2.1.2) p *Is Non-Enumerative*
Non-enumerativity implies definability by means of a procedure acting on a set of (enumerative) attribute values. One or more inverse procedures (processes) must be provided to effect the definition or change of certain attribute values so that (8.18) may become true. Such procedures not only depend on domain specific knowledge but reflect particular design practices or designers' intentions. To assert

$$intersect(l, pee_{+}(p, e))$$

for example, one might want to extend the edge $pee_+(p, e)$ (by changing the coordinate values of one of its vertices) or translate l, or take any of the possible measures for the two entities to intersect with each other.

Let us suppose the evaluation of predicate p actually uses enumerative functions $h_1, h_2, \ldots$ of its arguments. Two sub-cases are further distinguished.

(2.1.2.1) *All the h_i Are Defined*
A minimal extension of $\mathcal{I}$ with respect to p will be realized by treating one of the changeable functions h_i as a variable and determining its value from the current value and the rest of the arguments. In another expansion of MDL, the procedure for p and its inverses can be defined in some such form as the following:

$$
\begin{array}{l}
\textbf{predicate } p(u : sort_u, v : sort_v, \ldots) \equiv pp(h_1(u,v,\ldots), h_2(u,v,\ldots), \ldots); \\
\langle \text{block, the procedure body of } pp \text{ to determine the truth value of } p \rangle; \\
\quad \textbf{inverse } q_1 : qq_1(\ldots, ?h_{i_1}(u,v,\ldots), \ldots); \\
\quad \langle \text{block, the procedure body to determine the value of} \\
\quad i_1\text{-th argument prefixed by ? so that } qq_1 \text{ may be true} \rangle; \\
\quad \textbf{inverse } q_2 : qq_2(\ldots, h_{i_2}(u,v,\ldots), \ldots); \\
\quad \ldots
\end{array}
\tag{8.19}
$$

The inverse procedures are to be invoked selectively by the application agent through a statement of the form

$$\textbf{assert } p(u(x,y,\ldots), v(x,y,\ldots), \ldots) \textbf{ by } q_j[\textbf{or } q_k \ldots] \tag{8.20}$$

or automatically by the system when no q_j is specified.

(2.1.2.2) *One or More h_i Are Undefined*
In a way similar to the above, the inverses of pp should be defined to determime the values of the h_i. The differences from the above are that:

- all the h_i must be evaluated simultaneously by each version of the inverse procedures

$$rr(\ldots, ?h_i(u,v,\ldots), \ldots, ?h_j(u,v,\ldots), \ldots) \tag{8.21}$$

- their current values are non-existent, hence unavailable.

(2.1.3) *p Involves the Equality Sign '='*
The relation we want to consider is

$$f(h_1(u,v,\ldots), h_2(u,v,\ldots), \ldots) = g(h_1(u,v,\ldots), h_2(u,v,\ldots), \ldots) \tag{8.22}$$

which is a special case of the above in the sense that it is not defined enumeratively. The concept of equality is a purely abstract one, however, and as such, allows a general and more extensive treatment by the system. We assume that the value of either side of the equality, say the right-hand side, is defined. Otherwise, a failure will be reported.

(2.1.3.1) *f Is Enumerative*
If all the arguments of f have been evaluated, the functional assignment

$$f : (h_1(u,v,\ldots), h_2(u,v,\ldots),\ldots) \rightarrow g(h_1(u,v,\ldots), h_2(u,v,\ldots),\ldots) \quad (8.23)$$

is simply asserted. Otherwise, the currently undefined values are enumeratively determined by inverting f.

(2.1.3.2) *f Is Non-Enumerative*
If all the arguments of f have been evaluated and some are not contained in the right-hand side of (8.22), the value of each of these is put in turn into an adjustment by invoking the pertinent inverse function. The inverse functions will also be invoked if any arguments remain unevaluated.

A simple example of this type is the statement

$$\textbf{assert}\ \textit{r-of-curvature}(i, \theta(i), \theta(i-1), \textit{radius}(O)) = \textit{radius}(R)$$

within the example (3) of the preceding sub-section which is executed by changing the value of $\theta(i)$ with the help of an inverse function *angle*. It should be noted that *angle* returns only the increment of $\theta(i)$, hence the current value is necessary to determine its new value.

Two sub-cases are left out of our considerations, namely,

- case 1: neither side of (8.22) has been evaluated, and
- case 2: both sides are non-enumerative and contain the same set of arguments.

For these, the application agent is still responsible for resolving indeterminacies and potential inconsistencies by entering pertinent procedures.

(2.2) *One or More Arguments $u, v, \ldots$ of* (8.18) *Are Undefined*
We note first that those arguments that remain undefined must be of enumerative type, as $x, y, \ldots$ are all defined, hence can be determined independently of each other.

(2.2.1) *p Is Enumerative*
The currently undefined values of arguments are enumeratively determined by inverting p.

(2.2.2) *p Is Non-Enumerative and Depends on Enumerative Arguments h_i*
This is a generalization of (2.1.2.2) as at least one h_i will inevitably be undefined. If any inverse procedure is available to determine the values of h_i so that p may be true, however, the values of $u, v, \ldots$ can in turn be enumeratively determined from those of h_i.

(2.2.3) *p Involves the Equality Sign '='*
Quite similarly, the method of (2.1.3) is applied, following which the values of $u, v, \ldots$ can be obtained by inverting each h_i.

(3) *Identification of Primary Axioms*
By 'primary axioms' are meant those which the designer wants to have established as the immediate objectives at the present stage of his/her work.

What distinguishes these is that they stipulate the most essential properties or conditions of designated individual or classes of objects, or the existence of such objects. More specifically, each primary axiom will be a formula in the following form, or a conjunction of such formulas:

- The objects of primary interest are of sort ι and designated either
  - by individual constants, in which case α is considered as lacking a constraint part, or
  - by the (universally or existentially quantified) bound variable of the constraint part (ρ_1) of α.
- The assertion part (α_1) may itself have a constraint, whose bound variable designates the attributes or related objects.
- The nesting of assertions is repeated until the final one (α_n or β) is reached. The general form of α will thus be

$$\begin{aligned} &Q_1x_1(Q_2x_2(\ldots Q_nx_n(\beta(x_1,x_2,\ldots,x_n)\\ &\quad \textbf{under } \rho_n(x_1,x_2,\ldots,x_n))\ldots) \textbf{ under } \rho_1(x_1)) \end{aligned} \tag{8.24}$$

  where Q_i stands for either $\forall$ or $\exists$. Some Q_i together with ρ_i could be missing, in which case the corresponding x_i are replaced by a list of individual constants.
- β stipulates the conditions that the primary objects must meet in terms of attribute values, relations among themselves and with other objects. It is free of quantifiers, that is, it is a boolean combination of literals.

Associated with primary axioms are other axioms which stand as subordinates. These are such universally quantified formulas that their constraint parts contain literals which are unifiable with the ground instances of literals constituting the final assertions of primary axioms. Whenever a set of primary axioms are invoked, so should be their subordinates, subordinates' subordinates and so on. The idea behind this is that the establishment of immediate objectives must be accompanied by that of inevitable consequences or necessary conditions of them. It is possible and desirable for the system to recognize the primary-subordinate relationships among axioms and automatically invoke the subordinates in a propagatory manner whenever the primaries are invoked.

As an example, let us reformulate the axioms, *A1*, *A2* and *A4* for the problem of clipping a polygon (the first version) in the expanded MDL and see how these are related to each other through the primary-subordinate relationship.

$$\begin{aligned} &\textbf{axiom } E1 : \forall p \forall e \exists e_+(((e_+ = pee_+(p,e))\\ &\quad \textbf{under } pe(p,e_+) \land (ev_-(e_+) = ev_+(e))) \textbf{ or}\\ &\quad ((e_+ = pee_+(p,e)) \land pe(p,e_+)\\ &\quad \textbf{under } edge(e_+) \land (ev_-(e_+) = ev_+(e))) \textbf{ or}\\ &\quad ((e_+ = pee_+(p,e)) \land pe(p,e_+) \land (ev_-(e_+) = ev_+(e))\\ &\quad \textbf{under new}(e_+)) \textbf{ under } polygon(p) \land pe(p,e)); \end{aligned}$$

$$E2 : \forall p \exists e(((pe(p,e) \textbf{ under } edge(e)) \textbf{ or } \exists v(pe(p,e) \wedge (v = ev_-(e)) \textbf{ under new}(e) \wedge vertex(v))) \textbf{ under } polygon(p));$$

$$E4 : \forall p \forall l \exists \tilde{p}((polygon(\tilde{p}) \textbf{ and } clip\text{-}p(p,l,\tilde{p})) \textbf{ under new}(\tilde{p}) \wedge line(l) \wedge polygon(p));$$

$$clip\text{-}p(p,l,\tilde{p}) \equiv \forall \tilde{e}((eligible(\tilde{e}) \textbf{ and } sameside(l,\tilde{e},pee_+(\tilde{p},\tilde{e}))) \textbf{ under } pe(\tilde{p},\tilde{e}));$$

$$eligible(\tilde{e}) \equiv (\textbf{under } \neg intersect(l,\tilde{e}) \wedge pe(p,\tilde{e})) \textbf{ or } \exists e(clip\text{-}e(e,l,\tilde{e})) \textbf{ or } clip\text{-}l(l,p,\tilde{e});$$

$$clip\text{-}e(e,l,\tilde{e}) \equiv (((coord(ev_+(\tilde{e})) = intersection(l,e)) \textbf{ under } (ev_-(e) = ev_-(\tilde{e})) \wedge \textbf{new}(ev_+(\tilde{e}))) \textbf{ or } ((ev_+(\tilde{e}) = ev_+(e)) \textbf{ under } (coord(ev_-(\tilde{e})) = intersection(l,e)))) \textbf{ under } pe(p,e) \wedge intersect(l,e);$$

$$clip\text{-}l(l,p,\tilde{e}) \equiv \exists e1 \exists e2((coord(ev_+(\tilde{e})) = intersection(l,e2)) \textbf{ under } (e1 \neq e2) \wedge pe(p,e1) \wedge pe(p,e2) \wedge internal(p,(intersection(l,e1), intersection(l,e2))) \wedge (coord(ev_-(\tilde{e})) = intersection(l,e1)) \wedge \textbf{new}(ev_+(\tilde{e}))).$$

The primary axiom is *E4*. The relations between axioms are as depicted in Fig. 8.10. It should be noted that *E4* and *E1* will be invoked repeatedly, that *E1* which is self-invoking must be invoked in between the assertions of *eligible* and *sameside* to provide the latter with the adjacent edge of each eligible edge, and that backtracking will occur during the process of asserting *E1*, degrading efficiency if either of the sub-formulas of *clip-p* fails to be asserted. The system is responsible for controlling the invocation, synchronization and backtracking of interrelated processes. It should also be noted that the formations of *E1* and *E2* are quite general while that of *E4* reflects the specificity of the present problem. The generality of *E4* is the principal cause of the efficiency degradation.

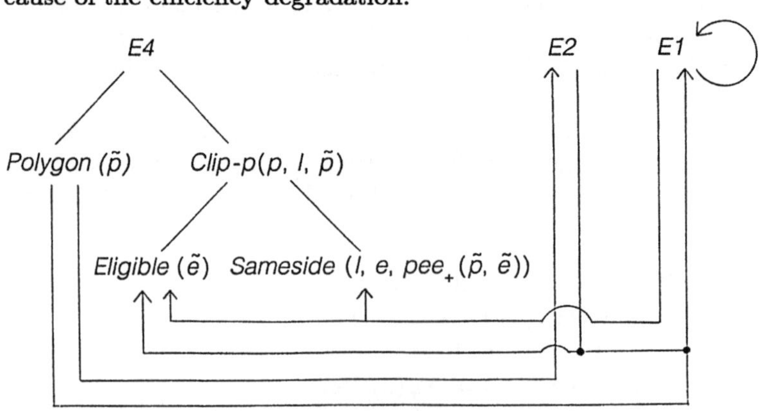

Fig. 8.10. Primary-subordinate relations between axioms for clipping a polygon

With the axioms reformulated as above, the task of clipping a given polygon will be accomplished simply by issuing the statement

assert $\widetilde{E4}$

where $\widetilde{E4}$ is the formula obtained by instantiating p and l in $E4$.

(4) *Logical Relationships between Formulas*
The preceding analysis can be generalized to include the relationships between axioms via the occurrences of unifiable sub-formulas which are open yet may contain either or both of universally or existentially quantified variables and serve as assertions or constraints. The establishment of a currently false axiom is accomplished by increasing or decreasing the extensions of one or more of its sub-formulas, which may adversely affect those of other axioms. Specifically, we are concerned about the danger of entering an infinite loop or coming to a dead end in trying to extend the current interpretation to satisfy the axioms in question.

To ease the task of analysis, every axiom should be turned into a clause by Skolemizing existentially quantified variables. At the same time, the assertion and constraint parts should be reconstructed according to the following rules:

- rule 1: $\forall x(\forall y(\beta(x,y)\ \textbf{under}\ \sigma(x,y))\ \textbf{under}\ \rho(x))$
  $\rightarrow \forall x \forall y(\beta(x,y)\ \textbf{under}\ \rho(x) \wedge \sigma(x,y))$
- rule 2: $\forall x(\exists y(\beta(x,y)\ \textbf{under}\ \sigma(x,y))\ \textbf{under}\ \rho(x))$
  $\rightarrow \forall x(\beta(x,s(x))\ \textbf{under}\ \rho(x))\ \textbf{and}\ \forall x(\textbf{under}\ \neg\rho(x) \vee \sigma(x,s(x)))$
- rule 3: $\exists x(\forall y(\beta(x,y)\ \textbf{under}\ \sigma(x,y))\ \textbf{under}\ \rho(x))$
  $\rightarrow \forall y(\beta(t,y)\ \textbf{under}\ \sigma(t,y))\ \textbf{and}\ \rho(t)$
- rule 4: $\exists x(\exists y(\beta(x,y)\ \textbf{under}\ \sigma(x,y))\ \textbf{under}\ \rho(x))$
  $\rightarrow \beta(t,u)\ \textbf{and}\ \sigma(t,u) \wedge \rho(t)$

where s, t and u are Skolem function symbols. By repeated applications of these rules, rearrangement of β as a formula in conjunctive normal form and distribution of the constraint part to each conjunct, an axiom will finally be transformed into an 'and' combination of sub-axioms (α_i) of the form

$$\begin{aligned}
&\forall x \forall y \ldots (\beta_i(x,y,\ldots)\ \textbf{under}\ \rho_i(x,y,\ldots)) \\
&\beta_i(x,y,\ldots) \equiv v(x,y,\ldots) \vee \cdots \vee \neg w(x,y,\ldots) \vee \cdots \qquad (8.25)\\
&v(x,y,\ldots) \equiv p(f(x,y,\ldots),\ldots),\ldots, w(x,y,\ldots) \equiv q(g(x,y,\ldots),\ldots),\ldots
\end{aligned}$$

where $p, q, \ldots$ are predicate symbols or names of defined formulas. One of the sub-axioms may lack the assertion part representing a pure constraint.

We note from (8.25) that the increase of extension of v (or $\neg w$) will contribute to that of β_i and write

$$\frac{\Delta\beta_i}{\Delta v} > 0 \quad \left(\text{or } \frac{\Delta\beta_i}{\Delta w} < 0\right) \tag{8.26}$$

to express this fact. Suppose two sub-formulas v_1 and v_2 which are unifiable occur in axioms α_1 and α_2 respectively. Assuming v_1 occurs in β_1, four cases are distinguished depending on whether v_2 is negated or not and whether it occurs in the assertion or constraint part of α_2.

Case 1 v_2 is positively contained in ρ_2, that is,

$$\frac{\Delta\beta_1}{\Delta v_1} > 0 \quad \text{and} \quad \frac{\Delta\rho_2}{\Delta v_2} > 0. \tag{8.26.1}$$

Asserting β_1 may strengthen the constraint ρ_2 making it more difficult to simultaneously establish the truth of β_2. If another pair of unifiable sub-formulas occur positively in ρ_1 and β_2, the danger of an infinite loop exists.

Case 2 v_2 is negatively contained in ρ_2, that is,

$$\frac{\Delta\beta_1}{\Delta v_1} > 0 \quad \text{and} \quad \frac{\Delta\rho_2}{\Delta v_2} < 0. \tag{8.26.2}$$

Asserting β_1 may weaken the constraint ρ_2 making it easier to simultaneously establish the truth of β_2.

Case 3 v_2 is positively contained in β_2, that is,

$$\frac{\Delta\beta_1}{\Delta v_1} > 0 \quad \text{and} \quad \frac{\Delta\beta_2}{\Delta v_2} > 0. \tag{8.26.3}$$

Asserting β_1 may increase the extension of β_2 making it easier to simultaneously establish the truth of β_2.

Case 4 v_2 is negatively contained in β_2, that is,

$$\frac{\Delta\beta_1}{\Delta v_1} > 0 \quad \text{and} \quad \frac{\Delta\beta_2}{\Delta v_2} < 0. \tag{8.26.4}$$

Asserting β_1 may contradict asserting β_2 and vice versa unless the extensions of ρ_is are disjoint.

The system should be capable of carrying out the above syntactical analysis of the set of all axioms automatically, and on that basis, of indicating the places where potential dangers of contradictions and infinite loops exist.

8.3.2 System Facilities at the Macroscopic and Intelligent Level

To begin with, let us summarize the functions required of the system for facilitating designers' and application programmers' work in accordance with the principles set forth in the preceding subsection. These included;

- expanding a simple **assert** α into a complex of **foreach/forsome** $\rho(x)$ **assert** $\beta(x)$ statements based on the definition of axiom α,
- invoking pertinent procedures to establish non-enumerative predicates or functional relations,
- recognizing the primary-subordinate relationship between axioms and propagating the assertions of axioms through this relationship, and

– recognizing more general relationships between axioms via the occurrences of unifiable sub-formulas and indicating potential dangers of contradictions and infinite loops.

Our hope was to bring the interpretation of a given theory under better control by making a static and structured description of its inherently dynamic and indeterminate aspect. One of the aims of the present subsection is to explore further possibilities along the same lines. The other aim is concerned with constructing the theory itself. For both of them, the exploitation of expert knowledge and experience will be pursued.

(1) *Interpretation*
(1.1) *Selection from Alternate Paths of a Process*
The multiplicity of different paths which is one of the principal causes for the indeterminacy of a process derives from existential quantification of variables, disjunction of sub-formulas and changeable attributes. Means must be provided to set and change priorities among competing alternatives in order to reflect designers' intentions as well as to reduce indeterminacy.

(1.1.1) *Instantiation of an Existentially Quantified Variable*
While there is generally no reason to choose any particular object or value as the instance of an existentially quantified variable, it is nevertheless possible to provide measures with which to screen the candidates. Suppose the formula

$$\alpha \equiv \exists x(\beta_1(x) \wedge \cdots \wedge \beta_n(x) \textbf{ under } \rho(x))$$

is to be satisfied and we have currently

$$R \equiv \{x : \rho(x)\} \subset \{x : \neg\beta_1(x) \vee \cdots \vee \neg\beta_n(x)\}.$$

The properties of sub-formulas ρ and β_i could be relevant.

(1.1.1.1) *Selection from R*
If the domain of definition S of ρ has a structure based on which functions $f, g, \ldots$ from 2^s to S can be defined, $f(R), g(R), \ldots$ may be natural selections for x. Examples are the average on a domain of arithmetic type, or the maximum and minimum on a totally ordered set. These functions should be prioritized and invoked in turn automatically by the system.

(1.1.1.2) *Priorities among Sub-Formulas β_i*
If some β_i are more difficult to assert than others, their assertion should be given priority. Objects should be selected from their common extension, if it is non-empty, and assertion of the easier sub-formulas with respect those objects should then be attempted. A simple way of specifying the priorities will be by the index numbers i.

The ultimate recourse should be to dialogue with the designers themselves. Pertinent pieces of information such as R, set functions $f, g, \ldots$, relative priorities and extensions of β_is should be presented to designers to support them in making evaluations and decisions, which should in turn be reflected in the path selection by the system.

(1.1.2) *Priorities among Disjuncts*
In order to facilitate asserting an axiom of the form

$$\forall x(\beta_1(x) \vee \cdots \vee \beta_n(x) \textbf{ under } \rho(x)),$$

it is relevant to prioritize the β_i with the easiest to assert or the most desired ranking the highest. While the final recourse is again to instructions from the designers, there remains room for a general treatment by the system. Suppose $\beta_i(x)(i = 1, \ldots, n)$ takes the form either

$$\forall y(\gamma_i(x, y) \textbf{ under } \rho_i(x, y)) \quad \text{or} \quad \exists y(\gamma_i(x, y) \textbf{ under } \rho_i(x, y)),$$

and let R and $R_i(x)$ be defined by

$$R \equiv \{x : \rho(x) \wedge \neg\beta_1(x) \wedge \cdots \wedge \neg\beta_n(x)\}$$
$$R_i(x) \equiv \{y : \rho_i(x, y) \wedge \neg\gamma_i(x, y)\} \quad (x \in R).$$

For each x in R, $\beta_i(x)$ should be treated as preceding $\beta_j(x)$ if and only if the following conditions are met.

- The bound variables (y) of both $\beta_i(x)$ and $\beta_j(x)$ are under the same kind of quantification ($\forall$ or $\exists$).
- $R_i(x)$ contains a smaller number of elements than $R_j(x)$.
- $\gamma_i(x, y)$ is a logical consequence of $\gamma_j(x, y)$.

Note that the predicate **new** should be considered as having an infinite extension ($\tilde{I} - I$, see Sect. 6.1 (2.1)), hence as ranking the lowest. Also to be noted is that the precedence relation is not a total order, hence the system may possibly be unable to select the unique highest ranking disjunct based on it.

(1.1.3) *Changeable Attributes*
Of each changeable attribute, its name, relative priority, and permissible range of values should be specified. Furthermore, a prioritized set of standard procedures should be provided and invoked automatically by the system to assert each predicate or defined formula of the form

$$w(f(x, y, \ldots), g(x, y, \ldots), \ldots)$$

where $f, g, \ldots$ are changeable attribute names. An example is the axiom

$$\textbf{min}_i(\textit{r-of-curvature}(cs(i))) \geq \textit{radius}(R)$$

for designing a cam contour (Sect. 8.2.2 (3)). These sets of procedures which are necessarily domain-dependent are understood to embody pieces of expert designers' knowledge of the domain in question.

(1.2) *Selection from Alternate Processes for a Theory*
The multiplicity of possible processes for a given theory is represented by the existence of different versions of asserting an axiom. Relative priorities among versions and the coherency of assortments of versions, one taken for each axiom, must be maintained and exploited by the system.

A service by the system at a deeper level is to propose the definition of new processes of asserting an axiom, that is, new divisions of an axiom into assertion and constraint parts. Let β_1 be a process for an axiom of the form either

$$\beta_1 \equiv \forall x(\gamma(x) \textbf{ under } \rho_1(x) \wedge \rho_2(x)) \quad \text{or} \quad \beta_1 \equiv \exists x(\gamma_1(x) \wedge \gamma_2(x) \textbf{ under } \rho(x)).$$

It will be preferable if a version β_2 is defined by

$$\beta_2 \equiv \forall x(\gamma(x) \vee \neg \rho_1(x) \textbf{ under } \rho_2(x)) \quad \text{or} \quad \beta_2 \equiv \exists x(\gamma_2(x) \textbf{ under } \rho(x) \wedge \gamma_1(x))$$

as β_2 precedes β_1 by the criteria in Sect. (1.1.2) above. This cannot be done arbitrarily, however. For one thing, $\rho_1(x)$ may not be negatable, as in the case of an enumerative predicate or relation between unchangeable attributes. For another, $\gamma_1(x)$ may be disqualified as a constituent of a constraint, with its assertion entailing the creation of new objects.

An example is the removal of the predicate *eligible* from the assertion part and the inclusion of it in the constraint part of an axiom for clipping a polygon. In the first version (Sect. 8.3.1 (3)), the conjunction of the primary axiom $\widetilde{E4}$ and a subordinate $E2$ will produce a theorem $\widetilde{E2}$:

$$\widetilde{E2} : \exists \tilde{e}(\widetilde{E2.1} \text{ or } \widetilde{E2.2})$$

$$\widetilde{E2.1} \equiv eligible(\tilde{e}) \wedge sameside(l, \tilde{e}, pee_{+}(\tilde{p}, \tilde{e})) \wedge pe(\tilde{p}, \tilde{e}) \textbf{ under } edge(\tilde{e})$$

$$\widetilde{E2.2} \equiv \exists v(eligible(\tilde{e}) \wedge sameside(l, \tilde{e}, pee_{+}(\tilde{p}, \tilde{e})) \wedge pe(\tilde{p}, \tilde{e}) \wedge (v = ev_{-}(\tilde{e})) \textbf{ under } \textbf{new}(\tilde{e}) \wedge vertex(v)).$$

It appears from the formation of $\widetilde{E2.1}$ that moving $eligible(\tilde{e})$ from the assertion to the constraint part is possible. Unfortunately, the existence of $\textbf{new}(\tilde{e})$ in $\widetilde{E2.2}$ denies this. A second thought on the nature of *eligible* suggests to us, however, that the difficulty might be overcome by creating in advance all the eligible edges and their related objects (vertices) and defining the values of attributes (coordinates). We may as well apply a similar treatment to the predicate *sameside*. We have actually done this, the result of which is the second version comprising axioms *C1* through *C3* (Sect. 8.2.2 (2) *Discussion*).

It should be noted that while the above reformulation of the process relies inherently on a knowledge of geometry, the system can at least find and propose possible alternatives by examining the syntactical formations of axioms and theorems. Designers (or rather database administrators) may accept or decline the proposal, and if they accept, must provide the definitions of (that is, the necessary and sufficient conditions for) new constituents of constraints.

(1.3) *Priorities among Primary Axioms*

In the purely logical or mathematical sense, there is no distinction of priorities among axioms. For constructing a model, however, it is vitally important to decide at each instant which of the primary axioms is to be asserted. In the second version of clipping a polygon, *C1*, *C2* and *C3* (which are all primary axioms) should be asserted in this order. Otherwise, unwarranted successes or failures will be reported. The system should be able to facilitate designers' making judgement by presenting to them the current state of interpretation (truth values of axioms, extensions of their constraint, etc.) and the logical relationships between axioms. In a way somewhat similar to the case of the primary-subordinate relationship, higher priorities should be given to the axioms whose assertion would adversely affect that of others by increasing the extensions of their constraint parts or decreasing that of their assertion parts.

(1.4) *Learning, or Memorizing Expert Designers' Actions and Decisions*
It will be one of the simplest, most fundamental, yet most practical ways of exploiting designers' knowledge and experience for the system to do the following:

- to memorize the traces of expert designers' instructions and responses to proposals by the system for design alternatives, and
- to have them ready for replay for novice designers to help them practice their professions in the manner of those more knowledgeable and experienced.

While the possible results of designers' actions and decisions and the specific structures to incorporate these depend largely on each application agent, we can define a general form and let the system manipulate the structures in that form. Each such structure is the pair of an MDL procedure and a set of 'standard' actual parameters. Parameters include the identifier of a local view (that is, the collection of theories this procedure is concerned with) and the instructions/responses from designers on crucial questions and decision points. The system should be capable of gathering one or more 'standard' cases and updating them through a series of sessions with expert designers on the same subject. The view identifier is the primary key to identify the procedure to be called during the dialogue with a novice to whom the system will present the accompanying set of actual parameters as the recommended decision.

(1.5) *Derivation of Theorems*
Although our prime concern is not with the formal manipulation of theories but with the construction of their models, it is obvious from the preceding discussions, in Sects. (1.1.2) and (1.2) in particular, that the system should have a certain degree of capability for logical inferences. Note that if we have all the axioms transformed into clauses, every model of either theory is also a model of the other (Lemma 3.16). In other words, a formula in the original theory will be a theorem if and only if the corresponding clause is a theorem in the Skolemized theory. Skolem constants and function symbols are treated as being unifiable to each other. Roughly speaking, PROLOG-like resolution by unification should suffice. Restricting the class of permissible formulas to that of definite clauses might appear a significant shortcoming. It will be so indeed if soundness and completeness of the proof procedure is a mandatory requirement. In our case, however, consequences which are only possible or

intermediate may contain valuable clues to help the system to search out and propose better alternatives. Besides, a forward reasoning starting at a limited number of specified axioms is more relevant than the usual backward process starting at the given goal. Suppose, for example, we infer

$$\alpha\theta \rightarrow \delta_1\theta \vee \delta_2\theta$$

from the axioms

$$\alpha \rightarrow \beta \quad \text{and} \quad \gamma \rightarrow \delta_1 \vee \delta_2$$

with the substitution θ unifying β and γ, and have the truth of $\alpha\theta$ established. While neither $\delta_1\theta$ nor $\delta_2\theta$ is a necessary consequence of $\alpha\theta$, the process must be programmed so that it may properly address both cases of asserting

$$\beta\theta \wedge \delta_1\theta \quad \text{and} \quad \beta\theta \wedge \delta_2\theta.$$

By examining the constructions of these, the system should be able to come up with new (and hopefully better) divisions of them into assertion and constraint parts.

(2) *Theory Construction*
In this subsection, we want to examine the possibility for the system to assist designers and application programmers in constructing new or revising existing theories. Specifically, we focus our attention on three different approaches, namely, metatheoretic reasoning, reuse of existing theories and organization of partial interpretations.

(2.1) *Metatheoretic Reasoning*
By this is meant a set of rules and conventions that enables inferences exceeding the normal level of FOPL in an attempt to reflect more or less 'common sense' knowledge. Circumscription and default reasoning, which we have touched on (Sects. 4.2.2 (1), 4.6.3 (4) and 4.6.4 (1)), are among the best-known examples. We want to see how these methods can be incorporated into our system to enhance its inference capability.

(2.1.1) *Circumscription*
As has been formulated in McCarthy [McC], the circumscription of a predicate p in a formula $\alpha(p)$ known to be a theorem is the formula schema

$$(\alpha(\phi) \wedge \forall x(\phi(x) \rightarrow p(x))) \rightarrow \forall x(p(x) \rightarrow \phi(x)) \tag{8.27}$$

where ϕ is an arbitrary formula and $\alpha(\phi)$ is the result of substituting all occurrences of p in α by ϕ. Each instance of (8.27) defines an axiom. The set of all ϕ that make its left side true constitutes the class of formulas which are equivalent to p.

The formula α is to be understood as a defining formula for p. By appropriately selecting one such, the system gives a precise description of the predicate in question, and at the same time, may possibly eliminate the need to postulate some axioms. Let us consider, for example, the axiom

$$\alpha(pe) \equiv \forall p \forall e(pe(p,e) \rightarrow \exists e^+(pe(p,e^+) \wedge (pee_+(p,e) = e^+))) \tag{8.28}$$

which relates a polygon to each pair of adjacent edges, and

$$\alpha(\phi) \equiv \forall p \forall e(\phi(p,e) \rightarrow \exists e^{+}(\phi(p,e^{+}) \wedge (pee_{+}(p,e) = e^{+}))) \tag{8.29}$$

which is the circumscription of p in α. By substituting

$$\tilde{\phi}(p,e) \equiv pe(p,e) \wedge \exists e^{-}(pe(p,e^{-}) \wedge (pee_{+}(p,e^{-}) = e)) \tag{8.30}$$

into $\alpha(\phi)$, we obtain

$$\begin{aligned}\alpha(\tilde{\phi}) \equiv\ & \forall p \forall e((pe(p,e) \wedge \exists e^{-}(pe(p,e^{-}) \wedge (pee_{+}(p,e^{-}) = e))) \\ & \rightarrow \exists e^{+}((pe(p,e^{+}) \wedge \exists e(pe(p,e) \wedge (pee_{+}(p,e) = e^{+}))) \wedge \\ & \qquad (pee_{+}(p,e) = e^{+}))) \\ \equiv\ & \forall p \forall e((pe(p,e) \wedge \exists e^{-}(pe(p,e^{-}) \wedge (pee_{+}(p,e^{-}) = e))) \\ & \rightarrow \exists e^{+}(pe(p,e^{+}) \wedge (pee_{+}(p,e) = e^{+}))) \\ \leftarrow\ & \forall p \forall e(pe(p,e) \rightarrow \exists e^{+}(pe(p,e^{+}) \wedge (pee_{+}(p,e) = e^{+}))) \equiv \alpha(pe)\end{aligned}$$

which together with the obvious truth of $\forall p \forall e(\tilde{\phi}(p,e) \rightarrow pe(p,e))$ gives

$$\forall p \forall e(pe(p,e) \rightarrow \tilde{\phi}(p,e)). \tag{8.31}$$

This is nothing but the axiom *A3* for clipping a polygon (Sect. 8.2.2 (2)). It should be noted that while our excluding this axiom from the program was based on our knowledge of geometry and the finiteness of interpretation, the present argument is purely formal-theoretic.

The same problem offers another example. First, let the axiom *A4* be slightly weakened and rearranged as the conjunction of two axioms defined by

$$\begin{aligned}A4.1: \quad & ((pe(p,\tilde{e}) \wedge \neg intersect(l,\tilde{e})) \vee \exists e(pe(p,e) \wedge clip\text{-}e(e,l,\tilde{e})) \vee \\ & clip\text{-}l(l,p,\tilde{e})) \rightarrow eligible(p,l,\tilde{e}) \\ A4.2: \quad & \exists \tilde{p}(polygon(\tilde{p}) \wedge \forall \tilde{e}(pe(\tilde{p},\tilde{e}) \rightarrow (eligible(p,l,\tilde{e}) \wedge \\ & eligible(p,l,pee_{+}(\tilde{p},\tilde{e})) \wedge sameside(l,\tilde{e},pee_{+}(\tilde{p},\tilde{e})))))\end{aligned}$$

and let ϕ denote the left side of the implication of *A4.1*. Circumscribing *eligible* in *A4.2* and instantiating the result with ϕ will produce

$$\begin{aligned}& \exists \tilde{p}(polygon(\tilde{p}) \wedge \forall \tilde{e}(pe(\tilde{p},\tilde{e}) \rightarrow (\phi(p,l,\tilde{e}) \wedge \phi(p,l,pee_{+}(\tilde{p},\tilde{e})) \\ & \qquad \wedge sameside(l,\tilde{e},pee_{+}(\tilde{p},\tilde{e}))))) \\ & \rightarrow \forall \tilde{e}(eligible(p,l,\tilde{e}) \rightarrow \phi(p,l,\tilde{e})).\end{aligned} \tag{8.32}$$

By actually providing a procedure to define a polygon clipped from another by a line, both arbitrarily given, we may conclude the truth of the left side, hence also the right side of the above implication. At the same time, this disqualifies certain classes of polygons generated from the original from being legitimate ones (Fig. 8.11).

In general, it will be beneficial if the system can make the following use of the concept of circumscription for supporting human designers and programmers in constructing theories:

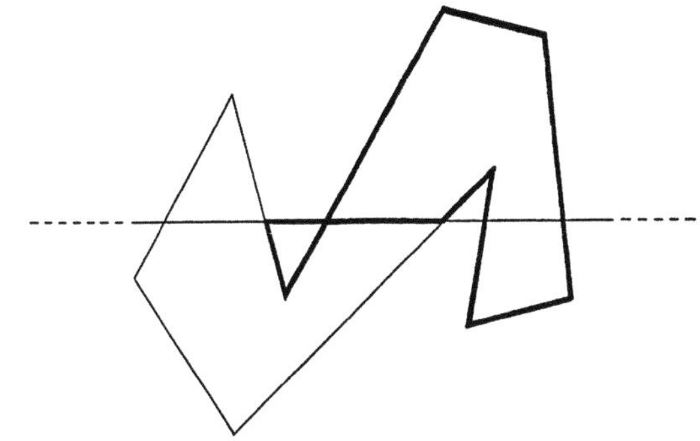

Fig. 8.11. Some illegitimate polygons generated from a polygon and a line

- to find the predicates, their circumscriptions, and possible instances which may contribute to simplifying the theory,
- to let these be known to designers and programmers, and take their instructions in turn during dialogue with them, and
- to circumscribe certain predicates in certain formulas, make inferences and present the consequences to designers and programmers.

(2.1.2) *Default Reasoning*
Default reasoning [Rei] is more general than circumscription in that it allows the formula schema (8.27) to be transcribed into the 'normal default'

$$(\alpha(\phi) \land \forall x(\phi(x) \to p(x)) \land \mathcal{M}(\forall x(p(x) \to \phi(x)))) \to \forall x(p(x) \to \phi(x)) \quad (8.33)$$

which conforms to the syntax defined by (4.53). Let us take a few examples discussed in McCarthy [McC] to see that the consequents are indeed likely to hold true.

The first and simplest example concerns a predicate p that is characterized by

$$\forall x(\psi(x) \to p(x)) \quad (8.34)$$

and its circumscription in the same formula

$$(\forall x(\psi(x) \to \phi(x)) \land \forall x(\phi(x) \to p(x))) \to \forall x(p(x) \to \phi(x)). \quad (8.35)$$

The left side of the implication becomes trivially true if ψ is substituted for ϕ, which results in

$$\forall x(p(x) \to \psi(x)). \quad (8.36)$$

The rationale for postulating (8.35) lies in the impossibility of logically concluding the falsehood of its consequent from the defining formula (8.34), which is the only knowledge we have concerning p. In fact, the negation of (8.36)

$$\exists x(p(x) \land \neg\phi(x))$$

demands the existence of elements lying outside the extension of ϕ, hence that of ψ, of which p is predicated contrary to the premise that as far as we know, only those elements that satisfy ψ may satisfy p. Thus, the same rationale supports the default

$$\mathcal{M}(\forall x(p(x) \rightarrow \phi(x))). \tag{8.37}$$

When ψ is enumeratively defined by

$$\psi(x) \equiv (x = a_1) \vee \cdots \vee (x = a_n), \tag{8.38}$$

the circumscription (8.35) will reduce to

$$(p(a_1) \wedge \cdots \wedge p(a_n)) \rightarrow \forall x(p(x) \rightarrow ((x = a_1) \vee \cdots \vee (x = a_n))). \tag{8.39}$$

The next example is obtained by replacing '$\wedge$' in the left side of the above implication by '$\vee$', namely

$$p(a_1) \vee \cdots \vee p(a_n) \tag{8.40}$$

or equivalently

$$\exists x(\psi(x) \wedge p(x)) \quad \psi(x) \equiv ((x = a_1) \vee \cdots \vee (x = a_n)). \tag{8.40}$$

Instantiating the circumscription of p in this by each of $x = a_i$ $(i = 1, \ldots, n)$ gives

$$p(a_i) \rightarrow \forall x(p(x) \rightarrow (x = a_i)) \quad (i = 1, \ldots, n) \tag{8.41}$$

which in conjunction with (8.40) produces

$$\forall x(p(x) \rightarrow (x = a_1)) \vee \cdots \vee \forall x(p(x) \rightarrow (x = x_n)). \tag{8.42}$$

For each $i = 1, \ldots, n$, the default must be

$$\mathcal{M}(\forall x(p(x) \rightarrow (x = a_i))) \tag{8.43}$$

which is a special case of (8.37).

Another familiar way of defining a predicate is by means of a recursive formula

$$\alpha(p) \equiv p(a) \wedge \forall x(p(x) \rightarrow p(f(x))). \tag{8.44}$$

The circumscription of p in this definition

$$(\alpha(\phi) \wedge \forall x(\phi(x) \rightarrow p(x))) \rightarrow \forall x(p(x) \rightarrow \phi(x)) \tag{8.45}$$

implies that the intended extension of p is the transitive closure comprising

$$a, f(a), f(f(a)), \ldots. \tag{8.46}$$

All the instances

$$p(a), p(f(a)), p(f(f(a))), \ldots \tag{8.47}$$

and only these are the logical consequences of $\alpha(p)$. In other words, the negation of the consequent of (8.45)

$$\exists x(p(x) \wedge \neg\phi(x))$$

where ϕ denotes the characteristic predicate for the set (8.46) cannot be logically inferred from all that is known. The default

$$\mathcal{M}(\forall x(p(x) \rightarrow \phi(x))) \tag{8.48}$$

should therefore be maintained.

In order to show the greater applicability of default reasoning, let us consider yet another problem which is of a different nature than those discussed above. Suppose we are given an axiom of the form

$$\forall x(p(f(x), g_1(x), \ldots, g_n(x))) \tag{8.49}$$

which stipulates a constraint on the value of function f in terms of those of $g_1, \ldots, g_n$. If we can think of a particular way of determining f. which is safer and more likely than others, and have no logical reason to refuse it, we may as well take it. To put it more formally, we may postulate

$$\begin{aligned}(\forall x(p(f(x), g_1(x), \ldots, g_n(x))) \wedge \mathcal{M}(\forall x(f(x) = F(g_1(x), \ldots, g_n(x)))))\\ \rightarrow \forall x(f(x) = F(g_1(x), \ldots, g_n(x))).\end{aligned} \tag{8.50}$$

For example, when (8.49) stipulates the maximum and minimum of permissible values of f,

$$\begin{aligned}(\forall x(g_1(x) \leq f(x) \leq g_2(x)) \wedge \mathcal{M}(\forall x(f(x) = (g_1(x) + g_2(x))/2)))\\ \rightarrow \forall x(f(x) = (g_1(x) + g_2(x))/2)\end{aligned}$$

defines the arithmetic mean of the two as the default f value.

If we provide an alternate method for assigning a value of f, taking the geometric mean for example, either method but not both can be included in the theory. The system should be capable of adding or deleting defaults to or from the theory according to the prescribed priorities or explicit instructions from designers and programmers. It will be profitable if the successes or failures of attempted defaults are recorded within each interpretation and certain summary types of information (e.g. statistics, trends, etc.) are stored within each theory. Consistencies must be observed among the defaults which can be simultaneously admitted, and for that purpose, defaults should be organized to form admissible clusters.

(2.2) *Reuse of Existing Theories*

Reuse of design objects is somewhat akin to that of computer programs. Each program, like a design object, is constituted of the specification (formal theory) and one or more realizations (interpretations). There are generally three categories of methods for constructing a program based on other (existing) programs, namely, horizontal activities (instantiations), vertical activities (refinements) and structural editing. In LIL (Library Interconection Language,

Goguen [Gog]) for example, the former two categories of activities are facilitated by defining 'views' between specifications and pointing to lower-level entities in the hierarchy of substructures. A 'view' is a mapping from one specification (T) to another (A), considered as the sets of non-logical symbols (constants, functions and predicates), which preserves types and arities of these symbols. The set of non-logical axioms of T will thus be imported to A. An example is the mapping $NATD$ from $POSET$, the formal theory of a partially ordered set, to $NATURAL$, that of the set of natural numbers defined by

$$NATD: \quad \leq \rightarrow |$$

which means $x \leq y$ if and only if $x \mid y$ (x divides y). Each specification is formed as a theory or generic package (a theory does not have an associated program body whereas a package has) which is to be instantiated by binding formal parameters with views as actual parameters. A generic package may contain a 'needs' clause pointing to another, and this can be repeated until a program body (a 'version', that is, an interpretation of a specification) is reached. For example,

generic package *SORT*[*x* :: *POSET*]
  needs *LISTP* :: *LIST*[*x*]
  ...
end *SORT*

make *SORT-NATD* **is** *SORT*[*NATD*]
  needs *LISTP* → *LIST.1*
end

in conjunction with the definition of $NATD$ will cause the version *LIST.1* of *LIST*[*NATD*] to be reused in the construction of *SORT*[*NATD*].

In our case, both horizontal and vertical organizations of a theory are formed through the inclusion of sub-theories as depicted in Fig. 6.4. Renaming of non-logical symbols (Sects. 6.2 (1.6) and 7.2.1 (24)) plays the role which is equivalent to that of view in LIL.

It is generally admitted that organizing a library of specifications is a difficult task. In a multi-level classification scheme, keys for retrieval might range from simple names to full theories. The central problem is how to provide the framework in which to express the pieces of information that characterize classes of theories at any level of detail precisely and most economically. One possible approach is to stratify each theory in a successive refinement, with each level given its own identifier. For example, a cam mechanism could be specified in a three-step refinement such as the following:

theory *rotary-to-linear*
begin
  constant *input*, *output* : **identifier**;
  predicate *rotary*(**identifier**), *linear*(**identifier**),
  in-out(**identifier**,**identifier**);
  axiom $rotary(input) \wedge linear(output) \wedge in\text{-}out(input, output)$
end

theory *rotary-to-linear.cam*
begin
  subtheory *rotary-to-linear*, *cam* **for** *input*, *tool-end* **for** *output*;
  constant *follower* : **identifier**;
  var *contact* : **identifier**;
  predicate *contact*(**identifier**,**identifier**), *part-of*(**identifier**,**identifier**);
  axiom $part\text{-}of(tool\text{-}end, follower) \wedge \forall contact(part\text{-}of(contact, follower))$;
  $\exists contact(contact(contact, cam))$
end

theory *rotary-to-linear.cam.planar-contour*
begin
  subtheory *rotary-to-linear.cam*, *roller* **for** *contact*;
  ⟨the fully detailed theory for a cam mechanism, such as in Sect. 6.3 (3)⟩;
end

The identifier could serve as the key to retrieve the theory that is to be reused if it carries an adequate meaning for designers and programmers. Otherwise, some higher level should first be entered at which the contents of the theories with the same identifier should be examined before proceeding further down the search path. A pictorial presentation of the tree of related theories and sub-theories such as shown in Fig. 8.12 should be made to aid human users in finding those that best suit their purposes.

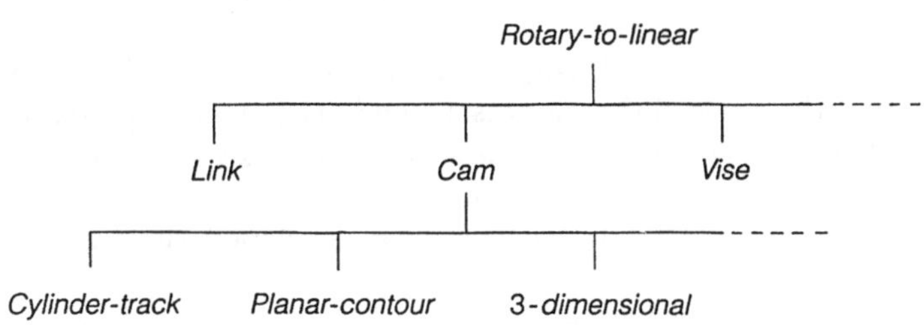

Fig. 8.12. Cam mechanism and related theories

It would also be beneficial if the schematic description of a typified and simplified design is associated with each theory and made readily available for inspection by designers and programmers. In the case of a cam mechanism, it should contain among others kinematic constraints on components, characteristics of possible tool end motions, animated movements of cam-and-follower arrangements, limits on pressures, frictions and other potentially crucial physical quantities, etc. Human users aided by these facilities should be able to edit the target theories in a manner which is quite natural to them, yet is translatable to a rigorous language such as defined in Sect. 7.2.

(2.3) *Organizing Partial Interpretations*
The concept of 'view' in LIL is nothing but a special case of that of an imbedding of a formal theory in a stronger one. Another important class of mappings are those in which a theory is made stronger by adding axioms that stipulate further constraints on constant symbols. In mathematics, science and engineering, many theories derive from others in this fashion. A very simple example is the theory for cylinders which can be obtained from that for truncated cones by equating the radii of top and bottom circles (Sect. 6.3 (1.1)). For finite cones, the radius of the top circle is given the value of zero (0). Such a theory is regarded as a partial interpretation of the original. We apply this term also to those cases where no constant expressions are given final interpretations (such as cylinders).

We may envisage the set of all partial interpretations of a given theory ($\mathcal{T}$) as a partially ordered set (denoted by $pi(\mathcal{T})$) with $\mathcal{T}$ being the minimum and each full interpretation being a maximal element. A family of standard parts of the same type may fit well into this structure. The basic services provided by the system should include the following:

- to admit entry at any point of $pi(\mathcal{T})$ by human users and the application agent (in theory construction mode),
- to move up and down the hierarchy retrieving pertinent pieces of information, and
- to insert a new entry into or delete an existing one from $pi(\mathcal{T})$ and update the structure accordingly.

Again, it would be beneficial if a schematic description associated with each element of $pi(\mathcal{T})$ is stored.

While the imbedding relates one theory as a sub-theory to another, this differs essentially from the reuse of theories discussed in the preceding subsection in that there will occur an increase of axioms only, and not of non-logical symbols (that is, *concepts*). In other words, in this subsection we have dealt with the specialization of a theory which could be developed through the reuse of other theories.

The exposition in this section has been more or less sketchy, and no rigorous definition of the language has been given. The reasons for this are threefold. First, such a language belongs to a different level than the MDL which is the principal point of interest for us throughout this book. Such a language must be capable of describing and manipulating such pieces of information as records of designers' actions and decisions, libraries of theories and partial interpretations, and so on, which will be structured in different ways than the design database. Second, the aspect of communications between human designers and the application agent which is outside of our concern has to be given serious consideration. Third, it will be unrealistic if we attempt to incorporate all the proposed concepts and their manipulations into a single, well-organized and well-balanced framework. Besides, there are other proposals which we have skirted, including automatic truth maintenance, qualitative reasoning, induction and abduction based logics, and so on, and more will come. There could be a multitude of possible combinations of alternate methods and approaches. We are yet to see which will prevail eventually.

References

[McC] McCarthy, J.: Circumscription – A Form of Non-monotonic Reasoning. Artificial Intelligence 13, 1980

[Rei] Reiter, R.: A Logic for Default Reasoning. Artificial Intelligence 13, 1980

[Gog] Goguen, J.A.: Reusing and Interconnecting Software Components. IEEE Computer, Feb. 1986

Chapter 9
Implementation and Remaining Problems

While a full implementation of the framework described in the preceding two chapters is admittedly not an easy task, a database management system called LMS (Logical Structure Management) which partially meets the requirements was developed in late 1979 and has been in service since then in practical application environments. As a matter of fact, the language specification presented in this book is the result of a further theoretical development of the idea of LMS. Applications of LMS include the design and manufacture of such diverse kinds of object as:

- outer shapes of auto-bodies
- die faces and structural parts of stamping devices,
- iron and steel frames,
- lead frames of integrated circuits,
- outer shapes of electric appliances,
- jig and fixures,
- various mechanical parts and products, and
- outer and inner parts of houses.

9.1 General Organization

LMS is organized in the traditional way for the construction of data management systems, and comprises a schema descriptor, a data manipulation language, execution modules, and utilities for logical and physical administrations. These subsystems are named respectively as follows after the corresponding subsystems of the CODASYL dbcs:

- LDL (Logical Structure Description Language)
- LML (Logical Structure Manipulation Language)
- LMR (Logical Structure Manipulation Routine)
- LMU (Logical Structure Manipulation Utility)
- LRU (Logical Structure Reorganization Utility).

Of these, we will give in the next section a description of LDL and LML, and a part of LMU, all of which are directly related to the modeling capability of LMS.

9.1.1 Structure of a Model

The same world view as presented in Sect. 5.3 underlies LMS, and hence in particular, the database is a model of a first order theory. The actual language constructs, however, are made under the following restrictions:

(1) Let I, A and N denote the set of identifiers, the union of data types permitted in the host language (e.g., FORTRAN) and the set of names respectively. The functions that can be defined over these sets are:

$$\begin{array}{ll} \textbf{name} & : I \to N \\ \textbf{id} = \textbf{name}^{-1} & : N \to I \\ \textbf{value} & I \to A. \end{array}$$

The range of **value** is type **record**. This means a compromise with the record-based modeling and reflects an early technical decision that the CODASYL dbcs is to be used as the internal data description and manipulation vehicle.

(2) Arbitrary kinds of predicate of arbitrary arities may be defined. However, their arguments must all be in the domain I.

(3) Every term (constant or variable) that occurs in a formula must denote either an identifier or a name, but not any value or any part thereof. This means that LMS does not 'see' attribute values or relations involving attribute values, and leaves the manipulation of these to application programs. In other words, LMS is dedicated to the construction and manipulation of explicit structures among objects represented in the domain I.

Figure 9.1 depicts what is said above. Comparing it with Fig. 6.1 will clarify the difference between the two structures of interpretation, one for the general first order theory and the other permissible under LMS.

9.1.2 Structure of the Data Storage

Except for the physical aspect, which is not our concern, knowledge to a certain extent of the storage structure is necessary for database administrators and application programmers to use LMS effectively.

(1) The database as a collection of pairs, each comprising an identifier and its value (hereafter called an object), is divided into mutually disjoint workunits each of which is in turn is the union of sections, also mutually disjoint. An application program is allowed to write into sections in one workunit and read from sections in an arbitrary number of workunits at any given instant. These sections must have been opened, and when access becomes no longer necessary, should be closed for efficiency (localizing the extent of accessibility.)

(2) The database is also divided into areas, each of which is further divided into pages. A section comprises a set of consecutive pages within an area. (See Fig. 9.2.)

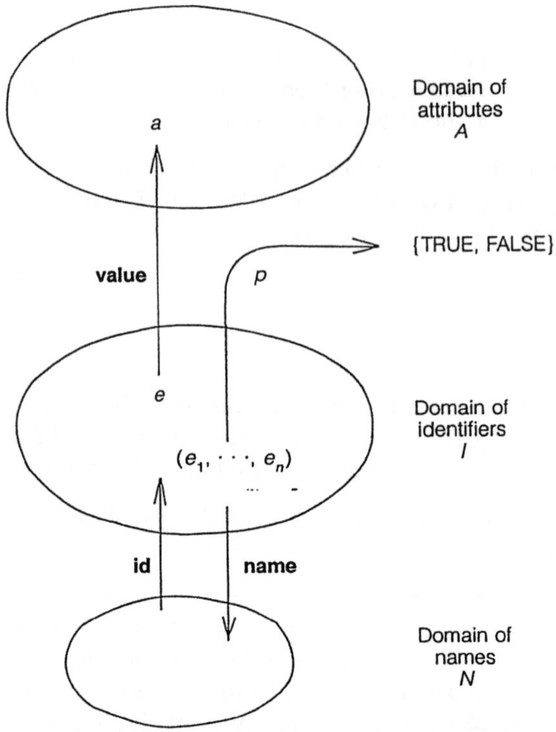

Fig. 9.1. The structure of an interpretation under LMS

(3) Each application program must declare one or more record-type areas (DDA (Data Delivery Area)) through which pieces of information such as identifiers, names, values, types of values, and names of workunits are transmitted to and from the database. Another kind of area (CA (Communication Area)) is to hold the control information generated by LMR during an execution of LML commands.

9.2 Logical Structure Desription Language (LDL)

(1) *Definition of Areas and Sections*
Each area definition is followed by one or more statements to define sections within that area. Their syntax is as follows:

area ⟨area name⟩ (⟨area code⟩, ⟨no. of pages⟩),
  ⟨no. of words per page⟩, ⟨no. of records per page⟩ ;
section ⟨section name⟩ (⟨starting page⟩, ⟨ending page⟩),
  ...
  ⟨section name⟩ (⟨starting page⟩, ⟨ending page⟩).

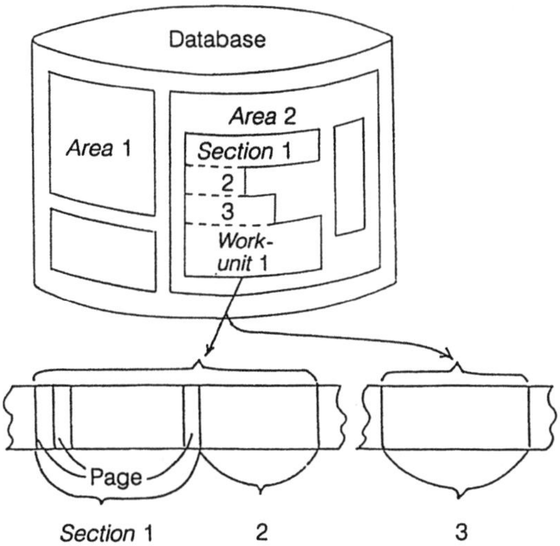

Fig. 9.2. The structure of the data storage

(2) *Registration of Predicate Names and Arities*

predicate ⟨predicate name⟩ ⟨number of arguments⟩,

...

⟨predicate name⟩ ⟨number of arguments⟩.

9.3 Logical Structure Manipulation Language (LML)

(1) *Definition of Logical Formulas*

fomula ⟨formula name⟩ (⟨variable name⟩, ... , ⟨variable name⟩)
= ⟨formula⟩

(1.1) A variable name must start with '?'.

(1.2) The names of formulas defined by the above statements may be used anywhere in an LML program in the same way as registered predicate names.

(1.3) Recursive definitions such as the following are permitted:

$$r(?) = p(?) \wedge \exists ?1 \ldots \exists ?n(r(?1) \wedge \cdots \wedge r(?n) \wedge q(?1, \ldots, ?n, ?))$$

where p and q are names of predicates or previously defined formulas.

(2) The permissible formulas are those constructed from atomic formulas including equalities and previously defined formulas using logical connectives and quantifiers ($\wedge, \vee, -, \rightarrow, \neg, \exists, \forall$) with the following restrictions:

(2.1) $\rightarrow$ and $\forall$ may only occur together in such a form as

$$\forall ?1 \ldots \forall ?n(p(?1,\ldots,?n) \rightarrow q(?1,\ldots,?n,?x1,\ldots,?xm)).$$

(2.2) $\neg$ (negation) can be used only to negate atomic formulas in structure modification (see (11) below).

(3) *Declarations*
Program variables of those types which are proper to LMS and do not belong to any host language must be declared in LML. These types are **genid** (for generalized identifier) and **genname** (for generalized name), and hold identifiers and names respectively.

(4) *Creation of Objects (Identifiers and Their Values)*

create ⟨DDA name⟩ [: ⟨structure modifier⟩] [: ⟨naming of objects⟩]

(4.1) The number and values of objects to be created must be given in the DDA by the application program prior to issuing this command. Upon execution, the identifiers (that LMS automatically creates) are returned to the DDA.

(4.2) In an optional structure modification and a naming of objects (for the functions of which, see (11) and (8) below respectively), the identifiers being created can be referred to as $*(1), *(2), \ldots$ with numbers indicating the order in which corresponding values are placed in the DDA.

(5) *Creation of Objects Accompanied by a Simultaneous Imbedding of These Objects in an Existing Set*

imbed : ⟨series name⟩ (**initial**) : ⟨range characterizer⟩
imbed ⟨DDA name⟩ : ⟨series name⟩ : $*(1)$.**to**.⟨term⟩, ... , $*(n)$.**to**.⟨term⟩
  [/⟨object selector⟩] [: ⟨structure modifier⟩] [: ⟨naming of objects⟩]
imbed : ⟨series name⟩ (**final**)

(5.1) The initial command determines the range of the imbedding mapping. The ⟨range characterizer⟩ is a formula which contains precisely one free variable.

(5.2) A series of commands with the same ⟨series name⟩ in which more than one intermediate command may appear creates a set of objects and imbeds it in the specified range. The correspondence is specified by $*(i)$.**to**. ⟨term⟩ where $*(i)$ denotes the i-th object being created (see (4.2) above, ordinals are local within each command) and ⟨term⟩ is an identifier in the range that satisfies the optional ⟨object selector⟩ when substituted for the free variable thereof. The object selector is a formula which contains precisely one free variable. An imbedding θ is a logical homomorphism, that is, $p(\theta(a_1),\ldots,\theta(a_n))$ holds whenever $p(a_1,\ldots,a_n)$ holds.

(5.3) The final command defines the relations in the newly created set of objects so that the above correspondence may indeed become an imbedding.

(6) *Deletion of Objects*

delete ⟨DDA name⟩ : ⟨object selector⟩

(6.1) Those objects selected by the object selector are deleted from the database, and their identifiers are returned to the DDA.

(6.2) The object selector must contain precisely one free variable.

(7) *Modification of Values*

modify ⟨DDA name⟩ : ⟨object selector⟩ ; ... ; ⟨object selector⟩
[: ⟨structure modifier⟩] [: ⟨naming of objects⟩]

(7.1) The values of selected objects are replaced by those given in the DDA in the same order as the object selectors separated by semicolons (;), and the identifiers of these objects are returned to the DDA.

(7.2) The object selector must contain precisely one free variable, and must uniquely determine its value.

(8) *Naming of Objects*

name ⟨term⟩ = ⟨name⟩, ... , ⟨term⟩ = ⟨name⟩ [/⟨object selector⟩]

(8.1) Names specified as values of **genname** type variables are given to those objects the identifiers of which are explicitly specified (in **genid** type variables) or uniquely determined by the optional object selector.

(8.2) The currently given name is deleted if ⟨name⟩ is equal to **denamed**.

(9) *Retrieval of Objects (Simple Mode)*

get ⟨DDA name⟩ :: ⟨object selector⟩

(9.1) The identifiers, names and values of the selected objects are retrieved from the database and placed in the specified DDA. These identifiers satisfy the object selector when substituted for the free variables thereof.

(9.2) The object selector may contain an arbitrary number of free variables.

(10) *Retrieval of Objects (Piped Mode)*

get ⟨series name⟩ (**initial**) : ⟨object selector⟩
get ⟨DDA name⟩ : ⟨series name⟩
get ⟨series name⟩ (**final**)

(10.1) While the function of these commands combined is essentially the same as that of a simple retrieval (as in (9)), this mode is useful for piecemeal retrieving and processing of a large number of objects. The initial command determines the set of objects to be retrieved. There may be more than one intermediate command, for each of which the number of objects to be received is specified in the DDA. The final command is only to indicate the end of a series.

(10.2) The object selector may contain an arbitrary number of free variables.

(10.3) **set**(series name) can be used as the name of the temporary predicate that characterizes the set of objects being retrieved.

(11) *Modification of Structure*

modify str ⟨structure modifier⟩ [/⟨object selector⟩]

(11.1) The logical structure of the database (more precisely, the set I) is modified so that the specified ⟨structure modifier⟩ becomes true.

(11.2) The structure modifier is such a logical formula that the truth value of every atomic formula contained therein is uniquely determined by that of the structure modifier itself. The general form it takes is

$$L_1 \wedge L_2 \wedge \cdots \wedge L_n$$

where L_i is a literal. When (free) variables occur in it, they must be uniquely determined by the optional object selector.

(12) *Evaluation of a Logical Formula*

get tval ⟨DDA name⟩ : ⟨closed logical formula⟩
get pred ⟨DDA name⟩ : (⟨n-tuple⟩), ... , (⟨n-tuple⟩)

(12.1) **get tval** evaluates the specified formula and returns its truth value to the DDA.

(12.2) **get pred** searches for such predicates as are satisfied by all the specified n-tuples and returns their names to the DDA. Note that this command (and only this) is of the second order in that it treats the predicate names as data.

9.4 Logical Structure Manipulation Utility (LMU)

Primary functions of LMU are:

(1) Registration and deletion of workunits,
(2) Definition, and initialization of sections, and their insertion to and release from workunits,
(3) Copying of objects from one or more sections of a workunit into a section of another workunit
(4) Compaction of workunits,
(5) Saving and restoration of workunits to and from archival storage, and
(6) Collection and summarization of statistical information on the use of database.

Copying between sections is the most important function from the viewpoint of modeling capability. Values of and relations among copied objects are the same as the originals, but identifiers are completely different since the copies are different entities from the originals. In other words, an isomorphic set of objects will be created. This point forms a significant contrast against the case of business applications where taking copies is primarily for operational security and convenience. (See the discussion in Sect. 1.2 (7).) Saving and restoration are rather similar to the latter. The same names are given to copies, but with different (user specified) qualifiers. In this way, the correspondences between the originals and the copies can be retained. If necessary, the user's own programs can be inserted to add, delete and/or change the data during copying. Though it is of course possible to carry out all the above tasks using the **get** and **imbed** commands of LML in a yet finer and more precise manner, this LMU function offers a very easy and efficient means to application programmers and database administrators.

9.5 Examples and Applications

9.5.1 Examples

(1) Given a face ($f1$) and one of its boundaries ($e1$) in the winged-edge topology of a polyhedron, find the surface opposite to $f1$ across $e1$ (Fig. 0.6).

(1.1) If the ternary predicate $boundary(f, e, f')$ is available,

get $dda :: boundary(f1, e1, ?f)$

(1.2) If only the binary predicate $boundary(f, e)$ is defined,

get $dda :: boundary(?f, e1) - (?f = f1)$

(2) Given a mesh of surface patches, a point (pt_0) on a boundary curve and a plane (pl), search for all the patches that intersect pl and calculate the intersection. The availability is assumed of predicate *cbdrys* (curve on the boundary of surface) between a patch and its boundaries, function *inters* to determine if a patch intersects pl, and procedure *calint* to calculate the intersection. (See Figs. 1.6 and 7.4.)

var $c0, s$: **genid**; $ssed, ssing$: **set of genid**;
{determine the curve c_0 on which pt_0 lies};
get $dda :: cbdrys(c0, ?s)$; $ssed := 0$;
{move id's of surface-patches containing c_0 to $ssing$};
foreach s **in** $ssing$ **do**
begin
  $ssed := ssed \cup [s]$; **if** $inters(s, pl) =$ **true then**
  begin
    $calint(s, pl)$;
    get $dda :: \exists ?c(cbdrys(?c, s) \wedge cbdrys(?c, ?s)) - ssed$;
    {move id's of $?s$ that share boundary with s to $ssing$}
  end;
  $ssing := ssing - [s]$
end.

The above program employs an algorithm and data structures which are much more straightforward than those in the case of Sect. 7.4 (2). It should be noted, however, that:

- for each patch s which intersects pl, every adjacent patch of s is examined one by one to see whether it intersects pl or not, and
- the entire attribute value of each patch is retrieved from the database, and hence
- the total time required to execute the program could be significantly greater than in the case of Sect. 7.4 (2).

(3) Let $use(x, y, t)$ represent the fact that part x uses part y applying transformation t. Starting with a given part a, search for all the parts directly or indirectly used by a, and arrange them in such an order that the user precedes the used.

formula $f(?) = (? = \mathbf{id}(a)) \vee \exists ?x(f(?x) \wedge \exists ?t(use(?x, ?, ?t)))$;
get $dda :: f(?1) \wedge \exists ?t(use(?1, ?2, ?t))$;
{sort $\{(?1(i), ?2(i)) : 1 \leq i \leq n\}$ so that the tuple of the form (y, z) never precedes those of the form (x, y).}

(4) Let $assy$ be the characteristic predicate of the set of parts directly or indirectly used by a in the above example. Create a copy headed by $aprime$ and imbed it in $assy$ (Fig. 9.3).

formula $assy(?) = \exists ?1(((?1 = \mathbf{id}(a)) \vee assy(?1)) \wedge \exists ?2(use(?1, ?, ?2) \vee use(?1, ?2, ?)))$;
get in(**initial**) : $(? = \mathbf{id}(a))$; $assy$ {initiates reading $assy$};
imbed out(**initial**) : **set**(in) {initiates creating a set of elements and imbedding it in **set**(in), the range of series in, that is, $assy$};
get $w0$: in {reads a into $w0$ placing its id in $oldid(0)$};
imbed $w0$: out : $(*.\mathbf{to}.oldid(0))$: **name** $*$ = $aprime$ {creates $aprime$};
{The following are made available to this program with each **get** statement;
n = the number of elements delivered with the current **get**
nt = the total number of elements in the range $assy$
np = the number of elements already delivered.}
while $np \leq nt$ **do**
begin
get w : in {reads $assy$ into w placing id's in array $oldid$};
imbed w : out: $(i = 1, n)(*(i).\mathbf{to}.oldid(i))$ {maps the i-th created element to $oldid(i)$ with the imbedding function}
end;
imbed out(**final**);
get in(**final**);

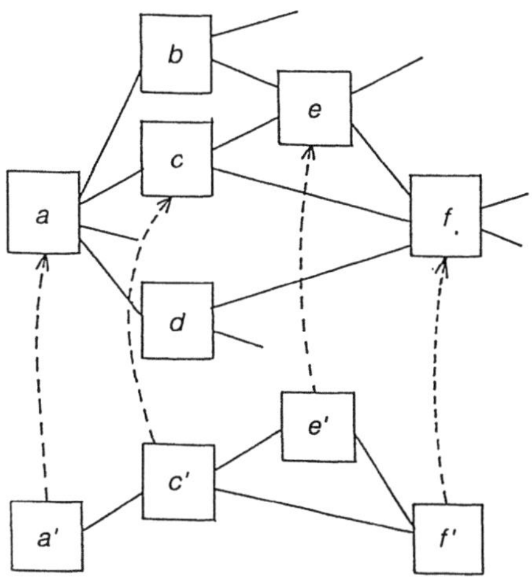

Fig. 9.3. An imbedding function

9.5.2 Applications

The purpose of this section is just to indicate the wide range of applicability of LMS by citing two practical and fairly large applications. It is not intended to give a detailed account of these applications or of how LMS is used in them.

(1) *Designing the Outer and Inner Constructions of a House*
Basic input operations of this system comprise the reading from a tablet of the main coordinate values of outer and inner walls of a house, the selection and placement of attachments, furniture, appliances, etc., and the specification of material and finishing. Roofs and foundations are mostly created automatically. The database is divided into two portions, one for storing standard parts and the other for containing the model of the individual house being designed. The latter represents a house as a collection of 3-dimensional objects having also attributes other than shape from which all the necessary drawings, perspectives, cost estimates and other kinds of document are automatically produced. Figure 9.4 shows a portion of the LMS structure of the model. Dependencies between objects make it very easy for a change at one place to propagate to other places. For example, the sizes of rooms and the positions of attachments will be automatically adjusted if the walls are moved (by changing the datum).

Figure 9.5 shows the hand-written planar sketch of a two-storey house which is input to the system to build the precise model, which takes approximately 200 interactive commands and 35 minutes of time. A relatively large consumption of time per command (average 10.5 seconds) is due to the fact that many commands perform complicated processing (e.g., automatic

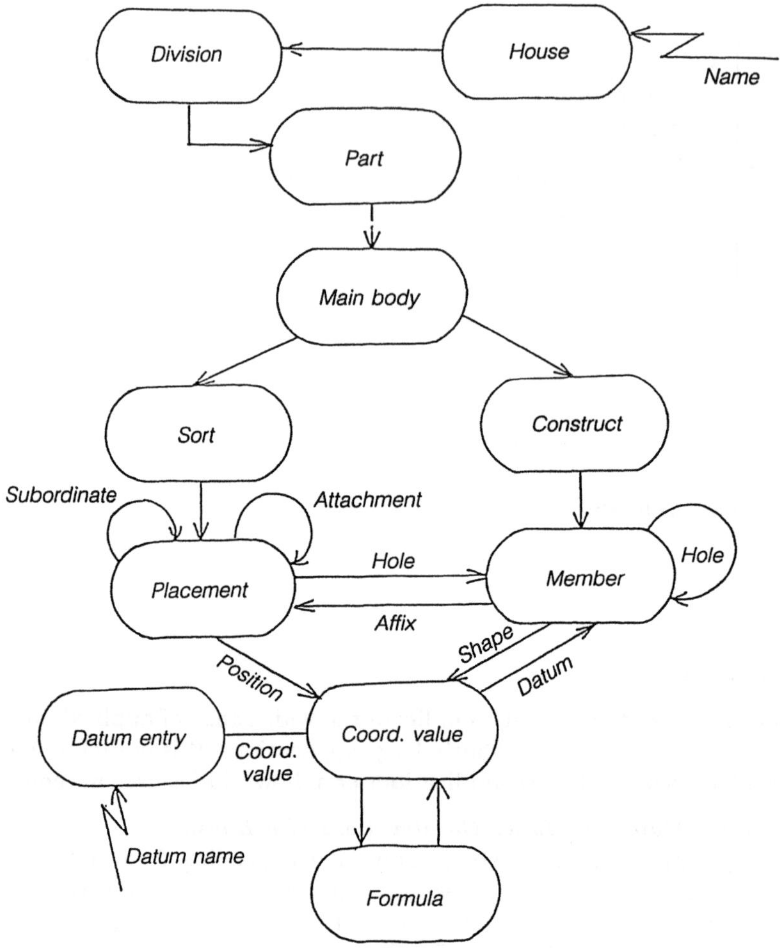

Fig. 9.4. The LMS structure of a house model (part)

creation of roofs) or repetitive operations (e.g., multiple placement of parts of the same kind) as opposed to conventional drafting-type systems. Following the design, the production of drawings (such as Figs. 9.6, 9.7 and many others) and the cost estimation are done in approximately 45 minutes.

It is quite possible and is actually being planned to extend the application to include downstream processes such as producing bills of materials, drawings for fabrication at construction sites, etc. It should be emphasized that the extensibility of applications can only be obtained by the generality or flexibility of the database schema that allows addition of any desired information.

Figures 9.4 through 9.7 are made available by the permission and courtesy of Sekisui House, Ltd.

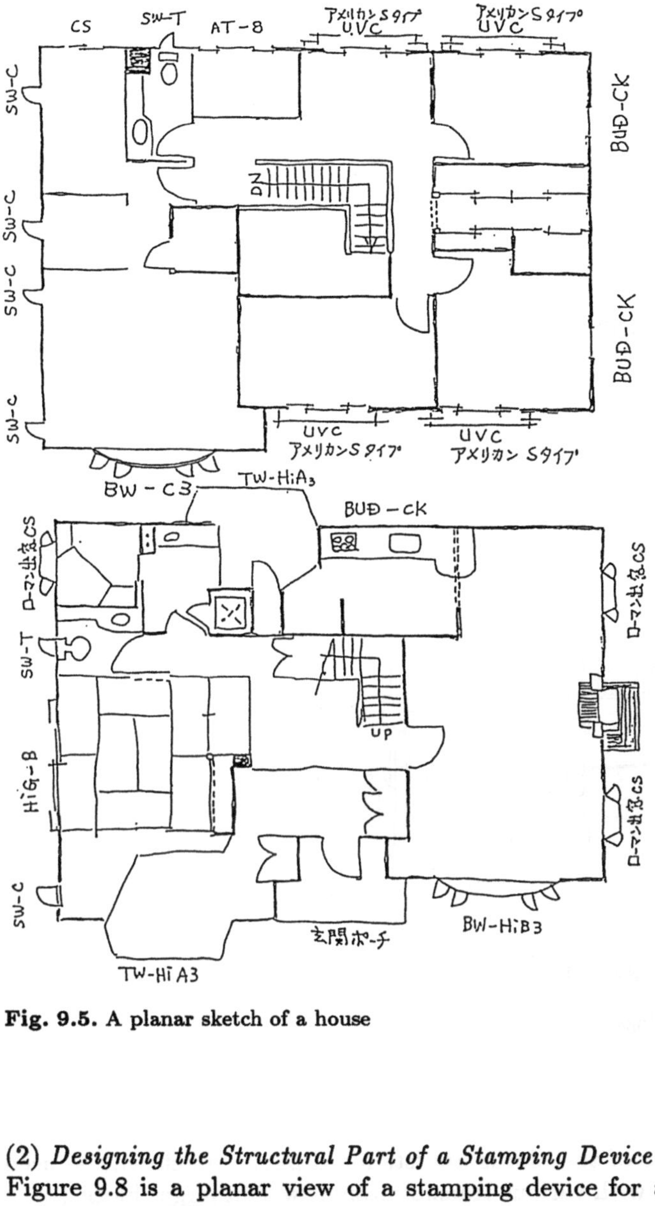

Fig. 9.5. A planar sketch of a house

(2) *Designing the Structural Part of a Stamping Device*
Figure 9.8 is a planar view of a stamping device for auto-body parts under process of design. We can observe that there exist three topmost level constituents or *blocks*, that is, the left, right and center portions. The former two are 'mirrors' to each other, and hence can be manipulated as copies or instances of the same prototypical block. The connections between these blocks are yet to be completed, as indicated by arrows. Each block has its own (local) coordinate axes and may contain references to (uses of) smaller blocks together with the transformation to fix the positions and orientations

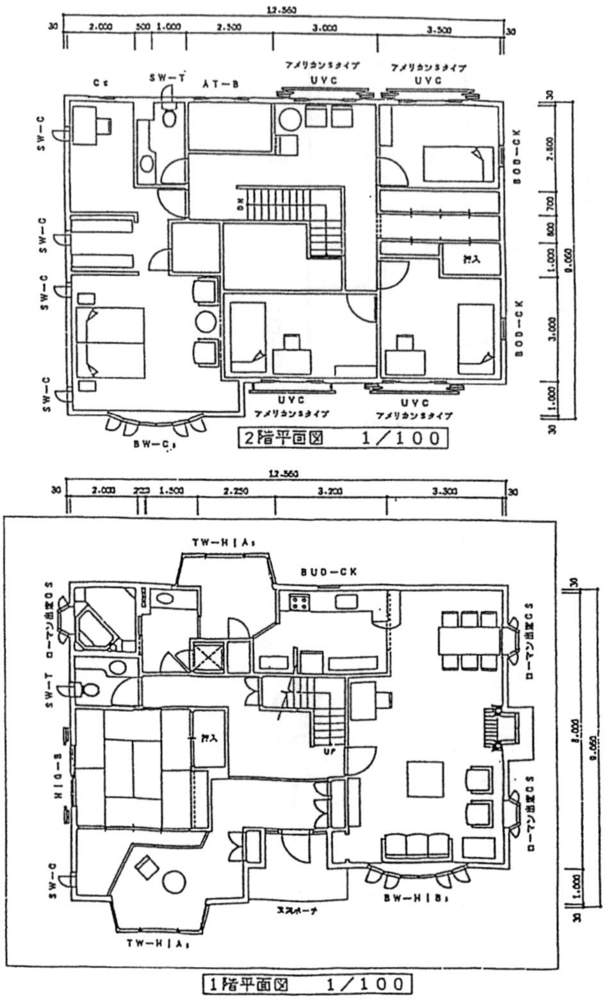

Fig. 9.6. Finished drawings

of these blocks. Some of the blocks are 'standard parts' which will be used at the various places with various transformations within larger blocks.

Figure 9.9 shows a cross section of the same design, in which the display of standard parts is suppressed. A perspective view is given in the much busier Fig. 9.10. There, it should be noted that standard parts are created and displayed as 3-dimensional objects while the remaining blocks are designed in $2\frac{1}{2}$-dimensional (or 'depth') mode. An advantage of designing in such a mixture of modes is to facilitate dimensioning and editing of drawings and yet to guarantee consistency between different views. Figure 9.11 depicts a

Fig. 9.7. A perspective view

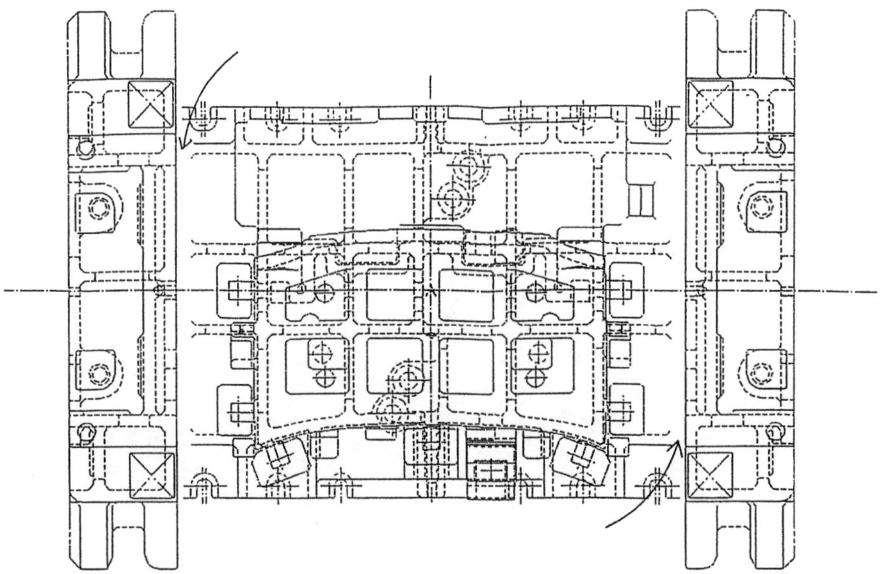

Fig. 9.8. A planar view of a stamping device

part of the tree-like data structure which is headed by the block for an entire die and ends with individual parts.

Figures 9.8 through 9.11 are made available by the permission and courtesy of Fuji Heavy Industries, Ltd.

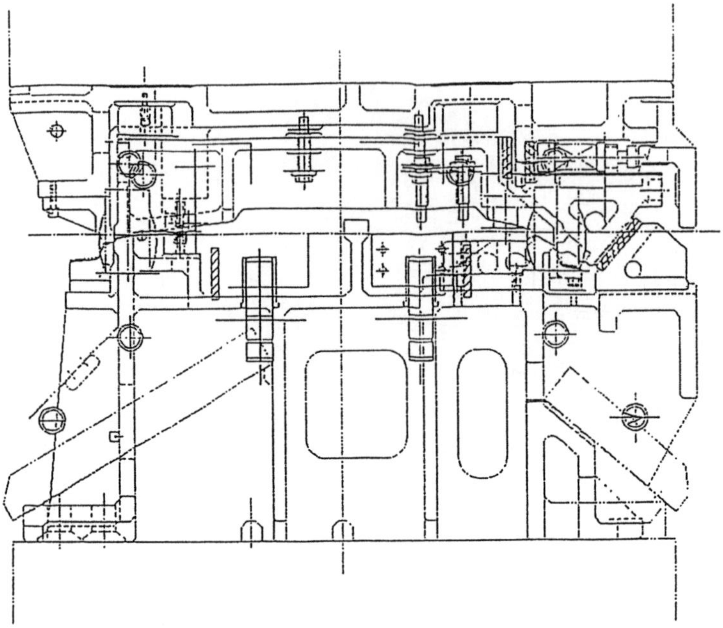

Fig. 9.9. A cross section with partial suppression of display

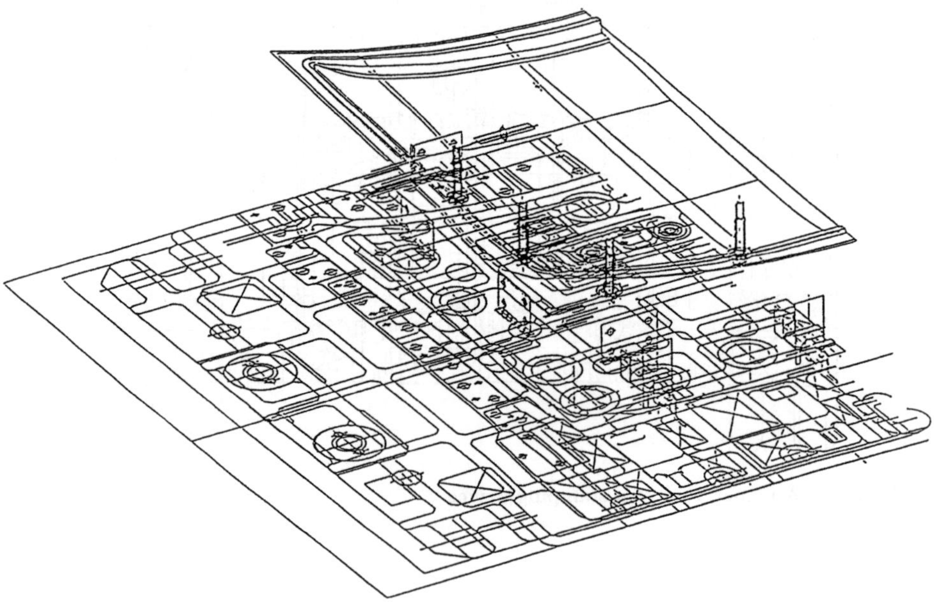

Fig. 9.10. A perspective view

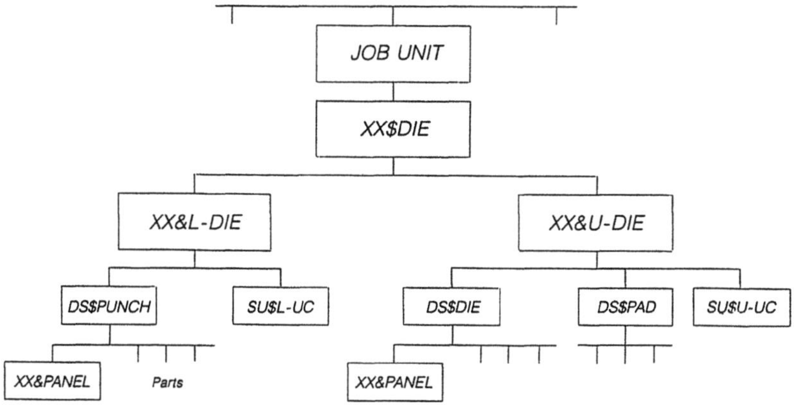

Fig. 9.11. The LMS structure (part)

9.6 Remaining Problems

The author feels confident of the methodological propriety of the conceptual framework presented in this book from the soundness of its theoretical foundation, small but non-trivial examples and the experiences with a partial implementation (that is, LMS). Several problems remain to be resolved, however, in order to make it truly applicable in practical environments. Some of the problems are concerned with the theory construction mode and others with the interpretation mode of operation. Roughly speaking, these are the problems concerning the external and internal realms respectively of a schema, and include the following.

(1) *Theory Construction*
(1.1) *User-Friendly Interface*
It is hardly expected that FOPL serves as a language in which designers can communicate with the system in a natural and smooth manner. An external, more user-friendly layer, which might look like a natural language and/or contain a domain-specific vocabulary, and yet allows a precise translation into FOPL is required.

(1.2) *Support of Metatheoretic Reasoning*
The man-system interface must support metatheoretic reasoning, such as circumscription or default reasoning, which enables designers to augment and strengthen theories by qualifying some of the undecidable (that is, neither provable nor negatable) formulas as axioms.

(1.3) *Templates for Theories*
It is desirable to develop an assortment of templates or pre-theories, each of which serves as a prototype, and can be refined and/or augmented to produce a family of small formal theories. These small theories in turn serve as sub-

theories or constituents of larger theories. An example: templates for form features.

(1.4) *Management of Theory-Base*
It must be possible to organize existing theories into a theory-base from which designers can retrieve the specifications which appear close to the ideas or pre-conceptions yet to be precisely formulated. A library of form features is a part of such a theory-base.

(2) *Interpretation*
(2.1) *Internal Schema Description Language*
The need to develop a language for describing internal schemas is twofold: first, to attain the highest possible efficiency for evaluating formulas in each given theory, and second, to define precisely the method of selection among alternate paths for asserting axioms.

(2.2) *Evaluation of Dynamic Characteristics of Application Agents*
It is important for the system to be able to collect and analyze the run-time performance data of each application agent, the result of which will be used to adjust the internal schemas.

Further problems present themselves at system level, in the solution of which it will be particularly important to facilitate:

- alternate designs,
- distributed processing, and
- control of versions and configurations in a concurrent engineering environment.

The technical core which is common to all these is the capability to create copies of an arbitrary portion of the database (object model) which are isomorphic to each other under a specified view and to detect any changes that may possibly affect the interface of the said portion with the rest of the design object.

It is hoped that our framework, when these problems as described above are solved, will provide a theoretically sound and practically well-equipped basis on which a wide range of CAD/CAM systems, conventional or intelligent, can be built and integrated in a natural and flexible manner. Moreover, it is expected to serve as a framework in which designers' behaviors can be formally described, thus opening the way to the establishment of the general theory of design as a rigorous scientific discipline.

References

[NUL] A User's Manual: Logical Structure Management System (available from Nihon Unisys Ltd., Tokyo, Japan)

Subject Index

Springer Series
Computer Graphics – Systems and Applications

Former Subseries of SYMBOLIC COMPUTATION

J. L. Encarnação, R. Schuster, E. Vöge (eds.):
Product Data Interfaces in CAD/CAM Applications.
Design, Implementation and Experiences.
IX, 270 pages, 147 figs., 1986

U. Rembold, R. Dillmann (eds.):
Computer-Aided Design and Manufacturing.
Methods and Tools. Second, revised and enlarged edition.
XIV, 458 pages, 304 figs., 1986

G. Enderle, K. Kansy, G. Pfaff:
Computer Graphics Programming. GKS – The Graphics
Standard. Second, revised and enlarged edition.
XXIII, 651 pages, 100 figs., 1987

Y. Shirai:
Three-Dimensional Computer Vision.
XII, 297 pages, 313 figs., 1987

D. B. Arnold, P. R. Bono:
CGM and CGI. Metafile and Interface Standards
for Computer Graphics.
XXIII, 279 pages, 103 figs., 1988

J. L. Encarnação, P. C. Lockemann (eds.):
Engineering Databases. XII, 229 pages, 152 figs., 1990

P. Wisskirchen:
Object-Oriented Graphics. From GKS and PHIGS
to Object-Oriented Systems.
XIII, 236 pages, 83 figs., 1990

J. L. Encarnação, R. Lindner, E. G. Schlechtendahl:
Computer Aided Design. Fundamentals and System
Architectures. Second, revised and extended edition.
XII, 432 pages, 240 figs., 1990

H. Hagen, D. Roller (eds.):
Geometric Modeling. Methods and Applications
VII, 286 pages, 140 figs., 1991

T. Yagiu:
Modeling Design Objects and Processes
X, 327 pages, 1991